44
AF758309

Structure and Action of Molecular Chaperones

Machines that Assist Protein Folding in the Cell

Series in Structural Biology

Series Editor: Anders Liljas *(Lund University, Sweden)*

Published

Vol. 1 Structural Aspects of Protein Synthesis, 2nd ed.
by Anders Liljas and Måns Ehrenberg

Vol. 2 Found in Translation: Collection of Original Articles on Single-Particle Reconstruction and the Structural Basis of Protein Synthesis
by Joachim Frank

Vol. 3 Selected Papers of Michael G. Rossmann with Commentaries: The Development of Structural Biology
by Michael G. Rossmann

Vol. 4 From a Grain of Salt to the Ribosome: The History of Crystallography as seen Through the Lens of the Nobel Prize
edited by Ivar Olovsson, Anders Liljas and Sven Lidin

Vol. 5 The Struggles and Dreams of Robert Langer
by Robert Langer

Vol. 6 Structure and Action of Molecular Chaperones: Machines that Assist Protein Folding in the Cell
by Lila M. Gierasch, Arthur L. Horwich, Christine Slingsby, Sue Wickner and David Agard

Vol. 7 Curiosity and Passion for Science and Art
by Uwe B. Sleytr

Forthcoming

Vol. 8 Textbook of Structural Biology, 2nd ed.
by Anders Liljas, Lars Liljas, Miriam-Rose Ash, Göran Lindblom, Poul Nissen and Morten Kjeldgaard

Series in Structural Biology — Vol. 6

Structure and Action of Molecular Chaperones

Machines that Assist Protein Folding in the Cell

Lila M. Gierasch
University of Massachusetts, Amherst, USA

Arthur L. Horwich
Yale School of Medicine, USA

Christine Slingsby
University of London, UK

Sue Wickner
National Cancer Institute, USA

David Agard
UC San Francisco, USA

World Scientific

EW JERSEY • LONDON • SINGAPORE • BEIJING • SHANGHAI • HONG KONG • TAIPEI • CHENNAI • TOKYO

Published by

World Scientific Publishing Co. Pte. Ltd.
5 Toh Tuck Link, Singapore 596224
USA office: 27 Warren Street, Suite 401-402, Hackensack, NJ 07601
UK office: 57 Shelton Street, Covent Garden, London WC2H 9HE

Library of Congress Cataloging-in-Publication Data
Names: Gierasch, Lila M., author. | Horwich, Arthur, author. | Slingsby, Christine, author. | Wickner, Sue, author. | Agard, David, author.
Title: Structure and action of molecular chaperones : machines that assist protein folding in the cell / Lila M. Gierasch, University of Massachusetts, Amherst, USA, Arthur L. Horwich, Yale School of Medicine, USA, Christine Slingsby, University of London, UK, Sue Wickner, National Cancer Institute, USA, David Agard, UC San Francisco.
Description: New Jersey : World Scientific, 2016. | Series: Series in structural biology | Includes bibliographical references and index.
Identifiers: LCCN 2016015384| ISBN 9789814749329 (hardcover : alk. paper) | ISBN 981474932X (hardcover : alk. paper)
Subjects: LCSH: Protein folding. | Molecular chaperones.
Classification: LCC QP551 .G54 2016 | DDC 572/.633--dc23
LC record available at https://lccn.loc.gov/2016015384

British Library Cataloguing-in-Publication Data
A catalogue record for this book is available from the British Library.

Copyright © 2016 by World Scientific Publishing Co. Pte. Ltd.

All rights reserved. This book, or parts thereof, may not be reproduced in any form or by any means, electronic or mechanical, including photocopying, recording or any information storage and retrieval system now known or to be invented, without written permission from the publisher.

For photocopying of material in this volume, please pay a copying fee through the Copyright Clearance Center, Inc., 222 Rosewood Drive, Danvers, MA 01923, USA. In this case permission to photocopy is not required from the publisher.

Printed in Singapore by Mainland Press Pte Ltd.

Contents

Preface

This book is part of a series covering fields where structural biology has made substantial contribution. The book reviews the beautiful architectures and varying mechanical actions of the set of specialized cellular proteins called molecular chaperones, which provide essential kinetic assistance to processes of protein folding and unfolding in the cell. Ranging from multi-subunit ring-shaped chaperonin and Hsp100 machines that use their central cavities to bind and compartmentalize action on proteins, to machines that use other topologies of recognition — binding cellular proteins in an archway or at the surface of a "clamp" or at the surface of a globular assembly — the structures show us the ways and means the cell has devised to assist its major effectors, proteins, to reach and maintain their unique active forms, as well as, when required, to disrupt protein structure in order to remodel or degrade. Each type of chaperone is beautifully illustrated by X-ray and EM structure determinations at near- atomic level resolution and its structure and action described by a leader in the study of the respective family. The beauty of what Mother Nature has devised to accomplish essential assisting actions for proteins *in vivo* is fully appreciable.

Arthur Horwich
New Haven, CT
2015

Chapter 1

Hsp70 Molecular Chaperones: Versatile Modular Nanomachines that Mediate Multiple Biological Functions

*Lila M. Gierasch**

Abstract

This chapter discusses the multitalented Hsp70 molecular chaperones, with particular emphasis on structural insights into their allosteric mechanism and how this mechanism enables them to perform a panoply of functions in nearly all organisms and cellular compartments. As our focus is the structural aspects of Hsp70 chaperone mechanisms, we call the interested reader's attention to a number of recent reviews for more extensive discussion of many other topics related to Hsp70s (Clerico and Gierasch, 2014; Mayer, 2010, 2013; Saibil, 2013; Zuiderweg *et al.*, 2013).

The Discovery of the Hsp70 Molecular Chaperone Family

The Hsp70 molecular chaperone family was first described as a set of 70 kDa proteins with greatly increased expression after heat shock, and hence the members were categorized as "heat shock proteins" (Lindquist, 1986; Lindquist and Craig, 1988). It was soon recognized that this family of proteins was widespread in nature, with the identification of Hsp70s from *Drosophila* (Tissieres *et al.*, 1974) to bacteria (Bardwell and Craig,

*Departments of Biochemistry & Molecular Biology and Chemistry, University of Massachusetts Amherst, Amherst, MA 01003, USA.

1984) to humans (Hunt and Morimoto, 1985). Constitutively expressed homologues were soon discovered in addition to the originally observed heat-inducible Hsp70s, and Hsp70 homologues were found in nearly all eukaryotic cellular compartments, including the cytoplasm, mitochondrion, chloroplast, endoplasmic reticulum (ER), and nucleus (Gething and Sambrook, 1992; McKay, 1993). The association of the Hsp70s with heat shock led to proposed functions in the protection of cells from heat-induced damage, particularly protein unfolding and aggregation. It soon became clear that the Hsp70 family performs a much broader array of cellular functions than simply protecting cells from heat shock damage. For example, the bacterial 70 kDa protein, DnaK, is indeed essential for viability and growth after heat shock (Paek and Walker, 1987), but it is also required for λ phage replication (Georgopoulos, 1977), a role that is based on the facilitation of complex disassembly (Alfano and McMacken, 1989). The ER Hsp70, also known as heavy-chain-binding protein or BiP, was first described as a facilitator of the folding and assembly of immunoglobulins (Munro and Pelham, 1986), but it is also a critical player in protein translocation across the ER membrane (Vogel *et al.*, 1990). In another eukaryotic system, a 70 kDa ATPase is implicated in the disassembly of clathrin coats from endocytic vesicles; originally called the clathrin-uncoating ATPase, it is identical to the constitutively expressed cytoplasmic Hsp70 molecular chaperone, Hsc 70 (Ungewickell, 1985). We now know that Hsp70 family members participate in cellular functions as diverse as early protein biogenesis, disaggregation, blocking misfolding, protection from heat stress, organellar import, disassembly of specific complexes (like clathrin cages), synaptic vesicle exocytosis, translocation into the ER lumen, and more (Clerico and Gierasch, 2014).

The striking reality is that Hsp70s all share a common biochemical mechanism and somehow this mechanism enables them to perform this array of seemingly disparate functions: essentially, they bind to unstructured regions of their polypeptide substrates with affinity that is modulated by nucleotide binding (McKay, 1993). The physiological context of Hsp70s confers their specialization: both their localization and the team of cochaperones (J-proteins; nucleotide exchange factors, or NEFs) and complicit chaperones they partner with (for example, disaggregases such as ClpB or upstream chaperones such as Hsp90s) permit Hsp70s to adapt to diverse cellular roles. Nonetheless, in all their roles, the underlying allosteric mechanism (Fig. 1.1A) is central: substrates bind with slow on/off-kinetics and high affinity to ADP-bound Hsp70s, and with rapid on/off kinetics and low affinity to ATP-bound Hsp70s (Schmid *et al.*, 1994). Substrate binding to the ATP-bound state, usually facilitated by a J-protein, leads to enhancement of the otherwise low basal Hsp70 ATPase activity, resulting in conversion to the high substrate affinity ADP-bound state. This state is cycled back to the ATP-bound state by the action of a NEF, and the substrate is released and a new one bound. While most of the current mechanistic understanding of Hsp70s derives from studies of *E. coli* DnaK, it is

(A)

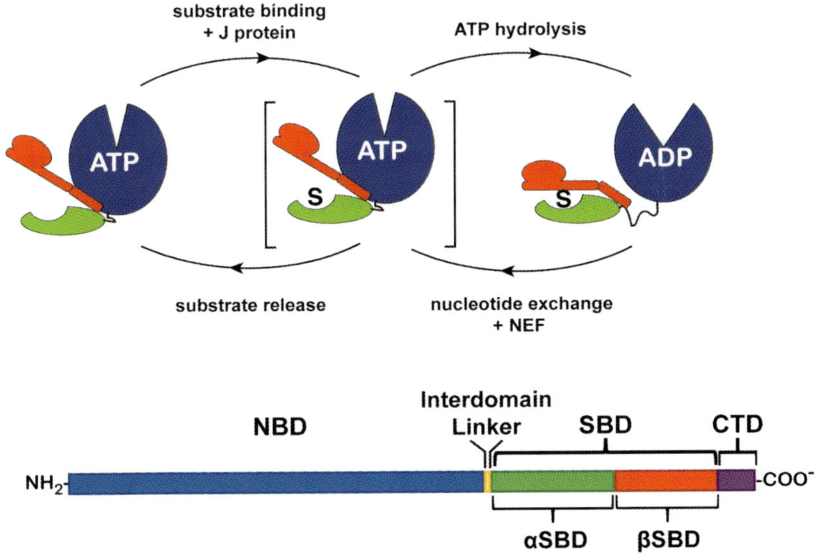

Fig. 1.1 **A**. Schematic of the allosteric mechanism of Hsp70 molecular chaperones, showing steps that are facilitated by J domain and nucleotide exchange factor (NEF) cochaperones. **B**. Sequence organization of Hsp70 molecular chaperones. [This figure was prepared by Eugenia Clerico.]

clear from many studies carried out in a number of laboratories that the same fundamental mechanism underlies the action of all Hsp70s (Clerico *et al.*, 2015; Mayer, 2013; Zuiderweg *et al.*, 2013). The fascinating structural basis for the allosteric mechanism of Hsp70s has only been elucidated quite recently, due to active research in a number of laboratories and a resulting array of structures determined by X-ray crystallography and NMR (Table 1.1). We now appreciate that the ability of Hsp70 chaperones to respond to ligands by a profound structural rearrangement is a result of the intrinsic properties of their constituent domains and is built into these molecular machines by their modular architecture, viz. how these domains are linked to one another. Moreover, this architecture and the resulting energy landscape enable Hsp70 actions to be modulated by key interactions with partner cochaperones.

Modular Structural Organization of Hsp70 Molecular Chaperones

The overall structural organization of Hsp70 molecular chaperones is highly conserved, consistent with the common mechanism of action that is at the heart of the multitude of Hsp70 physiological roles: Each family member consists of an N-terminal, 40 kDa nucleotide-binding domain (NBD) connected to a C-terminal approximately 30 kDa substrate-binding domain (SBD) (Fig. 1.1B). As implied by their names, the ligand-binding abilities of Hsp70 chaperones segregate to these domains. Based on its structure, the SBD can be further divided into the βSBD, the first 115 or so residues (390–502; unless otherwise noted, numbering will correspond to the *E. coli* DnaK sequence), which adopt predominantly β-structure, and αSBD, a helical region of about 100 amino acids (503–607) (see detailed discussion below). The functional collaboration between the NBD and the SBD in Hsp70 chaperones is made possible by an interdomain linker (approximately 12 residues long in prokaryotes and 16 residues long in eukaryotes). The crucial role of the linker in mediating interdomain communication is signaled

Table 1.1. Representative Atomic Resolution Structures of Hsp70 Chaperones*

Hsp70	Species	PDB ID	Method/ Resolution	Comments	Reference
Nucleotide-Binding Domain Structures					
Hsc70 (aka HspA8)	Bovine	3hsc	X-ray/1.93 Å	ADP-bound 1–386 Hsc70 NBD fragment; many structures of mutant and other nucleotide-bound states reported by this laboratory	Flaherty *et al.*, 1990
HspA1	Human	1s3x	X-ray/1.84 Å	ADP-bound 1–382 NBD fragment	Sriram *et al.*, 1997
HspA2	Human	3i33	X-ray/1.3 Å	ADP-bound 6–386 NBD fragment	Wisniewska *et al.*, 2010
HspA1L (HspA1-like)	Human	3gdq	X-ray/1.8 Å	ADP-bound 1–386 NBD fragment	Wisniewska *et al.*, 2010
HspA5 (aka BiP, Grp78)	Human	3iuc	X-ray/2.4 Å	ADP-bound 26–410 NBD fragment	Wisniewska *et al.*, 2010
HspA6	Human	3fe1	X-ray/2.2 Å	ADP-bound 6–385 NBD fragment	Wisniewska *et al.*, 2010
Mortalin (aka HspA9)	Human	4kbo	X-ray/2.8 Å	apo 47–431 NBD fragment	Amick *et al.*, 2014
Substrate-Binding Domain Structures					
DnaK	*E. coli*	1dkx (1dky, 1dkz)	X-ray/2.0 Å	389–607 fragment; peptide-bound; peptide ligand is NRLLLTG (NR)	Zhu *et al.*, 1996

(Continued)

Table 1.1. (*Continued*)

Hsp70	Species	PDB ID	Method/ Resolution	Comments	Reference
DnaK	*E. coli*	1bpr, 2bpr	NMR	386–561 fragment; C-terminal sequence intramolecularly bound to the substrate-binding site	Wang *et al.*, 1998
DnaK	*E. coli*	1dg4	NMR	393–507 fragment; βSBD with no helical lid; highly dynamic	Pellecchia *et al.*, 2000
DnaK	*E. coli*	1q5l	NMR	393–507 fragment; NR-bound; no helical lid; dynamic	Stevens *et al.*, 2003
DnaK	*E. coli*	4ezq, 4ezw, and many related	X-ray/ 2.0, 1.8 Å	389–607 fragment; many related structures reported, all bound to model peptides and small proline-rich antimicrobials; both forward and reverse binding modes (e.g., 4ezq) and alternative registers for NR (e.g., 4ezw) observed	Zahn *et al.*, 2013
Hsc70	Rat	7hsc, 1ckr	NMR	383–540 fragment; C-terminal sequence is bound intramolecularly in the substrate-binding site	Morshauser *et al.*, 1999
HspA1	Human	4po2	X-ray/2.0 Å	386–616 fragment; NR-bound	Zhang *et al.*, 2014
HspA1	Human	2lmg	NMR	537–610 fragment; corresponding to the helical bundle at the end of the helical lid	Gao *et al.*, 2012
HscA	*E. coli*	1u00	X-ray/1.95 Å	389–616 fragment; Fe,S center assembly dedicated Hsp70; bound to peptide ELPPVKIHC in reverse orientation	Cupp-Vickery *et al.*, 2004

BiP	Human	5e85	X-ray/2.57 Å	418-637 fragment with peptide substrate covalently attached at the C-terminus and bound to the SBD	Yang J *et al.*, 2015
Structures of Two-Domain Constructs					
DnaK	*E. coli*	2kho	NMR with modeling based on X-ray	1–605 fragment; bound to ADP and NR	Bertelsen *et al.*, 2009
DnaK	*E. coli*	4b9q	X-ray/2.4 Å	1–606 fragment; disulfide cross-linked between positions 47 and 529; T199A; bound to ATP	Kityk *et al.*, 2012
DnaK	*E. coli*	4jne, 4jn4	X-ray/1.96 Å	2–610 fragment; loop 3,4 shortened; T199A; bound to ATP	Qi *et al.*, 2013
DnaK	*Geobacillus kaustophilus*	2v7y	X-ray/2.37 Å	1–504 fragment; bound to ADP, Mg, and Pi; domains separate as in the ADP state; interdomain linker binds an adjacent molecule in the crystal	Chang *et al.*, 2008
Hsc70	Bovine	1yuw	X-ray/2.6 Å	1–554 fragment; no nucleotide; C-terminal segment intramolecularly bound to the substrate-binding site	Jiang *et al.*, 2005
BiP	Human	5e84	X-ray/2-99 Å	1-606 fragment with several mutations, bound to ATP	Yang J *et al.*, 2015

*The first structure of a particular Hsp70 domain is listed in the table. Frequently, additional structures have been reported; these are not included unless they added substantially new structural information.

by several sequence features highly conserved throughout the Hsp70 family, most notably a stretch of four hydrophobic residues, 389–392. A largely unstructured and overall poorly conserved C-terminal domain (CTD) is present in all Hsp70s; this segment is approximately 30 residues long but can vary in length.

How Does the Modular Structure of Hsp70 Molecular Chaperones Enable Ligand-Mediated Allostery?

The Nucleotide-Binding Domain (NBD)

The X-ray crystal structure of the NBD of Hsc70, the constitutive cytoplasmic Hsp70, was solved soon after the discovery of this chaperone family and has served as a paradigm for the structural features of the Hsp70 NBD. Inspection of the structure immediately pointed to a coarse resemblance to hexokinase, and a stronger homology to actin was soon recognized (Bork *et al.*, 1992; McKay, 1993). These similarities suggested that the Hsp70 NBD had an intrinsic capacity for allosteric conformational change triggered by ligand binding. Like the actin fold, the Hsc70 NBD has a bilobed structure (lobes termed I and II) with two α/β subdomains per lobe (designated IA, IB, IIA, and IIB); the nucleotide-binding site is located deep between the lobes (Fig. 1.2A) (Flaherty *et al.*, 1991). Beneath the nucleotide-binding site are the so-called crossing helices, a canonical feature of the actin fold. The scissors-like crossing helices offer the possibility of gating a hydrophobic binding site by dynamically switching the interhelical angle. In actin, this site is exploited by actin-binding proteins and regulated by allosteric ligands including nucleotide (Dominguez, 2004).

Interestingly, many structures of the Hsc70 NBD were solved for a number of mutant versions and various nucleotide-bound states without seeing any significant conformational change, even between ATP- and ADP-bound states (McKay, 1993). The first hint of actin-like subdomain rearrangements was provided by the structure of DnaK

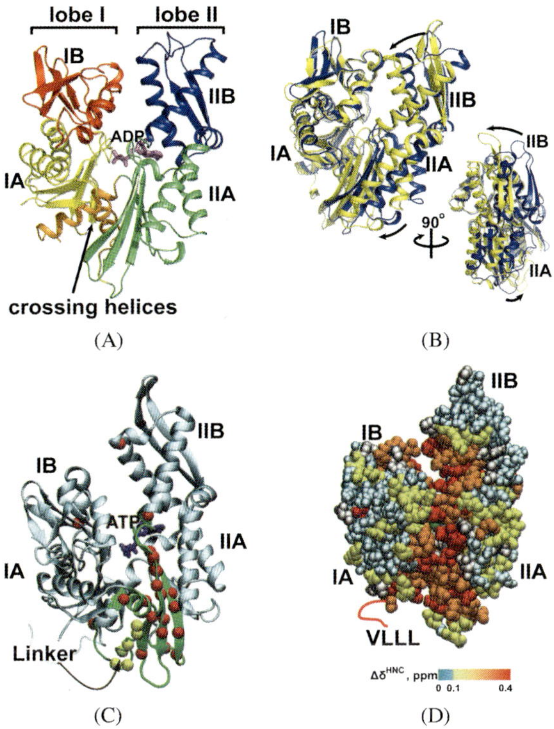

Fig. 1.2 **A**. Subdomain organization of the nucleotide-binding domain of Hsp70 molecular chaperones. The structure shown is ADP-bound Hsc70 (PDB 3hsc) (Flaherty *et al.*, 1990). **B**. Subdomain rearrangements in the Hsp70 NBD between ADP- and ATP-bound states. In blue is the structure of ADP-bound Hsc70 NBD (PDB 3hsc) (Flaherty *et al.*, 1990), and in yellow is the structure of ATP-bound DnaK taken from one of the two-domain crystal structures (PDB 4b9q) (Kityk *et al.*, 2012). **C**. Mapping of NMR chemical shift changes observed between ADP- and ATP-bound DnaK NBD, showing that the largest chemical shift differences (in red spheres) are located on the central β-sheet in subdomain IIA and along one of the crossing helices, features that constitute an allosteric pathway from the nucleotide-binding site to the binding site for the interdomain linker (Zhuravleva and Gierasch, 2011). **D**. The largest NMR chemical shift changes among a number of different ligand-bound states and constructs of the DnaK NBD ($\Delta\delta^{HNC}$) reveal the interfaces between the structural units of this domain, both the lobe interface and the interfaces between subdomains (Zhuravleva and Gierasch, 2011). [Panels a and b were prepared by Wenli Meng. Panels C and D are reproduced from: Zhuravleva and Gierasch, 2011.]

bound to a portion of the NEF GrpE, which showed one subdomain (IIB) twisted with respect to its position in Hsc70 (Harrison *et al.*, 1997). But it required a series of structural breakthroughs in recent years from both NMR and crystallography for the ATP-induced allosteric conformational changes within the NBD, as well as the mode of coupling to the SBD, to be revealed. One breakthrough was provided by an elegant NMR study of the Hsp70 NBD from *T. thermophilus* comparing ADP- and AMPPNP-bound states, which showed using a residual dipolar coupling analysis that subdomain rotations occurred between these two states (Bhattacharya *et al.*, 2009). The solution of crystal structures of an Hsp70 homologue, Hsp110, in the ATP-bound state (Liu and Hendrickson, 2007), and later two different ATP-bound DnaK forms (Kityk *et al.*, 2012; Liu and Hendrickson, 2007; Qi *et al.*, 2013), confirmed the model that had emerged from the NMR analyses (Bhattacharya *et al.*, 2009), and provided an opportunity for detailed inspection of ADP- and ATP-bound NBDs. Note that because no structure of the ADP-bound DnaK NBD has been reported, the ADP-bound structure of the DnaK NBD is generally modeled based either on that of Hsc70 (Flaherty *et al.*, 1994) or on the structure of the GrpE complex (Harrison *et al.*, 1997). Comparing structures of the NBD bound to ADP (PDB code 3hsc) and ATP (PDB code 4b9q) shows changes in the twist of the two lobes with respect to each other, and both altered angles of the subdomains and altered rotations with respect to each other (Fig. 1.2B). The concerted intradomain conformational change and subdomain rearrangement gates the hydrophobic binding site under the crossing helices and alters the presentation of residues on potential surfaces of interaction. Strikingly, a calculation of normal modes of the NBD fold using the web-based server ElNémo (Suhre and Sanejouand, 2004) revealed that the movements of subdomains between the ADP- and ATP-bound states observed in crystal structures are an intrinsic low frequency motion of the NBD (Zhuravleva and Gierasch, 2011). Similarly, an NMR analysis of 12 different ligand-bound states of the NBD yielded patterns of chemical shift perturbations (Fig. 1.2C) that showed how the rigid-body rearrangements yield a tunable, modular response unit at the heart of the Hsp70 molecular machine (Zhuravleva and Gierasch, 2011).

Both from this NMR analysis of different ligand-bound states (Zhuravleva and Gierasch, 2011) and from computational analysis of the DnaK NBD (General *et al.*, 2014), subdomain IIA emerges as a key mediator of the allosteric signal from ATP binding to the NBD. As can be seen in Fig. 1.2D, the top of the β-sheet in subdomain IIA is proximal to the γ-phosphate of bound ATP. Chemical shift changes support a downward movement of this entire small sheet in the ATP-bound state relative to the ADP-bound state. Moreover, this shift in subdomain IIA leads to opening of the angle between the crossing helices and thus the hydrophobic binding site beneath them, where communication with the interdomain linker and the SBD can occur. Direct evidence for the gating of this hydrophobic pocket was provided by comparative analysis of two DnaK NBD constructs, one ending at residue 388, before the hydrophobic tetrapeptide sequence from the interdomain linker, and the other ending at residue 392, thus including the hydrophobic VLLL sequence. The ^{388}VLLL^{392} sequence in the 1–392 NBD construct shows ATP-dependent binding to the hydrophobic binding site (Swain *et al.*, 2007; Zhuravleva *et al.*, 2012; Zhuravleva and Gierasch, 2011). Along with the binding of the interdomain linker sequence to the DnaK NBD comes activation of its ATPase activity by approximately 13-fold, comparable to the substrate-activated full-length DnaK molecule (Swain *et al.*, 2007; Vogel *et al.*, 2006). Thus, ATP-induced subdomain movements in the actin fold NBD of Hsp70s act as the initiator of transmission of an allosteric signal from the NBD, causing enhanced ATPase activity and binding of the interdomain linker. Key to interdomain allosteric communication in Hsp70 nanomachines is the proximity of the gated hydrophobic binding site under the crossing helices to the interdomain linker and therefore to the SBD.

The most conserved residues in the Hsp70 family are those that interact directly with bound nucleotides. Strikingly, nucleotide-binding residues are situated on all four subdomains of the NBD, thus providing a structural mechanism for the coordinated conformational responses of the NBD to nucleotide binding. In addition, both Mg^{2+} and K^{+} are integral ligands for Hsp70 action. Comparing structures of ADP- and ATP-bound NBDs sheds light on the roles of residues in direct contact with the bound nucleotide and suggests possible catalytic mechanisms. The subdomain rotations accompanying binding to ATP versus ADP position the catalytically essential Lys70 such that its ε-amino group hydrogen bonds with the γ-phosphate of ATP and is held by interactions with another essential residue, Glu171 (Fig. 1.3). The peptide segment 197–199 also coordinates the γ-phosphate of ATP, but via backbone amides. These structural details lead to a plausible mechanism for ATP hydrolysis invoking Glu171 as the general base catalyst, coordinating the hydrolytic water, together with Thr199, and accepting the released proton (Kityk *et al.*, 2012; Qi *et al.*, 2013), in contrast to an alternative mechanism of catalysis that had been proposed, in which Lys70 coordinated the attacking water (Flaherty *et al.*, 1994).

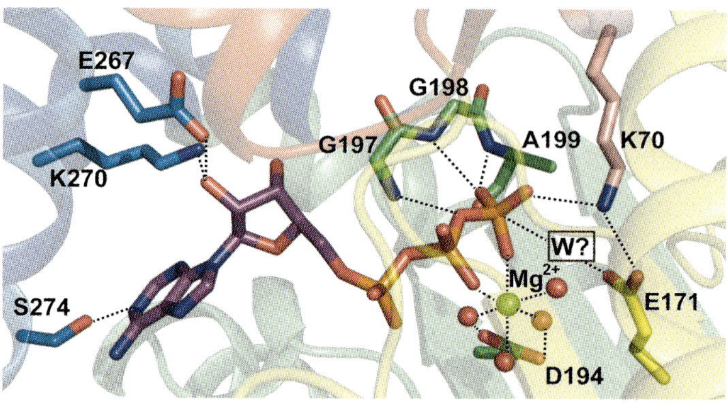

Fig. 1.3 Blowup of the ATP-binding site of DnaK as seen in the structure of a two-domain construct (PDB 4b9q) (Kityk *et al.*, 2012). Residues interacting with ATP and potentially involved in the catalytic mechanism are shown in sticks, with the color matching the subdomain to which they belong (subdomain coloring as in Fig. 1.2A). Crystallographically identified waters are shown as red spheres. As discussed in the text, the structure of ATP-bound DnaK suggests that E171 may bind to a catalytic water (not seen, but shown here schematically by "W?") which in turn is aligned to attack the γ-phosphate in the hydrolysis step (Kityk *et al.*, 2012). [This figure was prepared by Wenli Meng.]

The Substrate-Binding Domain (SBD)

As first seen in an X-ray crystal structure of the *E. coli* DnaK SBD bound to a model peptide (Zhu *et al.*, 1996), NRLLLTG, the SBD adopts a novel fold with two structural subdomains (Fig. 1.4A): βSBD, a compact β-sandwich domain that is made up of two sheets of four strands each arranged antiparallel to one another with loops between strands 1 and 2, strands 3 and 4, strands 4 and 5, strands 5 and 6, and strands 7 and 8 (called L1,2; L3,4; L4,5; L5,6; L7,8). L1,2 and L3,4 form a peptide-binding groove; and αSBD, a helical lid domain that covers the substrate-binding groove and forms a helical bundle at its C-terminal. The helical lid domain can be divided into three regions: helix A (from aa 507 to 517) and helix B (from aa 518 to 558) are connected by a 72° kink, and a three-helix bundle (comprising helices C, D, and E) from 559 to 603. Helix B is linked to the βSBD subdomain by an intricate pattern of salt bridges to the interstrand loops (Fig. 1.4B). The construct crystallized lacked the last 28 amino acids to improve crystal quality; this region is now known to be highly disordered.

The model substrate peptide is oriented perpendicular to the β-strands of the βSBD, with the side chain of the second Leu residue of NRLLLTG pointing deeply into a pocket, which is called the 0th position of the binding site (Fig. 1.4C). The bound peptide is extended, and seven hydrogen bonds are seen between the substrate backbone and residues from DnaK. Other interactions of the substrate peptide with DnaK are largely hydrophobic. The majority of contacts occur over a five-residue stretch of the peptide, which resides in sites −2 to +2 of the binding cleft. Met 404 from L1,2 and Ala 429 from L3,4 create an arch over the peptide-binding site (Fig. 1.4A).

It was speculated based on this structure of the SBD, and on the observation of an alternative form of the domain in the crystals, that the helical lid is dynamic and acts as a gate to allow substrate peptide entry and exit (Zhu *et al.*, 1996). Indeed, several experiments have demonstrated that the lid is mobile (Mapa *et al.*, 2010; Marcinowski *et al.*, 2011;

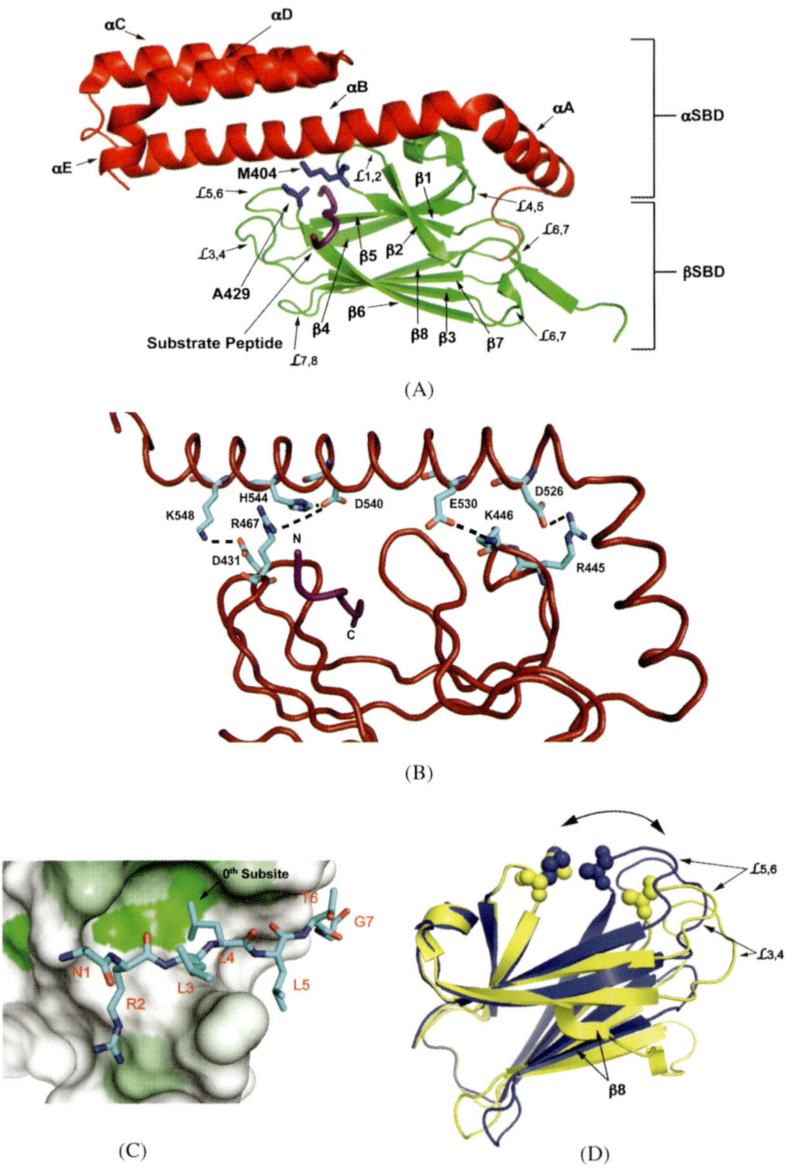

Fig. 1.4 **A**. Structure of the *E. coli* DnaK substrate-binding domain (SBD) bound to the peptide NRLLLTG (PDB 1dkz) (Zhu *et al.*, 1996). **B**. Detailed view from the same structure, showing the five salt bridges between the helical lid of the αSBD subdomain and the loops from the βSBD that surround the substrate-binding cleft. **C**. View from the top of the βSBD of the same structure, showing the hydrophobic pocket at the 0th position of the substrate-binding site interacting with the second Leu of the bound

Fig. 1.4 (*Continued*) NRLLLTG peptide. Hydrophobicity is scaled by the intensity of the green surface coloring. **D.** Overlay of the SBD structures from high affinity, ADP-bound DnaK (PDB 2kho) (Bertelsen *et al.*, 2009), in blue, and low affinity, ATP-bound DnaK (PDB 4b9q) (Kityk *et al.*, 2012). The arch residues, Met 404 and Ala 429, are shown in spheres to indicate the extent of opening of the substrate-binding cleft upon ATP binding. Note also that ATP binding leads to the detachment of strand $\beta 8$ and strand $\beta 7$. [Panels a, b, and d were prepared by Joseph Tilitsky. Panel c is reproduced with permission from: Clerico *et al.*, 2015.]

Schlecht *et al.*, 2011). Truncations of the helical lid lead to constructs that retain substrate-binding function but show faster association/dissociation rates for peptide substrates, as expected if the lid acts to trap the substrate (Buczynski *et al.*, 2001).

The isolated SBD displays increased dynamics, lower stability, and a higher aggregation propensity in the absence of a bound substrate. No significant conformational change was observed when substrate was removed from an isolated SBD construct that included the helical lid (Swain *et al.*, 2006). However, if devoid of the helical lid, the βSBD was found to be metastable and highly dynamic with the substrate-binding site partially collapsed, ablating the key hydrophobic pocket (Pellecchia *et al.*, 2000). This construct provided hints of the allosteric transition in the SBD upon conversion to the low affinity state, but detailed structural characterization of the alternative, low affinity allosteric conformation of the SBD requires the presence of the NBD.

The recent crystal structures of ATP-bound DnaK (Kityk *et al.*, 2012; Qi *et al.*, 2013) and its homologue Hsp110 (Liu and Hendrickson, 2007) have now provided insight into the mechanism by which the substrate affinity of the SBD is reduced relative to the high affinity, ADP-bound state (see details below). First, the helical lid detaches from the βSBD. Second, the substrate-binding groove opens and flattens as L1,2 and L3,4 separate. There are other rearrangements of the β-strands, in particular disruption of strand $\beta 8$ contacts with strand $\beta 7$, and the hydrophobic pocket at site 0 closes down (Fig. 1.4D). Characterization of subtle conformational changes throughout the βSBD, and application of a computational algorithm that identifies subdomain movements within a structure, showed that the observed changes within the βSBD could be grouped into scissoring motions around three hinge regions: residues 417–420, 438–440, and 454–457 (Zhuravleva and Gierasch, 2015). Strikingly, all three hinges regions have important roles in allostery, as evidenced by defects introduced upon their mutation, or structural connections they participate in. Again, as for the NBD, elNémo analysis (Suhre and Sanejouand, 2004) reveals that these subdomain movements in the βSBD are intrinsic motions. It is therefore plausible to posit that the βSBD samples (at least) two alternative conformations on its energy landscape and that its interaction with substrate or with other structural units (helical lid, NBD, linker) tips it toward one or the other of the alternative states.

The C-Terminal Domain (CTD)

We define the CTD as the segment of the chain from the helical bundle at the end of the αSBD to the C-terminal of the chaperone. CTDs of Hsp70s from both prokaryotes and eukaryotes display features of intrinsically disordered sequence in the first 15–25 amino acids, with few charges and many Gly, Ser, Ala, and Gln residues. A repeat motif of GGMP or GGLP occurs in bovine Hsc70 and some other eukaryotic Hsp70s (McKay, 1993). Toward the C-terminal, the sequences from prokaryotes and eukaryotes diverge. Prokaryotes have a short region that has a conserved motif from residues 624 to 638 (DnaK numbering) (Smock *et al.*, 2011). Mutations or truncations of this sequence in *E. coli* DnaK lead to phenotypic changes *in vivo*, viz. a severe growth defect in SecB$^-$ cells, and functional defects *in vitro*, viz. impaired refolding of luciferase. It was postulated that this region directly interacts with protein substrates as a secondary binding site (Smock *et al.*, 2011) — a role that had been proposed previously (Hu and Wang, 1996; Tsai and Wang, 1994). Eukaryotic Hsp70s harbor an EEVD sequence at their extreme C-terminal that is a specific TPR (tetratricopeptide repeat) motif recognition site utilized for cochaperone binding (see below) (Freeman *et al.*, 1995), and the ER Hsp70 has a KDEL (or HDEL in yeast) retention signal (Munro and Pelham, 1987).

Structural Insights into the Allosteric Mechanism of Hsp70s

The ADP-Bound State

In the ADP-bound Hsp70 structure as exemplified by *E. coli* DnaK, the NBD and SBD take up conformations that are essentially identical to those they adopt as isolated domains, and the interdomain linker is flexible, tethering the NBD and SBD to each other in a relatively unconstrained mutual arrangement (Bertelsen *et al.*, 2009). Apo-DnaK behaves similarly. This "beads on a string" structure is supported by NMR spectra showing overlap of ADP-bound DnaK resonances with those of the separate NBD and SBD (Swain *et al.*, 2007). The dynamics of this state, however, preclude its crystallization. The deepest structural insights have derived from knowledge of the isolated domain structures, and an NMR study using residual dipolar coupling and paramagnetic relaxation enhancements from spin labels at sites on the SBD and NBD, which led to a model for ADP-bound DnaK (PDB 2kho) in which the two domains do not interact and rotate relatively freely about the linker — restricted only to a 32° cone (Bertelsen *et al.*, 2009) (Fig. 1.5A). In terms of the allosteric mechanism, the ADP-bound state possesses biochemical properties that correspond to the sum of the two separate domains. In ADP-bound DnaK, the helical lid subdomain adheres strongly to the SBD via a number of salt bridges, as it does in the isolated domain, which is consistent with slow on- and off-kinetics for substrate binding. Indeed, the substrate affinity of the SBD is high in ADP-bound DnaK and is essentially the same as the separate domain. Similarly, the NBD in this construct adopts the same structure as the ADP-bound isolated domain of Hsc70, and has low ATPase activity. [The only structure available to date for the ADP-bound DnaK NBD

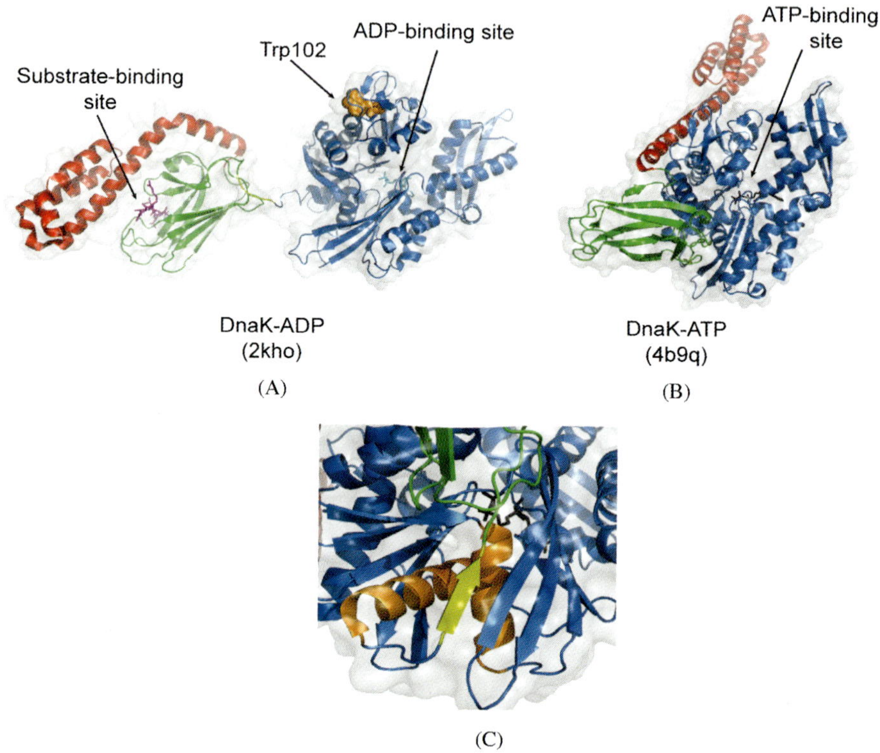

Fig. 1.5 **A**. Structure of full-length DnaK bound to ADP (PDB 2kho) (Bertelsen *et al*., 2009). Trp102, which becomes sequestered from solvent upon ATP binding to the chaperone, is shown in orange. **B**. Structure of full-length DnaK bound to ATP (PDB 4b9q) (Kityk *et al*., 2012). **C**. Blowup showing the interaction of the interdomain linker with the NBD in the ATP-bound state of DnaK. The conserved hydrophobic VLLL sequence is in yellow. Note the formation of a strand–strand interaction with subdomain IIA. The crossing helices are shown in orange. [This figure was prepared by Eugenia Clerico.]

is for the complex with GrpE (PDB 1dkg), and the cochaperone interaction leads to intradomain conformational changes (noted above, and described in greater detail below) (Harrison *et al*., 1997).] The interdomain linker is solvent-exposed in ADP-bound DnaK, consistent with its susceptibility to cleavage by a variety of proteases (Buchberger *et al*., 1995; Kamath-Loeb *et al*., 1995).

The ATP-Bound State

ATP binding to DnaK leads to a dramatic structural rearrangement including structural changes within each domain, intimate docking of the NBD and SBD, and sequestration of the interdomain linker. The interdomain linker, which is flexible, solvent-exposed, and proteolytically labile in the ADP-bound or apo-form, becomes sequestered and protease-resistant when ATP binds, and simultaneously the SBD becomes more dynamic (Rist *et al.*, 2006; Swain *et al.*, 2007) and proteolytically labile (Buchberger *et al.*, 1995; Kamath-Loeb *et al.*, 1995). A large packing interface is created between the NBD and the SBD, as revealed by spectral changes in a diagnostic tryptophan (Trp102) in DnaK (Fig. 1.5A), which resides near a solvent-exposed surface of its NBD and displays a fluorescence blueshift and quenching upon ATP binding (Banecki *et al.*, 1992; Montgomery *et al.*, 1999; Palleros *et al.*, 1992). These changes in DnaK show that Trp102 is buried away from solvent in the ATP-bound form of the chaperone. Importantly, the two domains of DnaK, which act like flexibly tethered "balls on a string" in either the ADP or no-nucleotide states, tumble as one globular entity in the ATP-bound state (Swain *et al.*, 2007; Zuiderweg *et al.*, 2013).

Structural descriptions of the ATP-bound state were elusive for many years, presumably because of the dynamic nature of this state. Early hopes that the first crystal structure reported for a two-domain Hsp70 construct, in this case Hsc70 (Jiang *et al.*, 2005), would provide insight into how interdomain allosteric communication worked were disappointed, because this structure had several features that raised questions about its relevance to the allosteric state in the presence of ATP. For example, SBD residues 539–544 of the C-terminal region of this truncated construct were bound intramolecularly to the substrate-binding cleft, and the interdomain linker, which had been shown to be proteolytically labile in ADP-bound or apo-Hsp70s and sequestered from protease in ATP-bound forms, was fully solvent-exposed. Moreover, there was in fact no indication that ATP was bound to the

NBD in these Hsc70 crystals. Fortunately, in the last decade, three crystal structures have broken the structural logjam and provided images of the ATP-bound states of Hsp70s. The resulting structures are consistent with other biochemical data and shed light on the allosteric mechanism of this family of molecular chaperones. The first structure to provide a glimpse of the ATP-bound state of an Hsp70 was that of a homologue, Hsp110 from *S. cerevisiae*, also known as Sse1 (Liu and Hendrickson, 2007) (PDB 2qxl). Hsp110s are clearly recognizable as being related to Hsp70s, although they have sequence insertions in their SBDs and at their C-terminals (Easton *et al.*, 2000). Hsp110 NBDs are highly homologous to those of Hsp70s, although they lack the capacity to hydrolyze ATP and, in keeping with this biochemical difference, they have substitutions in key nucleotide-interacting residues. The Hsp110 family SBD is more divergent from that of the Hsp70 family, consistent with their altered substrate binding properties (Xu *et al.*, 2012). Hsp110s have been shown to act as nucleotide exchange factors for Hsp70s (Andreasson *et al.*, 2008a; Andreasson *et al.*, 2008b; Andreasson *et al.*, 2010), some chaperone activity has been reported (Mattoo *et al.*, 2013), and more recently they have been ascribed activity as disaggregation partners working with Hsp70s (Rampelt *et al.*, 2012). Because of the high overall homology between these two chaperones, the structure elucidated for ATP-bound Sse1 yielded a plausible model for the ATP-bound state of an Hsp70, DnaK (Liu and Hendrickson, 2007). The NBD is highly structurally homologous between Hsp70 and Hsp110, and despite the sequence divergence the SBD secondary structural units could be overlaid in a plausible manner on the Hsp110 structure. As supported by previous biochemical data, the monomeric structure of ATP-bound Sse1 had the SBD forming a large interaction interface with the NBD, with the interdomain linker forming an integral part of the packing interface. The highly conserved hydrophobic sequence within the linker — VLLL in DnaK, but FKFE in Sse1 — formed a β-strand with strand 14 of the NBD in a pocket that is created under the crossing helices. The helical lid subdomain of the SBD was detached from the SBD and swiveled to wrap around the NBD. The SBD itself underwent significant conformational change relative to the peptide-bound domain structure that had been reported earlier, with the substrate-binding groove much shallower and L1,2 and L3,4, which had enclosed the peptide substrate, now rearranged to a more open structure. The Hsp110 chaperone crystallized as a dimer stabilized by contacts predominantly formed between the NBDs, and biochemical evidence suggested that it is indeed dimeric in solution.

Recently, structures have been reported for ATP-bound DnaK (Kityk *et al.*, 2012; Qi *et al.*, 2013) and for ATP-bound BiP (Yang *et al.*, 2015), and the results have shown that the use of the Hsp110 structure to model ATP-bound DnaK was valid (Fig. 1.5B). Interestingly, all of the recent structures required some special strategies to obtain diffraction quality crystals, most likely because of dynamics intrinsic to the ATP-bound state of Hsp70s. In one approach, two residues were mutated to Cys, one in the helical lid and one in the NBD, such that a disulfide bond formed between them and immobilized the otherwise dynamic helical lid (Kityk *et al.*, 2012). In the other Dnak structure solution, mutations were introduced into L3,4 of the SBD both to minimize aggregation (L3,4 was shortened by four residues from TAEDNQS to MGG, mimicking the loop of Sse1 (Xu *et al.*, 2012)) and to favor the ATP-bound state (Qi *et al.*, 2013). In the structure of ATP-bound Bip, several mutations were incorporated to favor crystallization (Yang *et al.*, 2015).

Concerns about any artifacts that might be introduced by the strategies used to favor crystallization of ATP-bound DnaK are allayed by the finding that the modified SBD adopts essentially an identical structure to the unmodified parent (Qi *et al.*, 2013), and the two new structures of ATP-bound DnaK are fully compatible with each other (Kityk *et al.*, 2012; Qi *et al.*, 2013). Both confirm the intimate docking of the NBD and the SBD onto each other, with the α-helical lid that covers the peptide-binding groove in the isolated domain and in the ADP/apo state lifting off by breakage of a number of salt bridges to the βSBD and forming new interactions with the NBD. As described above, the NBD subdomains experience rotations, tilting, and shearing relative to their positions in ADP-bound Hsc70, the paradigm for this nucleotide-bound state, which causes shifting of the crossing helices and opening-up of a hydrophobic pocket beneath these helices, into which is inserted the hydrophobic segment of the interdomain linker. As anticipated from work with the isolated NBD with linker attached, the highly conserved hydrophobic tetrapeptide VLLL sequence of the linker forms an additional strand of the central β-sheet within subdomain IIA of the NBD (Fig. 1.5C).

The helical lid lifts and rotates from its position in ADP-bound DnaK, and helices A and B straighten as the lid binds onto subdomains 1A and IB of the NBD. Neither of the crystal structures of ATP-bound DnaK, nor that of Hsp110, has the C-terminus, unstructured, approximately. 30-amino-acid segment present, as the constructs behave more tractably with this region truncated. Nonetheless, it is noteworthy that the attachment point of this region at the end of the helical lid three-helix bundle would move substantially upon the shift from ADP binding to ATP binding.

The intradomain conformational changes in the βSBD create complementary surfaces for docking with the NBD in ATP-bound DnaK (Zhuravleva and Gierasch, 2015) (Fig. 1.6A). Specifically, the conformational change in the βSBD involves repacking of a hydrophobic cluster near the interface with the NBD, and this repacking positions key residues for steric matching with three partner NBD residues: Lys414 and Asn415 from the βSBD form interactions with Asp326 from subdomain IIA of the NBD, Gln442 from the βSBD forms interactions with Asp148 from subdomain IA of the NBD, and Asp481 from the βSBD forms interactions with Ile168 from subdomain IA of the NBD. In the high affinity structure of the βSBD, the geometry of these residues precludes favorable interaction with the complementary NBD surface. Hence, dynamic sampling of alternative conformations with remodeled interfaces underlies the interdomain coupling of DnaK allostery.

It is striking to compare the changes in interdomain interacting interfaces between ADP- and ATP-bound DnaK (Fig. 1.6B). In the ADP-bound chaperone, the major interdomain contact surface is between the helical lid subdomain of the SBD and the βSBD, with approximately

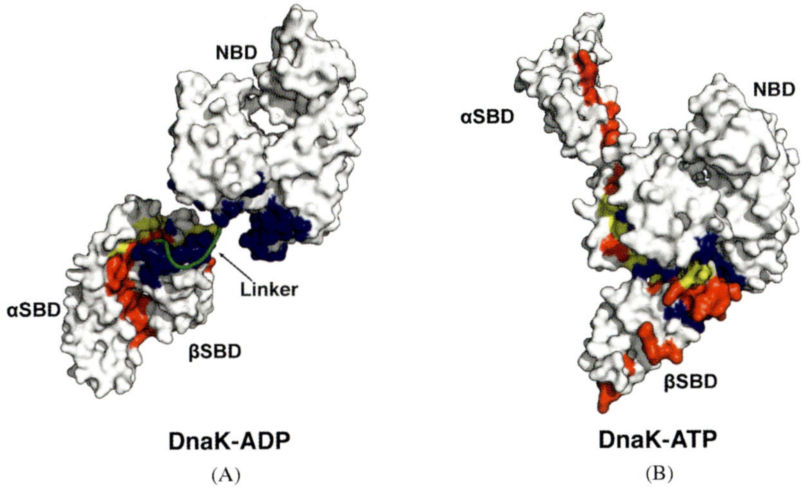

Fig. 1.6 Interaction interfaces formed between the αSBD, βSBD, and NBD in the ADP- (**A**) and ATP- (**B**) bound states of DnaK. In red are shown residues that belong to the interaction surface between the αSBD and the βSBD; in blue, residues that belong to the interface between the NBD and either the βSBD or the αSBD. In yellow are shown contact surfaces that participate in both interfaces. The interdomain linker is shown schematically in green in the ADP-bound state. Note the orthogonality of the contact surfaces: the same residue interactions cannot be formed in the two states, underlying the energetic competition in the ligand-mediated allostery of Hsp70s. [This figure is reproduced from: Zhuravleva *et al.*, 2012.]

1620 Å^2 of buried surface. The linker, which is flexible in ADP-bound DnaK, has relatively small calculated interaction surface areas with the NBD and the SBD in this state (336 Å^2 and 828 Å^2, respectively), but it is important to note that the linker conformation was modeled in the ADP-bound NMR structure, and so quantitative conclusions about its interactions are not warranted. By contrast, in ATP-bound DnaK, the surface area of interaction between the NBD and the SBD is large — 2909 Å^2, including 1739 Å^2 between the NBD and the βSBD, and 1170 Å^2 between the helical lid and the NBD. The linker interacts to a substantial extent with both the NBD and the SBD, with a surface area of interaction of 1208 Å^2 between the linker and the NBD, and 288 Å^2 between the linker and the SBD. Peptide binding to the SBD buries about 470 Å^2. These surfaces of interaction represent energetic components of the allosteric conformational change. By themselves, they cannot be used straightforwardly to estimate the free energy changes in DnaK's conformational ensemble upon ATP binding, because both solvation and dynamics also change and contribute to both enthalpic and entropic terms. Nonetheless, it is valid to draw the conclusion that binding of ATP shifts the conformation of the NBD so that surfaces favorable for docking of the linker and the SBD are presented, and the formation of new buried surfaces of interaction "pays" for the conformational change and detachment of the helical lid in the SBD. To make this coupled set of rearrangements feasible relies on the fact that the helical lid is dynamically docked onto the βSBD, and the βSBD has an intrinsic capacity to sample alternative conformations, one of which presents a surface complementary to the NBD in its ATP-bound conformation, as described above.

The lessons learned from studies of DnaK offer a paradigm of the Hsp70 allosteric mechanism with a central role for conformational sampling and evolved capacities for domain and subdomain interactions. Nonetheless, dynamics are also crucial to the allosteric mechanism of Hsp70s. In particular, the allosteric activity of an Hsp70 requires the presence of both a substrate (which activates the ATPase) and ATP (which modulates substrate affinity). The interfaces described above are associated with a high affinity, ADP- and substrate-bound state and a low affinity, ATP-bound state that lacks substrate. What is the nature of the state that has both ligands present? The interfaces between the helical lid and the βSBD, and between the NBD and both the βSBD and the helical lid, are orthogonal; it is not possible for them to be present simultaneously. Therefore, conformational fluctuations will be a feature of the allosterically active form of DnaK. NMR evidence supported this model, arguing that the allosterically active state was characterized by linker binding to the ATP-bound NBD, but dynamic interaction of the helical lid with the βSBD and a high affinity conformation of the βSBD (Zhuravleva *et al.*, 2012). This model explains how the modular Hsp70 nanomachine has the ability to serve as a ligand-modulated binder of unstructured regions of protein substrates, and to be tuned in its allosteric cycle by interactions with cochaperones. In addition to the awareness of dynamics that is needed to understand DnaK functions, one must be wary of applying all the lessons learned from DnaK to Hsp70s more generally. Further experiments are urgently needed to explore the behavior of other Hsp70s, in particular eukaryotic homologues.

Substrate Interactions with Hsp70s

Our current understanding of the basis of substrate recognition by Hsp70 chaperones is based on experimental data from peptide scanning arrays, libraries, and phage display analyses together with the information offered from inspection of crystal structures of peptide complexes (Clerico *et al.*, 2015). While the binding is clearly very promiscuous, there are adequate biases toward arrangements of hydrophobic and charged residues to enable the development of predictive algorithms for finding potential DnaK binding sites in the sequence of a protein (Fourie *et al.*, 1994; Rudiger *et al.*, 1997; Srinivasan *et al.*, 2012; Van Durme *et al.*, 2009). Roughly stated, the pattern of residues that enables binding to the canonical substrate-binding groove in the SBD is a single large hydrophobic residue with additional hydrophobic neighbors and positive charges in the flanking regions, either N- or C-terminal to the primary hydrophobe. Most proteins in the *E. coli* proteome are predicted to have at least one potential DnaK-binding site (Srinivasan *et al.*, 2012). It is clear from analysis of DnaK interactors in *E. coli*, which number about 700, that additional factors obtain in the specificity of DnaK interaction *in vivo* (Calloni *et al.*, 2012).

Much of the work on recognition and binding of substrates to DnaK and other Hsp70s has been carried out using short model peptides. New efforts are being ramped up to tackle the interaction of Hsp70s with protein clients. Understanding Hsp70–substrate interactions is challenging because of the intrinsic dynamics of the system and has required the use of powerful biophysical techniques. Recent work on protein substrates indicates that the sites of DnaK binding predicted from peptide studies are indeed used in the context of the protein (Lee *et al.*, 2015; Sekhar *et al.*, 2015; Xing *et al.*, 2010). However, the exposure and "unfoldedness" of a potential binding region will also govern its likelihood of interaction with an Hsp70.

Experiments using single-molecule fluorescence (Marcinowski *et al.*, 2011; Mashaghi *et al.*, 2014) and electron spin resonance probes (Schlecht *et al.*, 2011) have shed light on the interaction of Hsp70s with their substrates, most importantly showing that the helical lid domain may remain detached from the βSBD when a protein substrate is bound, which allows the bound substrate to adopt a compact nonnative state with a segment of the extended chain bound to the substrate-binding cleft. Single-molecule FRET studies of the mitochondrial Hsp70 showed that it had a more heterogeneous set of conformations in the ADP-bound state than anticipated, but that a peptide substrate pushed the ensemble toward the fully undocked state (Mapa *et al.*, 2010).

The impact of Hsp70 binding on substrates is a physiologically crucial question, and an area of active research. Many assays show that DnaK can inhibit aggregation and facilitate folding, both activities that could be accomplished by binding to incompletely folded or misfolded states and releasing the substrate in an unfolded state. Many believe that this is the likely mode of action of Hsp70s in their generic housekeeping functions: the "holdase" function, to protect aggregation-prone species, and the "foldase" function, to facilitate folding by undoing unproductive interactions (Mattoo and Goloubinoff, 2014). Nonetheless, few studies provide direct experimental evidence for these models. Two papers that used NMR methods to examine the influence of DnaK binding on the conformational ensembles of model substrate proteins have recently appeared (Lee *et al.*, 2015; Sekhar *et al.*, 2015). While details vary to a minor extent, both studies found that DnaK binding to model substrate proteins (in one case a three-helix bundle from human telomere repeat binding factor 1 and in the other an SH3 domain) was mediated by a single region, that the binding site corresponded to a canonical binding motif, and that the rest of the substrate sampled conformations little changed from the unfolded ensemble in the absence of a chaperone. By contrast, a single-molecule FRET study of DnaK action on a larger substrate (rhodanese) led to a model in which multiple binding sites were used by the chaperone and the substrate expanded significantly due to chaperone action, which is consistent with an "unfoldase" activity (Kellner *et al.*, 2014).

Modulation of Hsp70 Action by Cochaperones

In essence, Hsp70 molecular chaperones are like the engine of a car: they perform the central function, but without the body, transmission, wheels, steering and brakes, they can do nothing. For Hsp70s, these other essential components of the chaperone machine are the partner cochaperones. It is the cochaperones that confer on a given Hsp70 the specialization to perform a particular function. The major bacterial Hsp70 chaperone system, consisting of DnaK, the J-protein DnaJ, and the NEF GrpE, is quite streamlined, and no alternative NEFs of J-proteins partner with DnaK. Note that there is one other major Hsp70 system in prokaryotes with homologues in eukaryotes, which is dedicated to the assembly of Fe,S proteins and has its own cochaperones (Vickery and Cupp-Vickery, 2007).

In contrast to this picture in *E. coli*, eukaryotes have evolved a vast array of J-proteins that partner with a smaller set of different Hsp70s, with partnerships restricted to those that occupy the same cellular compartments, creating a mix-and-match diversity to confront diverse physiological functions (Kampinga and Craig, 2010). Additionally, several NEFs exist and also contribute to functional diversity (Bracher and Verghese, 2015).

There are a growing number of structures that shed light on the way Hsp70s work with their cochaperones (Table 1.2), but many questions remain. Still, it is clear that the modular architecture that enables

Table 1.2. Atomic Resolution Structures of Hsp70 Complexes with Cochaperones

Hsp70	Species	PDB ID	Cochaperone	Method/ Resolution	Comments	Reference
Hsc70	Bovine	2qwn, 2qwo, 2qwp, 2qwq, 2qwr	Auxilin	X-ray/1.7 to 2.4 Å	Several forms of complex with auxilin J domain 810–910 fragment disulfide linked to 1–394 NBD fragment; ADP-, ATP-bound; different NBD constructs also reported	Jiang *et al.*, 2007
DnaK	*E. coli*	1dkg	GrpE	X-ray/ 2.8 Å	NBD 1–383 fragment in complex with 1–197 fragment of GrpE	Harrison *et al.*, 1997
DnaK	*Geobacillus kaustophilus*	4ani	GrpE	X-ray	1–509 DnaK construct with 1–213 GrpE fragment; 2:2 stoichiometry in crystal	Wu *et al.*, 2012
Hsc70	Bovine	1hx1	Human BAG1	X-ray/1.9 Å	NBD 4–381 fragment bound to BAG domain 150–263 fragment	Sondermann *et al.*, 2001
Hsc70	Human	3fzf	BAG1	X-ray/2.2 Å	ATP-bound 1–381 NBD fragment; 222–334 fragment of BAG1	Williamson *et al.*, 2009
Hsc70	Mouse	3cqx	BAG2	X-ray/2.3 Å	NBD 1–388 fragment bound to BAG domain 102–189	Xu *et al.*, 2008
HspA1	Human	3a8y	BAG5	X-ray/2.3 Å	NBD 1–388 fragment bound to 5th BAG domain of BAG5 (334–475)	Arakawa *et al.*, 2010
Hsc70	Human	4hwi	*Arabidopsis thaliana* BAG1	X-ray/2.3 Å	NBD 5–381 fragment bound to BAG domain 65–242 fragment	Fang *et al.*, 2013

Hsc70	Bovine	3c7n	*S. cerevisiae* Hsp110/ Sse1p	X-ray/3.1 Å	Sse1 1–666 bound to Hsc70 1–554 fragment	Schuermann *et al.*, 2008
HspA1	Human	3d2f, 3d2e	*S. cerevisiae* Hsp110/ Sse1p	X-ray/2.3 Å	Sse1 1–675 bound to HspA1 NBD 1–382 fragment	Polier *et al.*, 2008
HspA1	Human	1xqs	HspBP1	X-ray	HspBP1 84–359 fragment complexed with NBD fragment 184–371	Shomura *et al.*, 2005
BiP	*S. cerevisiae*	3qml	Sil1/HspBP1	X-ray/2.3 Å	NBD 43–426 complexed with Sil1 113–421 fragment	Yan *et al.*, 2011
HspA1	Human	4j8f	Rat Hip	X-ray/2.7 Å	Fusion protein containing an NBD fragment 1–382 of HspA1 and the middle domain of Hip, aa 77–247	Li *et al.*, 2013
Hsc70	Human	4kbq	CHIP TPR domain	X-ray/2.9 Å	Helical lid and tail (541–646 fragment with 626–638 deletion) of Hsc70 complexed with the TPR domain of CHIP	Zhang *et al.*, 2015

the Hsp70 allosteric cycle to respond with exquisite sensitivity to its ligands also poises it to be tuned by cochaperone interactions. For example, NEFs interact primarily with the Hsp70 NBD so as to enhance the dissociation and association rates of nucleotides. The structural basis for this enhancement is the opening-up of the interlobe cleft, thus easing access to the nucleotide-binding site deep at the bottom of the cleft, and therefore catalyzing nucleotide exchange. Beyond this common strategy, there appear to be alternative interaction modes for different classes of NEFs with Hsp70s; recently, the structures and functions of the NEF classes were thoroughly reviewed (Bracher and Verghese, 2015). The GrpE family of prokaryotes and eukaryotic organelles binds to subdomain IIB of DnaK and causes this subdomain to twist 14° relative to the rest of the NBD, which opens the interlobe cleft (Harrison *et al.*, 1997) (Fig. 1.7A). It should be noted that the structure solved for *E. coli* GrpE bound to DnaK contains only a fragment of the GrpE dimer. Binding to the SBD in addition to the NBD cannot be excluded, and indeed was suggested by these authors based on modeling of the full-length GrpE into their structure. Consistent with this possibility, a recent structure of a *Geobacillus* GrpE/DnaK complexed with a full-length GrpE shows a wide array of interactions, including some between the NEF and the SBD (Wu *et al.*, 2012). Interpretation of this structure is complicated, however, as there is a 2:2 stoichiometry with two different types of dimer interactions present.

Eukaryotes have three major families of Hsp70-interacting NEFs: the functionally diverse Bcl-2 associated athanogene, or BAG family, which have a modular architecture and a common C-terminal, approximately 100-amino-acid BAG domain; the Hsp110/Grp170 family, which, as noted above, share homology with Hsp70s but lack the ATPase catalytic residues and the ability to undergo an allosteric transition; and the Sil1/HspBP1 family, which are composed of a variable N-terminal segment and a conserved C-terminal domain that has NEF activity (Bracher and Verghese, 2015). Structures of complexes of examples from all three families have been reported (Table 1.2). Both the BAG domain and the Hsp110 NEFs, like GrpE, interact primarily with NBD subdomain IIB,

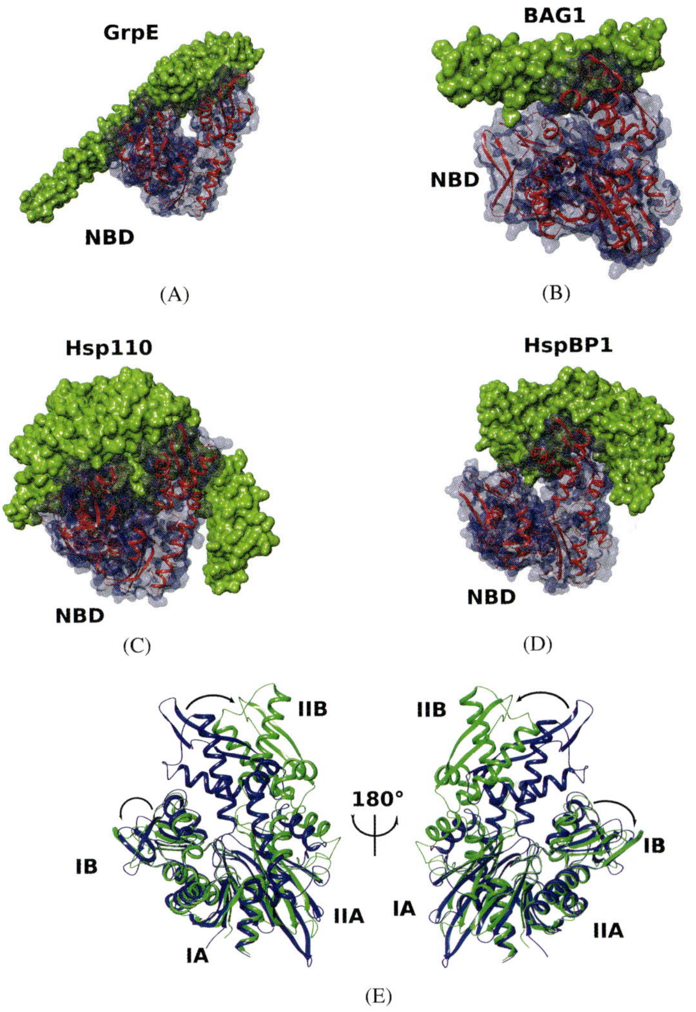

Fig. 1.7 A–D. Structures of Hsp70/nucleotide exchange factor (NEF) complexes. The NBDs are shown in a red cartoon with transparent surfaces, and the NEFs in lime green. (A) Complex of the DnaK NBD and a fragment of GrpE (PDB 1dkg) (Harrison *et al*., 1997). **B**. Complex of bovine Hsc70 NBD with the BAG domain of human BAG1 (PDB 1hx1) (Sondermann *et al*., 2001). **C**. Complex of the bovine Hsc70 NBD bound to *S. cerevisiae* Hsp110 (PDB 3c7n) (Schuermann *et al*., 2008). **D**. Complex of *S. cerevisiae* BiP NBD with *S. cerevisiae* HspBP1/Sil1 (PDB 3qml). **E**. Superposition of the NBD from ATP-bound DnaK (PDB 4jn4) (Qi *et al*., 2013) (in blue) and the NBD of Hsc70 in the complex with Hsp110 (PDB 3c7n) (in green). The structures were overlaid by aligning subdomains IB, IA, and IIA. Note the substantial movement of subdomain IIB upon NEF binding. Similar movements are observed upon binding of GrpE to DnaK, and BAG domains to other Hsp70s; HspBP1 causes a comparable rotation plus additional distortions to subdomain IIA. [This figure was prepared by Charles English.]

causing it to rotate away from the interlobe cleft (Figs. 1.7B and 1.7C). HspBP1 has extensive interactions with the NBD, including binding at subdomain IIB and causing rotation, but also causing displacement of lobe 1 (Fig. 1.7D).

Structural data on the interaction of J-proteins with Hsp70s are considerably sparser than those on NEFs. The diverse types of J-proteins would seem to challenge the development of general structural principles, although they are defined because of the presence of an approximately 40-residue J-domain. However, functional conclusions will be difficult, because J-proteins may mediate a wide array of functions, from delivery of substrate to anchoring at sites of Hsp70 action (Kampinga and Craig, 2010). In all cases, they are postulated to interact with the Hsp70 NBD in a cooperative manner with ATP binding, to activate the Hsp70 ATPase and thus drive the Hsp70 to a stronger interaction with its substrate, and to utilize a common His–Pro–Asp motif on the J-domain for binding to the Hsp70 NBD. Two signal studies of the atomic structure of a J-domain–Hsp70 interaction have appeared, but stand in contrast to each other: a crystal structure was determined for Hsc70 bound to the J-domain of its J-protein cochaperone, auxilin; to facilitate crystallization, Cys substitutions were introduced into both the Hsc70 and the J-domain and a disulfide bond was formed between them (Jiang *et al.*, 2007). The resulting structure shows the J-domain interacting with the site below the crossing helices where the hydrophobic interdomain linker binds. The linker was present in these structures, and it adopts either of two alternative conformations, one interacting directly with the J-domain. Binding of the J-domain as indicated in this structure, however, would be inconsistent with the ATP-bound, domain-docked conformation of the Hsp70.

Another study drew different conclusions about the J-domain–Hsp70 (here DnaK) interaction. In this work a noncovalent complex between a two-domain construct of DnaK (1–605, i.e. lacking only the unstructured CTD) and DnaJ 1–70 was analyzed by NMR. Interestingly, these authors compared the binding of a longer DnaJ construct that contains the GF-rich region and concluded that this latter part of DnaJ bound weakly to the SBD, but did not alter the functional interaction of the canonical J-domain. Using the 1–70 fragment and a combination of chemical shift perturbations and paramagnetic broadening, these authors concluded that the binding of DnaJ to ADP-bound DnaK is characterized by a highly dynamic interaction they called "tethered binding," but is incompatible with the binding observed in the crystal structure of the Hsc70/auxilin fragment. The authors argued for the relevance of their work on the interaction of the J-domain with the ADP-bound state of DnaK, but one might question this, since binding of DnaJ occurs to ATP-bound DnaK.

A recent report described an elegant reconstitution of the partnership between DnaJ and DnaK, in which the interaction of a substrate protein with DnaJ enhanced the probability of the substrate folding. However, only when the substrate also interacted with DnaK in the presence of ATP was optimal folding, i.e. full chaperone activity, observed (Nunes *et al.*, 2015).

Many classes of J-protein lack substrate-binding capacity (Kampinga and Craig, 2010). Nonetheless, they conserve the J-domain, which defines them, and thus interact with their partner Hsp70 to enhance its ATPase activity. In some cases, it is likely that the role of the J-protein is simply to tether the Hsp70 partner to a given site of action. It is safe to say that the full details of how J-domain cochaperones interact with their Hsp70 partners and, importantly, how this interaction facilitates the functional roles of this chaperone–cochaperone pair, have yet to be determined.

Hsp70 Oligomeric States

Observations from many labs have suggested that most Hsp70s tend to oligomerize both *in vitro* and *in vivo*, and a variety of functional implications of this behavior have been discussed but without consensus regarding either the physiological implications or the structural mechanism. In many cases, oligomerization is reversed by ligand binding, whether substrate, nucleotide, or cochaperones. The literature on this subject is well summarized in a recent report that sought to identify the mode of oligomerization for the human constitutive cytoplasmic Hsp70, known as HspA1 (Aprile *et al.*, 2013). These authors concluded that oligomerization occurred via an interaction between the interdomain linker and the helical lid of another molecule. Others have postulated interactions between the interdomain linker and the canonical substrate-binding site (Preissler *et al.*, 2015), and indeed this interaction has been observed in crystal structures (Chang *et al.*, 2008). The potential of Hsp70s to experience binding to the highly promiscuous substrate-binding site and thus self-associate or in other cases to form intramolecular interactions between accessible unstructured regions (Wang *et al.*, 1998), seems to be an unavoidable attribute of this system. The questions that remain to be addressed are to what extent these interactions occur *in vivo*, and if they do, what the physiological consequences are.

The Strikingly Diverse Physiological Roles of Hsp70 Molecular Chaperones: Relating Structure to Function

A question that underlies all Hsp70 functions is how the energy of ATP binding and hydrolysis is exploited to perform the many roles of these chaperones. The roles are diverse, suggesting that mechanistic diversity will be a characteristic of Hsp70s. However, as discussed at length in this chapter, the allosteric mechanism of Hsp70 chaperones is common to the whole family. The ability to bind unfolded regions of their substrates and release them with an appropriate timing, presumably optimized through evolution to work synergistically with folding rates, can adequately explain protection from aggregation, reversing misfolding, and facilitating folding to native. But what about other functions? And what are the detailed mechanisms whereby folding is favored and aggregation inhibited? Then there are the more specialized functions, such as facilitation of protein translocation across membranes and disassembly of clathrin-coated pits.

The main issue is whether the passive "holding" of an extended region of the polypeptide chain and the resulting steric impact on available conformations of the substrate is adequate for some functions. In the case of protecting aggregation-prone species, likely yes. In the case of disaggregation, perhaps not. Here it seems that some active energetic contribution may be required to disrupt favorable interactions within the aggregate. On the other hand, in many disaggregation reactions, Hsp70s partner with other chaperones like the ClpB family, which are capable of active force generation. For protein translocation, there are two alternative ways of thinking about the Hsp70 role on the distal side of a membrane (the role on the proximal side is generally to maintain a translocation-competent state): there is the "pulling," mechanically active role, or a passive "holding" that works to favor the chain's trajectory to the trans side of the membrane by a kind of ratchet function.

Recently, a model called "entropic pulling" has been described, initially for cases where multiple Hsp70s may bind simultaneously to the same substrate (De Los Rios *et al.*, 2006; Goloubinoff and De Los Rios, 2007; Mattoo and Goloubinoff, 2014). In this cooperative binding mechanism, the relatively prolonged binding of adjacent chaperones and steric clashes between them through Brownian motion exerts a force on the chain segment between them, causing it to be unfolded or extended. In the case of substrates held to a membrane pore or a partner chaperone, the action of a single Hsp70 could again lead to a force on the chain in between by entropic pulling while the substrate is held by an adjacent clamp, the pore or partner chaperone.

To decipher these mechanistic details and better understand how Hsp70s perform their cellular tasks will require deeper knowledge of the components and interactions active in any particular functional role. Promising results have been obtained in a few cases, and some examples are described here.

Clathrin uncoating ATPase. The clathrin-coated pit that is formed during endocytosis must be disassembled before the contents are released into the cytoplasm. Clathrin triskelions are acted on by a specialized J-protein, auxilin, which binds to vertices where trimers of clathrin hold the clathrin coat together. Auxilin recruits Hsc70, the constitutive cytoplasmic Hsp70, where it binds to a recognition motif QLMLT in a disordered and accessible region near the C-terminal of the clathrin triskelion. Through a combination of powerful electron microscopy, X-ray crystallography, and single molecule fluorescence imaging, a model has emerged for how this binding leads to disassembly of the clathrin coat (Bocking *et al.*, 2011; Xing *et al.*, 2010). Conformational selection of states for the clathrin molecules that are incompatible with the assembled coats tips the balance toward disassembly and uncoating of the vesicle. The action of multiple Hsc70s is required in order to push the system toward disassembly; thus, this function can be viewed as exploiting a combination of Hsc70's holdase, unfoldase, and entropic pulling functions.

Hsp70 as part of a chaperone team in steroid receptor maturation. Hsp70s have been implicated as part of a complex chaperone machinery in maturation of steroid hormone receptors, such as the glucocorticoid receptor (GR), for many years. The biochemical characterization and reconstitution of this five-component system paved the way for biophysical studies of the structures, interactions, and functional mechanisms of the various team members that work with the Hsp70: Hsp40 (the J-protein), Hop (the Hsp70/Hsp90 organizing protein), Hsp90, and p23. Recent application of state-of-the-art structural approaches applicable to large multimolecular complexes including cryoelectron microscopy (cryo-EM) has painted an increasingly detailed picture of this system (Alvira *et al.*, 2014; Kirschke *et al.*, 2014). In the presence of Hsp40, Hsp70 binds to GR (in both studies, the ligand-binding domain was used) and holds it in a nonnative conformation that is incapable of ligand binding. Interaction with Hop via binding of its TPR domain to the EEVD of Hsp70 prepares this complex for handoff to the Hsp90, whereupon the action of p23 and ATP hydrolysis yield functional GR. Cryo-EM images were consistent with Hp70 binding between dimeric Hsp90 molecules, and with GR binding to one monomer of Hsp90 and Hop interacting with the other. While the resolution of cryo-EM images does not allow definitive fits to known atomic structures of the component chaperones and cochaperones, one of these studies saw complexes that were both extended and compact and postulated a mechanism of substrate transfer, one of the central mysteries of the Hsp70/Hsp90 partnership (Alvira *et al.*, 2014). Validation of this proposal awaits higher resolution images. In overview, the Hsp70 role deduced from this recent work was largely as an unfoldase, preparing the GR substrate for reactivation by Hsp90 in a dynamic cycle that allows the cell to respond to changing demands for the receptor and enabling quality control simultaneously.

Hsp70 action as part of disaggregation machinery. Two examples of disaggregation machinery in which Hsp70s play a central role have been characterized in great detail, including structural aspects: the ClpB/Hsp70 system and the Hsp70/Hsp110 system (for recent reviews see: Aguado *et al.*, 2015; Mogk *et al.*, 2015). In both cases, the Hsp70 role commences with binding of accessible, unfolded regions of aggregates. The partnership with the Hsp100 family of AAA+ ATPases, most notably ClpB, has been extensively studied but not without controversy. Recent structural studies are elucidating the nature of the Hsp70 chaperone interaction with the ringlike ClpB partner. An insertion domain called M is characteristic of the disaggregase AAA+ family members, and serves as an essential part of the Hsp70 interaction site. The M-domain position in the hexameric ring has been poorly determined, most likely due to heterogeneity. But a consistent model is emerging: the M-domain gates between a conformation wherein the M-domains of adjacent protomers interact, inhibiting interaction with the partner Hsp70, DnaK (Carroni *et al.*, 2014). Tilting of the M domains breaks the contacts between adjacent domains and allows Hsp70 binding and transfer of substrate, where it is threaded and unfolded by the action of ClpB. An NMR description of the binding details shows that DnaK is bound via its NBD to a region of the M domain and part of the ClpB N domain in such a way that the SBD-bound substrate could be transferred to the ClpB substrate-binding site (Rosenzweig *et al.*, 2013). This is an active area of research and a promising system to yield molecular mechanistic details about the partnership between an Hsp70 and the system that specializes it for a cellular disaggregation function.

Similarly, the partnership of Hsp70s with Hsp110 as a disaggregation machine is receiving considerable attention (Rampelt *et al.*, 2012). The structural basis for cooperation between Hsp70 and Hsp110 as an NEF is well established (Polier *et al.*, 2008; Schuermann *et al.*, 2008). While this system has relatively poor disaggregation activity *in vitro*, new discoveries of cochaperone partnerships with J-proteins are pointing to this machinery as a key metazoan disaggregation machinery, and as such it is of major biomedical importance (Nillegoda *et al.*, 2015). The missing link in reconstituting disaggregation activity by this partnership was found to be a reliance on transient but specific interactions with two different J-proteins — one from class A and one from class B. Biochemical reconstitution of the disaggregation activity and biophysical studies of the modes of interaction of the players led to a model where one J-protein (class B) targets large aggregates and recruits the Hsp70, which exerts a pulling force and disaggregation ensues. For smaller aggregate sizes, the class A J-protein would play a similar role. The novel aspect of the model, however, is that class A and B J-proteins interact with one another, and higher order assemblies are proposed to be implicated in aggregate disassembly, with the diversity of J-proteins having been evolutionarily exploited for this important cellular function as well as other more specialized roles.

Outstanding Questions and Challenges

There are many unanswered questions regarding Hsp70 molecular chaperones and how they perform their cellular functions. In many cases, current understanding is limited by the methods available for studying these chaperones in action. The systems in which they act are complex, with many species present. All elements of these systems are dynamic, including conformational dynamics (for example of the substrates) as well as dynamics of transient interactions, and of course the domain and subdomain dynamics associated with these nanomachines. Nonetheless, meteoric technical advances over the past few years have led to optimism that the next decade will provide deep mechanistic insights into questions such as: How are substrates affected by the binding to Hsp70? Is there active remodeling of the substrate as a consequence of the interaction? How do Hsp70s accept substrates from cochaperones, particularly in systems where the J-protein possesses chaperone-binding capability? How does the handoff from the Hsp70 to downstream partners, such as Hsp90, happen? How does the tuning of a given Hsp70 by its interaction with partner J-proteins and NEFs allow it to be specialized for a given cellular role? Are all Hsp70 allosteric mechanisms truly the same, or does evolution select attributes that are specific to the functional spectrum of a given Hsp70 class? Lastly, what are the prospects for therapeutic modulators of Hsp70s? Is it possible to take advantage of their tunability and specialization, as opposed to their mechanistic commonalities, and develop small molecule modulators with specificity for particular functions? Stay tuned to this exciting field and watch for new developments in understanding this ubiquitous, abundant, and physiologically diverse class of multitalented molecular chaperone nanomachines.

Acknowledgements

The work in the author's laboratory on the structure and mechanism of Hsp70s is supported by a grant from the National Institutes of Health (GM027616).

References

Aguado A, Fernandez-Higuero JA, Moro F, Muga A (2015). Chaperone-assisted protein aggregate reactivation: different solutions for the same problem. *Arch Biochem Biophys* **580:** 121–134.

Alfano C, McMacken R (1989). Heat shock protein–mediated disassembly of nucleoprotein structures is required for the initiation of bacteriophage lambda DNA replication. *J Biol Chem* **264:** 10709–10718.

Alvira S, Cuellar J, Rohl A, *et al.* (2014). Structural characterization of the substrate transfer mechanism in Hsp70/Hsp90 folding machinery mediated by Hop. *Nat Commun* **5:** 5484.

Amick J, Schlanger SE, Wachnowsky C, *et al.* (2014). Crystal structure of the nucleotide-binding domain of mortalin, the mitochondrial Hsp70 chaperone. *Protein Sci* **23:** 833–842.

Andreasson C, Fiaux J, Rampelt H, *et al.* (2008a). Insights into the structural dynamics of the Hsp110–Hsp70 interaction reveal the mechanism for nucleotide exchange activity. *Proc Natl Acad Sci USA* **105:** 16519–16524.

Andreasson C, Fiaux J, Rampelt H, *et al.* (2008b). Hsp110 is a nucleotide-activated exchange factor for Hsp70. *J Biol Chem* **283:** 8877–8884.

Andreasson C, Rampelt H, Fiaux J, *et al.* (2010). The endoplasmic reticulum Grp170 acts as a nucleotide exchange factor of Hsp70 via a mechanism similar to that of the cytosolic Hsp110. *J Biol Chem* **285:** 12445–12453.

Aprile FA, Dhulesia A, Stengel F, *et al.* (2013). Hsp70 oligomerization is mediated by an interaction between the interdomain linker and the substrate-binding domain. *PLoS One* **8:** e67961.

Arakawa A, Handa N, Ohsawa N, *et al.* (2010). The C-terminal BAG domain of BAG5 induces conformational changes of the Hsp70 nucleotide-binding domain for ADP–ATP exchange. *Structure* **18:** 309–319.

Banecki B, Zylicz M, Bertoli E, Tanfani F (1992). Structural and functional relationships in DnaK and DnaK756 heat-shock proteins from *Escherichia coli*. *J Biol Chem* **267:** 25051–25058.

Bardwell JC, Craig EA (1984). Major heat shock gene of *Drosophila* and the *Escherichia coli* heat-inducible DnaK gene are homologous. *Proc Natl Acad Sci USA* **81:** 848–852.

Bertelsen EB, Chang L, Gestwicki JE, Zuiderweg ER (2009). Solution conformation of wild-type *E. coli* Hsp70 (DnaK) chaperone complexed with ADP and substrate. *Proc Natl Acad Sci USA* **106:** 8471–8476.

Bhattacharya A, Kurochkin AV, Yip GN, *et al.* (2009). Allostery in Hsp70 chaperones is transduced by subdomain rotations. *J Mol Biol* **388:** 475–490.

Bocking T, Aguet F, Harrison SC, Kirchhausen T (2011). Single-molecule analysis of a molecular disassemblase reveals the mechanism of Hsc70-driven clathrin uncoating. *Nat Struct Mol Biol* **18:** 295–301.

Bork P, Sander C, Valencia A (1992). An ATPase domain common to prokaryotic cell cycle proteins, sugar kinases, actin, and Hsp70 heat shock proteins. *Proc Natl Acad Sci USA* **89:** 7290–7294.

Bracher A, Verghese J (2015). GrpE, Hsp110/Grp170, HspBP1/Sil1 and BAG domain proteins: nucleotide exchange factors for Hsp70 molecular chaperones. *Subcell Biochem* **78:** 1–33.

Buchberger A, Theyssen H, Schroder H, *et al.* (1995). Nucleotide-induced conformational changes in the ATPase and substrate binding domains of the DnaK chaperone provide evidence for interdomain communication. *J Biol Chem* **270:** 16903–16910.

Buczynski G, Slepenkov SV, Sehorn MG, Witt SN (2001). Characterization of a lidless form of the molecular chaperone DnaK: deletion of the lid increases peptide on- and off-rate constants. *J Biol Chem* **276:** 27231–27236.

Calloni G, Chen T, Schermann SM, *et al.* (2012). DnaK functions as a central hub in the *E coli* chaperone network. *Cell Rep* **1:** 251–264.

Carroni M, Kummer E, Oguchi Y, *et al.* (2014). Head-to-tail interactions of the coiled-coil domains regulate ClpB activity and cooperation with Hsp70 in protein disaggregation. *Elife* **3:** e02481.

Chang YW, Sun YJ, Wang C, Hsiao CD (2008). Crystal structures of the 70-kDa heat shock proteins in domain disjoining conformation. *J Biol Chem* **283:** 15502–15511.

Clerico EM, Gierasch LM (2014). Structure and function of Hsp70 molecular chaperones. In: T Machajewski, Z Gao (eds), *Inhibitors of Molecular Chaperones as Therapeutic Agents*, Royal Society of Chemistry, Cambridge, pp. 65–125.

Clerico EM, Tilitsky JM, Meng W, Gierasch LM (2015). How Hsp70 molecular machines interact with their substrates to mediate diverse physiological functions. *J Mol Biol* **427:** 1575–1588.

Cupp-Vickery JR, Peterson JC, Ta DT, Vickery LE (2004). Crystal structure of the molecular chaperone HscA substrate binding domain complexed with the IscU recognition peptide ELPPVKIHC. *J Mol Biol* **342:** 1265–1278.

De Los Rios P, Ben-Zvi A, Slutsky O, *et al.* (2006). Hsp70 chaperones accelerate protein translocation and the unfolding of stable protein aggregates by entropic pulling. *Proc Natl Acad Sci USA* **103:** 6166–6171.

Dominguez, R (2004). Actin-binding proteins — a unifying hypothesis. *Trends Biochem Sci* **29:** 572–578.

Easton DP, Kaneko Y, Subjeck JR (2000). The Hsp110 and Grp1 70 stress proteins: newly recognized relatives of the Hsp70s. *Cell Stress Chaperones* **5:** 276–290.

Fang S, Li L, Cui B, *et al.* (2013). Structural insight into plant programmed cell death mediated by BAG proteins in *Arabidopsis thaliana. Acta Crystallogr D Biol Crystallogr* **69:** 934–945.

Flaherty KM, DeLuca-Flaherty C, McKay DB (1990). Three-dimensional structure of the ATPase fragment of a 70K heat-shock cognate protein. *Nature* **346:** 623–628.

Flaherty KM, McKay DB, Kabsch W, Holmes KC (1991). Similarity of the three-dimensional structures of actin and the ATPase fragment of a 70-kDa heat shock cognate protein. *Proc Natl Acad Sci USA* **88:** 5041–5045.

Flaherty KM, Wilbanks SM, DeLuca-Flaherty C, McKay DB (1994). Structural basis of the 70-kilodalton heat shock cognate protein ATP hydrolytic activity. II. Structure of the active site with ADP or ATP bound to wild type and mutant ATPase fragment. *J Biol Chem* **269:** 12899–12907.

Fourie AM, Sambrook JF, Gething MJ (1994). Common and divergent peptide binding specificities of Hsp70 molecular chaperones. *J Biol Chem* **269:** 30470–30478.

Freeman BC, Myers MP, Schumacher R, Morimoto RI (1995). Identification of a regulatory motif in Hsp70 that affects ATPase activity, substrate binding and interaction with HDJ-1. *EMBO J* **14:** 2281–2292.

Gao XC, Zhou CJ, Zhou ZR, *et al.* (2012). The C-terminal helices of heat shock protein 70 are essential for J-domain binding and ATPase activation. *J Biol Chem* **287:** 6044–6052.

General IJ, Liu Y, Blackburn ME, *et al.* (2014). ATPase subdomain IA is a mediator of interdomain allostery in Hsp70 molecular chaperones. *PLoS Comput Biol* **10:** e1003624.

Georgopoulos CP (1977). A new bacterial gene (groPC) which affects lambda DNA replication. *Mol Gen Genet* **151:** 35–39.

Gething MJ, Sambrook J (1992). Protein folding in the cell. *Nature* **355:** 33–45.

Goloubinoff P, De Los Rios P (2007). The mechanism of Hsp70 chaperones: (entropic) pulling the models together. *Trends Biochem Sci* **32:** 372–380.

Harrison CJ, Hayer-Hartl M, Di Liberto M, *et al.* (1997). Crystal structure of the nucleotide exchange factor GrpE bound to the ATPase domain of the molecular chaperone DnaK. *Science* **276:** 431–435.

Hu SM, Wang C (1996). Involvement of the 10-kDa C-terminal fragment of Hsc70 in complexing with unfolded protein. *Arch Biochem Biophys* **332:** 163–169.

Hunt C, Morimoto RI (1985). Conserved features of eukaryotic Hsp70 genes revealed by comparison with the nucleotide sequence of human Hsp70. *Proc Natl Acad Sci USA* **82:** 6455–6459.

Jiang J, Maes EG, Taylor AB, *et al.* (2007). Structural basis of J cochaperone binding and regulation of Hsp70. *Mol Cell* **28:** 422–433.

Jiang J, Prasad K, Lafer EM, Sousa R (2005). Structural basis of interdomain communication in the Hsc70 chaperone. *Mol Cell* **20:** 513–524.

Kamath-Loeb AS, Lu CZ, Suh WC, *et al.* (1995). Analysis of three DnaK mutant proteins suggests that progression through the ATPase cycle requires conformational changes. *J Biol Chem* **270:** 30051–30059.

Kampinga HH, Craig EA (2010). The Hsp70 chaperone machinery: J proteins as drivers of functional specificity. *Nat Rev Mol Cell Biol* **11:** 579–592.

Kellner R, Hofmann H, Barducci A, *et al.* (2014). Single-molecule spectroscopy reveals chaperone-mediated expansion of substrate protein. *Proc Natl Acad Sci USA* **111:** 13355–13360.

Kirschke E, Goswami D, Southworth D, *et al.* (2014). Glucocorticoid receptor function regulated by coordinated action of the Hsp90 and Hsp70 chaperone cycles. *Cell* **157:** 1685–1697.

Kityk R, Kopp J, Sinning I, Mayer MP (2012). Structure and dynamics of the ATP-bound open conformation of Hsp70 chaperones. *Mol Cell* **48:** 863–874.

Lee JH, Zhang D, Hughes C, *et al.* (2015). Heterogeneous binding of the SH3 client protein to the DnaK molecular chaperone. *Proc Natl Acad Sci USA* **112:** E4206–E4215.

Li Z, Hartl FU, Bracher A (2013). Structure and function of Hip, an attenuator of the Hsp70 chaperone cycle. *Nat Struct Mol Biol* **20:** 929–935.

Lindquist S (1986). The heat-shock response. *Annu Rev Biochem* **55:** 1151–1191.

Lindquist S, Craig EA (1988). The heat-shock proteins. *Annu Rev Genet* **22:** 631–677.

Liu Q, Hendrickson WA (2007). Insights into Hsp70 chaperone activity from a crystal structure of the yeast Hsp110 Sse1. *Cell* **131:** 106–120.

Mapa K, Sikor M, Kudryavtsev V, *et al.* (2010). The conformational dynamics of the mitochondrial Hsp70 chaperone. *Mol Cell* **38:** 89–100.

Marcinowski M, Holler M, Feige MJ, *et al.* (2011). Substrate discrimination of the chaperone BiP by autonomous and cochaperone-regulated conformational transitions. *Nat Struct Mol Biol* **18:** 150–158.

Mashaghi A, Kramer G, Lamb DC, *et al.* (2014). Chaperone action at the single-molecule level. *Chem Rev* **114:** 660–676.

Mattoo RU, Goloubinoff P (2014). Molecular chaperones are nanomachines that catalytically unfold misfolded and alternatively folded proteins. *Cell Mol Life Sci* **71:** 3311–3325.

Mayer MP (2010). Gymnastics of molecular chaperones. *Mol Cell* **39:** 321–331.

Mayer MP (2013). Hsp70 chaperone dynamics and molecular mechanism. *Trends Biochem Sci* **38:** 507–514.

McKay DB (1993). Structure and mechanism of 70-kDa heat-shock-related proteins. *Adv Protein Chem* **44:** 67–98.

Mogk A, Kummer E, Bukau B (2015). Cooperation of Hsp70 and Hsp100 chaperone machines in protein disaggregation. *Front Mol Biosci* **2:** 22.

Montgomery DL, Morimoto RI, Gierasch LM (1999). Mutations in the substrate binding domain of the *Escherichia coli* 70kDa molecular chaperone, DnaK, which alter substrate affinity or interdomain coupling. *J Mol Biol* **286:** 915–932.

Morshauser RC, Hu W, Wang H, *et al.* (1999). High-resolution solution structure of the 18kDa substrate-binding domain of the mammalian chaperone protein Hsc70. *J Mol Biol* **289:** 1387–1403.

Munro S, Pelham HR (1986). An Hsp70-like protein in the ER: identity with the 78 kd glucose-regulated protein and immunoglobulin heavy chain binding protein. *Cell* **46:** 291–300.

Munro S, Pelham HR (1987). A C-terminal signal prevents secretion of luminal ER proteins. *Cell* **48:** 899–907.

Nillegoda NB, Kirstein J, Szlachcic A, *et al.* (2015). Crucial Hsp70 co-chaperone complex unlocks metazoan protein disaggregation. *Nature* **524:** 247–251.

Nunes JM, Mayer-Hartl M, Hartl FU, Muller DJ (2015). Action of the Hsp70 chaperone system observed with single proteins. *Nat Commun* **6:** 6307.

Paek KH, Walker GC (1987). *Escherichia coli* dnaK null mutants are inviable at high temperature. *J Bacteriol* **169:** 283–290.

Palleros DR, Reid KL, McCarty JS, *et al.* (1992). DnaK, Hsp**73:** and their molten globules: Two different ways heat shock proteins respond to heat. *J Biol Chem* **267:** 5279–5285.

Pellecchia M, Montgomery DL, Stevens SY, *et al.* (2000). Structural insights into substrate binding by the molecular chaperone DnaK. *Nat Struct Biol* **7:** 298–303.

Polier S, Dragovic Z, Hartl FU, Bracher A (2008). Structural basis for the cooperation of Hsp70 and Hsp110 chaperones in protein folding. *Cell* **133:** 1068–1079.

Preissler S, Chambers JE, Crespilllo-Casado A, *et al.* (2015) Physiological modulation of BiP activity by trans-protomer engagement of the interdomain linker. eLife 4: e08961.

Qi R, Sarbeng EB, Liu Q, *et al.* (2013). Allosteric opening of the polypeptide-binding site when an Hsp70 binds ATP. *Nat Struct Mol Biol* **20:** 900–907.

Rampelt H, Kirstein-Miles J, Nillegoda NB, *et al.* (2012). Metazoan Hsp70 machines use Hsp110 to power protein disaggregation. *EMBO J* **31:** 4221–4235.

Rist W, Graf C, Bukau B, Mayer MP (2006). Amide hydrogen exchange reveals conformational changes in Hsp70 chaperones important for allosteric regulation. *J Biol Chem* **281:** 16493–16501.

Rosenzweig R, Moradi S, Zarrine-Afsar A, *et al.* (2013). Unraveling the mechanism of protein disaggregation through a ClpB–DnaK interaction. *Science* **339:** 1080–1083.

Rudiger S, Germeroth L, Schneider-Mergener J, Bukau B (1997). Substrate specificity of the DnaK chaperone determined by screening cellulose-bound peptide libraries. *EMBO J* **16:** 1501–1507.

Saibil H (2013). Chaperone machines for protein folding, unfolding and disaggregation. *Nat Rev Mol Cell Biol* **14:** 630–642.

Schlecht R, Erbse AH, Bukau B, Mayer MP (2011). Mechanics of Hsp70 chaperones enables differential interaction with client proteins. *Nat Struct Mol Biol* **18:** 345–351.

Schmid D, Baici A, Gehring H, Christen P (1994). Kinetics of molecular chaperone action. *Science* **263:** 971–973.

Schuermann JP, Jiang J, Cuellar J, *et al.* (2008). Structure of the Hsp110:Hsc70 nucleotide exchange machine. *Mol Cell* **31:** 232–243.

Sekhar A, Rosenzweig R, Bouvignies G, Kay LE (2015). Mapping the conformation of a client protein through the Hsp70 functional cycle. *Proc Natl Acad Sci USA* **112:** 10395–10400.

Shomura Y, Dragovic Z, Chang HC, *et al.* (2005). Regulation of Hsp70 function by HspBP1: structural analysis reveals an alternate mechanism for Hsp70 nucleotide exchange. *Mol Cell* **17:** 367–379.

Smock RG, Blackburn ME, Gierasch LM (2011). Conserved, disordered C terminus of DnaK enhances cellular survival upon stress and DnaK *in vitro* chaperone activity. *J Biol Chem* **286:** 31821–31829.

Sondermann H, Scheufler C, Schneider C, *et al.* (2001). Structure of a Bag/Hsc70 complex: convergent functional evolution of Hsp70 nucleotide exchange factors. *Science* **291:** 1553–1557.

Srinivasan SR, Gillies AT, Chang L, *et al.* (2012). Molecular chaperones DnaK and DnaJ share predicted binding sites on most proteins in the *E coli* proteome. *Mol Biosyst* **8:** 2323–2333.

Sriram M, Osipiuk J, Freeman B, *et al.* (1997). Human Hsp70 molecular chaperone binds two calcium ions within the ATPase domain. *Structure* **5:** 403–414.

Stevens SY, Cai S, Pellecchia M, Zuiderweg ER (2003). The solution structure of the bacterial Hsp70 chaperone protein domain DnaK(393–507) in complex with the peptide NRLLLTG. *Protein Sci* **12:** 2588–2596.

Suhre K, Sanejouand YH (2004). ElNemo: a normal mode web server for protein movement analysis and the generation of templates for molecular replacement. *Nucleic Acids Res* **32:** W610–W614.

Swain JF, Dinler G, Sivendran R, *et al.* (2007). Hsp70 chaperone ligands control domain association via an allosteric mechanism mediated by the interdomain linker. *Mol Cell* **26:** 27–39.

Swain JF, Schulz EG, Gierasch LM (2006). Direct comparison of a stable isolated Hsp70 substrate-binding domain in the empty and substrate-bound states. *J Biol Chem* **281:** 1605–1611.

Tissieres A, Mitchell HK, Tracy UM (1974). Protein synthesis in salivary glands of *Drosophila melanogaster*: relation to chromosome puffs. *J Mol Biol* **84:** 389–398.

Tsai MY, Wang C (1994). Uncoupling of peptide-stimulated ATPase and clathrin-uncoating activity in deletion mutant of Hsc70. *J Biol Chem* **269:** 5958–5962.

Ungewickell E (1985). The 70-kd mammalian heat shock proteins are structurally and functionally related to the uncoating protein that releases clathrin triskelia from coated vesicles. *EMBO J* **4:** 3385–3391.

Van Durme J, Maurer-Stroh S, Gallardo R, *et al.* (2009). Accurate prediction of DnaK-peptide binding via homology modelling and experimental data. *PLoS Comput Biol* **5:** e1000475.

Vickery LE, Cupp-Vickery JR (2007). Molecular chaperones HscA/Ssq1 and HscB/Jac1 and their roles in iron–sulfur protein maturation. *Crit Rev Biochem Mol Biol* **42:** 95–111.

Vogel JP, Misra LM, Rose MD (1990). Loss of BiP/GRP78 function blocks translocation of secretory proteins in yeast. *J Cell Biol* **110:** 1885–1895.

Vogel M, Mayer MP, Bukau B (2006). Allosteric regulation of Hsp70 chaperones involves a conserved interdomain linker. *J Biol Chem* **281:** 38705–38711.

Wang H, Kurochkin AV, Pang Y, *et al.* (1998). NMR solution structure of the 21kDa chaperone protein DnaK substrate binding domain: a preview of chaperone–protein interaction. *Biochemistry* **37:** 7929–7940.

Williamson DS, Borgognoni J, Clay A, *et al.* (2009). Novel adenosine-derived inhibitors of 70kDa heat shock protein, discovered through structure-based design. *J Med Chem* **52:** 1510–1513.

Wisniewska M, Karlberg T, Lehtio L, *et al.* (2010). Crystal structures of the ATPase domains of four human Hsp70 isoforms: HspA1L/Hsp70-hom, HspA2/Hsp70-2, HspA6/Hsp70B', and HspA5/BiP/GRP78. *PLoS One* **5:** e8625.

Wu CC, Naveen V, Chien CH, *et al.* (2012). Crystal structure of DnaK protein complexed with nucleotide exchange factor GrpE in DnaK chaperone system: insight into intermolecular communication. *J Biol Chem* **287:** 21461–21470.

Xing Y, Bocking T, Wolf M, *et al.* (2010). Structure of clathrin coat with bound Hsc70 and auxilin: mechanism of Hsc70-facilitated disassembly. *EMBO J* **29:** 655–665.

Xu X, Sarbeng EB, Vorvis C, *et al.* (2012). Unique peptide substrate binding properties of 110-kDa heat-shock protein (Hsp110) determine its distinct chaperone activity. *J Biol Chem* **287:** 5661–5672.

Xu Z, Page RC, Gomes MM, (2008). Structural basis of nucleotide exchange and client binding by the Hsp70 cochaperone Bag2. *Nat Struct Mol Biol* **15:** 1309–1317.

Yan M, Li J, Sha B (2011). Structural analysis of the Sil1–Bip complex reveals the mechanism for Sil1 to function as a nucleotide-exchange factor. *Biochem J* **438:** 447–455.

Yang J, Nune M, Zong Y, *et al.* (2015) Close and allosteric opening of the polypeptide-binding site in a human Hsp70 chaperone BiP. *Structure* **23:** 2191–2203.

Zahn M, Berthold N, Kieslich B, *et al.* (2013). Structural studies on the forward and reverse binding modes of peptides to the chaperone DnaK. *J Mol Biol* **425:** 2463–2479.

Zhang H, Amick J, Chakravarti R, *et al.* (2015). A bipartite interaction between Hsp70 and CHIP regulates ubiquitination of chaperoned client proteins. *Structure* **23:** 472–482.

Zhang P, Leu JI, Murphy ME, *et al.* (2014). Crystal structure of the stress-inducible human heat shock protein 70 substrate-binding domain in complex with peptide substrate. *PLoS One* **9:** e103518.

Zhu X, Zhao X, Burkholder WF, *et al.* (1996). Structural analysis of substrate binding by the molecular chaperone DnaK. *Science* **272:** 1606–1614.

Zhuravleva A, Clerico EM, Gierasch LM (2012). An interdomain energetic tug-of-war creates the allosterically active state in Hsp70 molecular chaperones. *Cell* **151:** 1296–1307.

Zhuravleva A, Gierasch LM (2011). Allosteric signal transmission in the nucleotide-binding domain of 70-kDa heat shock protein (Hsp70) molecular chaperones. *Proc Natl Acad Sci USA* **108:** 6987–6992.

Zhuravleva A, Gierasch LM (2015). Substrate-binding domain conformational dynamics mediate Hsp70 allostery. *Proc Natl Acad Sci USA* **112:** E2865–E2873.

Zuiderweg ER, Bertelsen EB, Rousaki A, *et al.* (2013). Allostery in the Hsp70 chaperone proteins. *Top Curr Chem* **328:** 99–153.

Chapter 2

Structural Analysis of GroEL/GroES Chaperonin-Mediated Protein Folding

*Arthur Horwich**

Abstract

This chapter concerns structural aspects of the GroEL/GroES chaperonin-mediated protein folding reaction. It first reviews the overall cellular actions of bacterial and organellar chaperonins (the so-called type I chaperonins), as well as the discovery and early EM imaging of these ring assemblies. It then covers X-ray crystallographic and cryoEM analyses of bacterial GroEL and its complexes with nucleotide, non-native substrate protein, and cochaperonin GroES and their implications for action in assisting protein folding (see Table of Structures on pp. 114, 115). The final section deals briefly with current models of action for GroEL/GroES-mediated protein folding.

Overall Physiologic Action of Chaperonins

Anfinsen's Speculations About Assistance in Protein Folding

The classic studies of Anfinsen and his coworkers in the late 1950's observed spontaneous *in vitro* refolding of ribonuclease, demonstrating that polypeptide chains contain within their amino acid sequence all of the information required for folding to the native state, presumed to lie at a thermodynamic minimum (Anfinsen, 1973). Yet Anfinsen himself recognized that some proteins could not spontaneously achieve

*Yale School of Medicine, Howard Hughes Medical Institute, 145 Boyer Center for Molecular Medicine, 295 Congress Ave., New Haven, CT 06510, USA.

the native state, under conditions either *in vitro* or *in vivo*, typically forming off-pathway aggregates such as cellular inclusion bodies. Such proteins apparently required assistance of a kinetic nature. His discovery of protein disulfide isomerase identified one type of assistance, involving acceleration of the acquisition of the proper arrangement of covalent disulfide bonds in secreted proteins (Golberger *et al.*, 1963). But he apparently recognized that, even for proteins residing in reducing compartments (where such disulfides are generally not present), there must be some form of assistance provided. He included a speculative drawing — in a piece in the "Cold Spring Harbor Quantitative Biology" series in 1963 — of a folded native multisubunit "machine" that could bind proteins at its outer aspect and properly template their folding (Epstein *et al.*, 1963). Yet he opined that he could not understand what property would be shared by the diverse array of non-native protein structures that could allow them all to be properly templated by such a machine. This question was only settled by two discoveries.

Discovery of Heat Shock Proteins and Their Role in Binding Other Proteins to Prevent Aggregation During Stress

First, in the mid-to-late 1970's, several groups identified a class of specialized proteins, known as heat shock proteins, that are induced in response to thermal stress and provide a protective role to the cell (see e.g. Schedl *et al.*, 1978; LeMaux *et al.*, 1978; Kelley and Schlesinger, 1978), although their role in protecting the cell was not immediately recognized. Then, in work in the mid-1980's, Pelham established that such specialized proteins act by maintaining the proper fold of other proteins through the common action of binding to exposed hydrophobic surfaces (Pelham, 1984, 1986). This, as amply borne out by analyses over the next decade, was the sought-for shared property of non-native conformations. Such hydrophobic surfaces become buried in the native state, and thus this constitutes a unique property of non-native proteins that allows their specific recognition.

Chaperone Action and Chaperonin Ring Assemblies

The action of heat shock proteins — and, more generally, molecular chaperones — in binding exposed hydrophobic surfaces of non-native "substrate" proteins via their own proffered hydrophobic surfaces acts to prevent such surfaces in the substrate proteins from forming intermolecular hydrophobic interactions that lead to multimolecular aggregation. [Note that "hydrophobic interactions" are directed by the entropically favored exclusion of water (e.g. Creighton, 1993).] This property of binding hydrophobic surfaces in non-native forms is indeed exhibited by the so-called chaperonin ring assemblies, which bind proteins through hydrophobic surfaces at the inside aspect of their rings. Chaperonins ultimately release substrate proteins into a large encapsulated hydrophilic cavity, a so-called "Anfinsen cage," where they are free to use their primary structure to find the native state.

Genetic Identification of a Cellular Function, *GroE*, in Bacteriophage Biogenesis

Chaperonins were first identified as a cellular function that assists bacteriophage biogenesis. Work presented in 1972 by two groups of investigators (Georgopoulos and Kaiser; Takano and Kakefuda) studying T4 and λ phages infecting bacterial cells, respectively, identified a cellular function whose mutational deficiency prevented the progression of infection to bacterial lysis and formation of plaques in a lawn of bacteria. The block of phage biogenesis was associated with the presence inside the infected mutant bacterial cells of phage heads with defective morphology, including aggregated phage head "monsters" and tubular forms (Fig. 2.1). The affected cellular locus was named *groE*, because aggregation of λ phage heads could be overcome by second site mutations in the phage gene λE, encoding the major capsid protein, lowering the level of this protein and preventing its aggregation (Georgopoulos *et al.*, 1973). The possibility of a cellular role for the *groE* locus was also raised, because the *groE*-deficient strains grew poorly, even in the absence

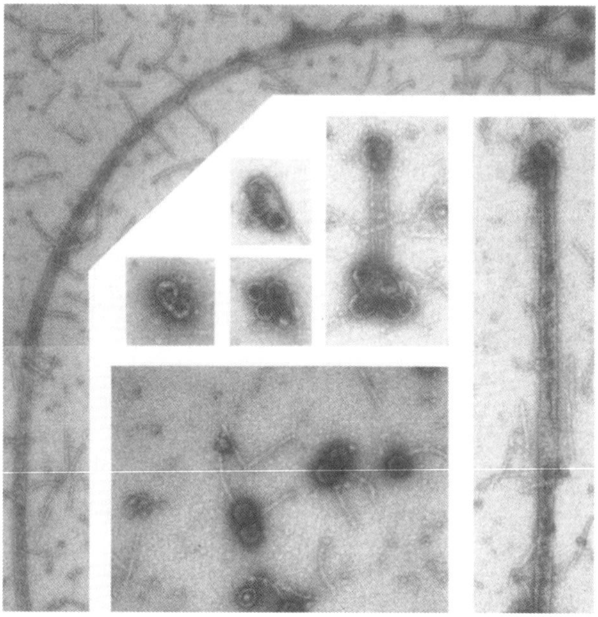

Fig. 2.1 Phage particles in wild-type and GroEL-deficient cells by scanning transmission EM analysis. *Bottom center*: Wild-type phage, with normal head and connected tail structures. *Center*: Aggregated phage heads ("monsters"), lacking tail structures. *Right and left*: Aberrant tubular structures. [From Georgopoulos *et al.*, 1972.]

of phage. Thereafter, it became clear that two cytosolic proteins are encoded by the *E. coli groE* operon (Tilly *et al.*, 1981): one an ~60 kDa protein, GroE Large, and one a 10–15 kDa protein, GroE Small — the proteins respectively called GroEL and GroES. The work of Hendrix and of Hohn *et al.* (described below) established in 1979 that GroEL is an oligomeric double-ring assembly (Fig. 2.2).

Rubisco Binding Protein in Chloroplasts, Related to GroEL in Sequence and Double-Ring Structure, Implicated in Assembly of CO_2-Fixing Enzyme Rubisco

In 1980, a second multisubunit complex was imputed by Barraclough and Ellis in the oligomeric assembly of the CO_2-fixing enzyme Rubisco. Called the Rubisco large subunit binding protein, it was identified as

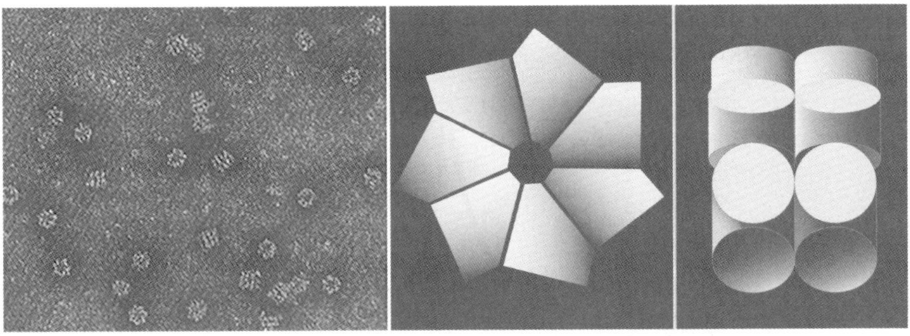

(A) (B) (C)

Fig. 2.2 Early image and model of GroEL. A. Negatively stained GroEL, raw image. Rings and four-stripe "bricks." [From Hendrix, 1979.] B, C. Early model of GroEL, end and side views, with seven-member stacked rings. [Adapted from Hohn *et al.*, 1979.]

associating with newly synthesized Rubisco large subunits encoded and synthesized inside the chloroplast stroma. The subunit binding protein was not found, however, with mature Rubisco formed upon association of Rubisco large subunits with small subunits imported from the cytosol. This led to the idea that the binding protein functioned in assembly of the large and small subunits into the final functional Rubisco heterooligomer. In 1983, a complex purified from pea leaf was identified that shared the quaternary structural property of two stacked sevenfold symmetric rings that had been observed for GroEL (Pushkin *et al.*, 1983); this was established as being the Rubisco large subunit binding protein. In 1988, the coding sequences for the subunit of GroEL and Rubisco large subunit binding protein were isolated and sequenced and found to encode proteins with greater than 50% identity across the predicted amino acid sequences (Hemmingsen *et al.*, 1988). John Ellis dubbed these ring assemblies "chaperonins."

A Third Related Component in Mitochondria Mediates Folding of Newly Imported Proteins to the Native State

Then, also in 1988, a third such double-ring component, whose subunit had been identified as a mildly heat-inducible protein inside the matrix

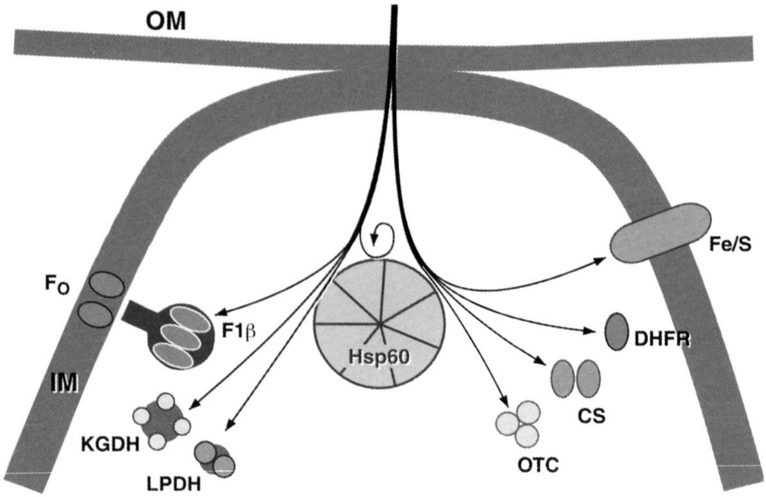

Fig. 2.3 The role of Hsp60 in the mitochondrial matrix in mediating proper folding of monomeric Rieske Fe/S protein and imported DHFR as well as subunits of a number of oligomeric proteins. [Model from work of Cheng *et al.*, 1989, and Ostermann *et al.*, 1989 as well as Glick *et al.*, 1992.]

of *Tetrahymena thermophila* mitochondria, was identified by Hallberg and coworkers, exhibiting two seven-fold symmetric stacked rings (McMullen and Hallberg, 1987, 1988). A year later, experiments with a temperature-sensitive mutation in the yeast version of this component indicated that its role was to mediate proper folding of newly imported mitochondrial proteins (Fig. 2.3; Cheng *et al.*, 1989). In particular, the yeast gene rescuing the temperature-sensitive folding-deficient mutant proved to be identical to the yeast homologue of the *T. thermophila* mitochondrial heat shock protein (Cheng *et al.*, 1989; Reading *et al.*, 1989). The mitochondrial component was called Hsp60 for the ~60 kDa size of the subunit, and sequence analysis predicted the subunit of this assembly to exhibit the same high degree of identity to the subunits of GroEL and Rubisco large subunit binding protein (Cheng *et al.*, 1989; Reading *et al.*, 1989). The functional role of Hsp60 as analyzed *in vivo* (Cheng *et al.*, 1989), along with *in organello* reconstitution of folding of

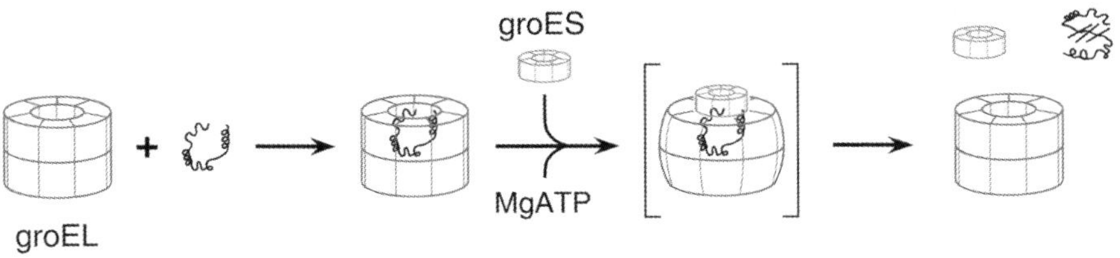

Fig. 2.4 The first reconstitution of GroEL/GroES–ATP-mediated protein folding, in sequential steps of non-native polypeptide binding to standalone GroEL, producing a binary complex, followed by addition of ATP–GroES, which caused release of folded protein. [See Goloubinoff *et al.*, 1989, and Martin *et al.*, 1991.] This model incorporates later topological understanding (see pp. 49–62).

an imported monomeric DHFR by Hsp60 (Ostermann *et al.*, 1989) and *in vitro* reconstitution of refolding of the subunit of a dimeric Rubisco by purified bacterial GroEL and GroES (Fig. 2.4; Goloubinoff *et al.*, 1989), provided demonstration that the chaperonins mediate assistance for the step of polypeptide chain folding, as distinct from oligomeric assembly. In the absence of such action, both *in vivo* and *in vitro*, the proteins on which these machines act are fated to misfolding and aggregation.

Essential Kinetic Assistance for Newly Imported Mitochondrial Proteins and for Newly Translated Cytosolic Proteins by Chaperonins

Both Hsp60 and GroEL were found to be essential for cell viability at all temperatures in genetic disruption experiments (Fayet *et al.*, 1989; Cheng *et al.*, 1989; Reading *et al.*, 1989), indicating that these machines provide an essential action to kinetically assist polypeptide chain folding even under non-stress conditions. This implies that many proteins can face kinetic difficulties of folding to native form even at physiological temperature in the milieu of an intact cell. This is likely a function of the high concentration of total solute (including macromolecules) in the cell, and of the notion that physiological temperature, such as 37°C, can already be relatively high enough to promote misfolding and aggregation.

Protein Substrates of Hsp60 and GroEL

Subsequent studies have sought to identify particular essential substrate proteins that depend on Hsp60 and GroEL for their proper folding. In mitochondria, it would appear from proteomic studies that many species are prone to aggregation in the absence of Hsp60 function (Dubuquaié *et al.*, 1998). This includes Hsp60 itself, with pre-existent Hsp60 required for the proper folding of newly imported Hsp60 subunits (see Cheng *et al.*, 1990). Recent studies of *E. coli* indicate that the class of TIM barrel proteins appears to have an absolute dependence on GroEL/GroES for their proper folding (Kerner *et al.*, 2005, Fujiwara *et al.*, 2010), whereas other structural classes of proteins employ the chaperonin system to a varying extent (Ewalt *et al.*, 1997; Chapman *et al.*, 2006). Under heat stress conditions, however, where GroEL/GroES rise to greater than 10% of total cell protein (Neidhardt *et al.*, 1984), many more protein species appear to show greater dependence on this system to enable binding/protection and proper folding. This finds a parallel *in vitro*, where purified GroEL could bind a large portion of the *E. coli* proteome following dilution from denaturant (Viitanen *et al.*, 1992).

Early Structural Studies — Negative Stain EM Observations

Contaminant of RNA Polymerase Preparations

Images of contaminants of early preparations of *E. coli* RNA polymerase are likely the first observed negative stain images of GroEL. For example, Lubin observed rosettes with seven bumps in a polymerase preparation analyzed by EM in 1969 (see image IX therein). In 1976, Ishihama *et al.* noted that there was ATPase activity contaminating the polymerase preparations and proceeded to dissociate polymerase into its subunits by high salt incubation. This treatment retained the ATPase in intact form, allowing its separation by further purification steps. The ATPase migrated in equilibrium sedimentation as a 930 kDa particle, and in SDS gels it was composed of a single polypeptide of ~70 kDa, suggesting 13 or 14 subunits per molecule. In negative stain EM, 7–9-member rings were observed. At this time, a connection to a particular molecule or cellular function could not be made.

Purified GroEL — Two Stacked Seven-Member Rings

In 1979, working independently, Hohn *et al.* and Hendrix overproduced GroEL from temperature-inducible prophage and transducing phage, respectively, purified the protein using glycerol gradients, anion exchange, and gel filtration, and carried out negative stain EM. Both groups observed seven-fold radially symmetric rings of 12.5–13 nm diameter with a central "hole," with side views showing rectangles with four stripes and dimensions of ~12.5–13 nm × 10–11.6 nm (Fig. 2.2). Both groups misinterpreted the stripes to run parallel to the long axis of the particle, but both correctly concluded that GroEL must be composed essentially of two stacked seven-member rings.

Purified GroES — A Single Seven-member Ring

In 1986, the GroES cochaperonin was first imaged by negative stain EM by Georgopoulos' group after overproduction from a plasmid at heat shock temperature and chromatographic purification steps (Chandrasekhar *et al.*, 1986). It behaved on a sizing column as an 80 kDa protein despite exhibiting a 10–15 kDa subunit in SDS gel analysis, indicating a multimeric form. In EM, a ring structure with rotational symmetry was observed, with a diameter of ~8 nm. A total of 6–8 subunits per ring was suggested. GroES was observed to inhibit GroEL ATPase activity by ~50% and to physically associate with GroEL in the presence of ATP.

Hsp60 EM Analysis

In 1988, as mentioned, McMullin and Hallberg reported on the heat-inducible protein found in a particle inside *T. thermophila* mitochondria as revealed by negative stain EM images, observing rings with seven-fold symmetry and side stripes resembling GroEL. Remarkably, an antibody against the 58 kDa subunit from *T. thermophila* cross-reacted with GroEL in a Western blot, anticipating the high degree of identity that would be revealed upon sequencing the Hsp60 and GroEL genes.

Resolution of Two Major Domains of GroEL

In 1989, a further EM study by Hutchinson and coworkers of the Hsp60 complex from *N. crassa* mitochondria, employing various staining procedures and specimen tilting, made clear that the four stripes observed in side views lie perpendicular, not parallel, to the seven-fold symmetry axis of the chaperonin particle (Hutchinson *et al.*, 1989). They correctly interpreted each stripe to represent a collective of a major globular domain from the seven subunits of a ring, with each subunit thus containing two major globular domains and the collective of seven subunits of a ring comprising two stripes. We now know these domains to be the so-called apical domains, at either end of the chaperonin cylinder, and the equatorial domains, lying back to back at its waistline.

GroEL/GroES Complexes

In 1991, Saibil and coworkers presented the first negative stain EM images of GroEL/GroES complexes, formed in the presence of ATP (Fig. 2.5). Asymmetric particles were observed, with one of the terminal stripes of GroEL "bowed out, forming a dome in the side view." This indicated a structural change in one of the GroEL rings that occurred upon GroES binding. Dome formation was subsequently shown to be at least in part a function of direct physical association of GroES by Ishii *et al.* (1992), who carried out immunoEM with an anti-GroES antibody on an asymmetric bacterial GroEL/GroES complex directly purified from *Thermus thermophilus*. The antibody was observed to bind exclusively to the domed end of the bullet-shaped complexes. Thus GroES binding

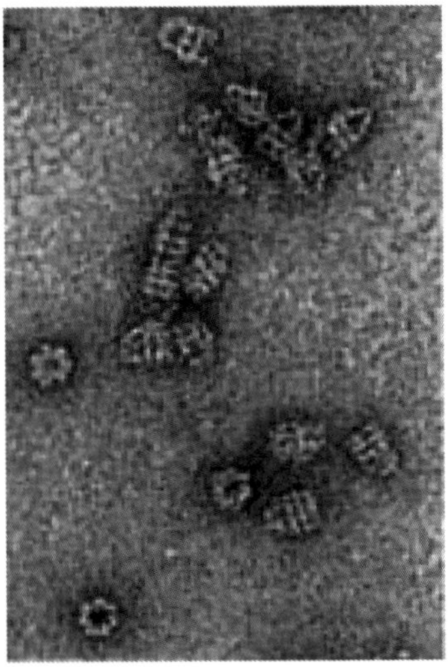

Fig. 2.5 The first EM image of GroEL/GroES–ATP complexes, appearing as bullet-shaped structures; see text. [From Saibil *et al.*, 1991.]

both presents mass at that terminal end of GroEL to which it binds and elevates the apical domains in the GroEL ring to which it binds.

Densities in the Central Cavity

In 1993, Saibil and coworkers provided a more detailed negative stain EM study of native GroEL purified from *Rhodobacter spheroides*. They used a tilt series to generate a 3D reconstruction, imposing seven-fold symmetry. The cutaway side view was particularly informative (Fig. 2.6), revealing the two major domains of each GroEL subunit connected by a slender density between them at the outside aspect of the cylinder. Along the seven-fold symmetry axis, each ring contained a large density at the equatorial level, these now known to be the collective of flexible C-terminal tails of the subunits of a ring, amounting to nearly 20 kDa of mass per ring. Small densities were also observed at the seven-fold symmetry axis at the apical domain level, speculated to be an average of density from the collective of GroEL-bound polypeptide substrates.

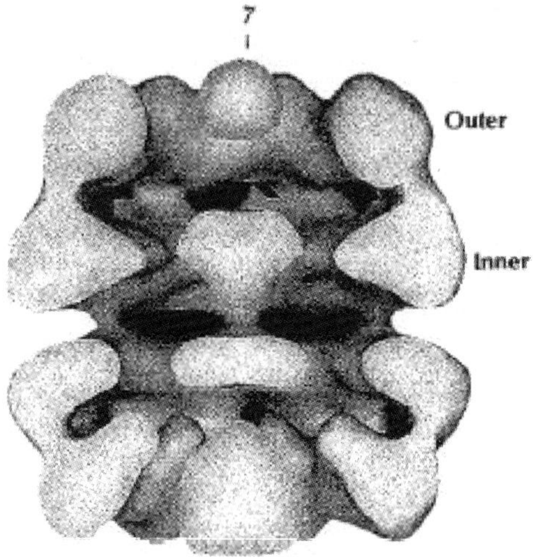

Fig. 2.6 Reconstruction of GroEL from negative stain images with a tilt series, showing the two major domains of a subunit, outer (apical) and inner (equatorial) and axial cavity masses at the outer (apical) level, later shown to be bound substrate polypeptide, and the inner (equatorial) level, later shown to be the collective of disordered subunit C-termini. [From Saibil *et al.*, 1993.]

GroEL/GroES Form an "Anfinsen Cage"

In further analysis of *E. coli* GroEL/GroES complexes in ATP, a density for GroES itself was appreciated at the rounded end of the asymmetric complex (Saibil *et al.*, 1993). This agreed with an earlier negative stain study of GroEL/GroES asymmetric complexes by Langer and coworkers (1992; Fig. 6 therein). GroES binding was proposed to form a so-called "Anfinsen cage," in which a polypeptide could fold without the possibility of aggregation.

Binary Complexes of GroEL with Non-native Protein

Three EM studies addressed the location of substrate proteins at GroEL. Langer *et al.* (1992) compared negative stain EM views of GroEL associated with non-native protein (the latter diluted from denaturant into a solution containing GroEL) with EMs of GroEL alone and observed the acquisition of stain-excluding density in the central "hole" of the GroEL rings in end views. The same density was observed in end views of Gro-

EL/GroES complexes, but the location of substrate protein, whether in the ring underneath GroES (in *cis*) or in the opposite, so-called *trans* ring, could not be gauged, because polypeptide substrate could not be located in side view images. In further studies of substrate localization in the absence of GroES, Braig *et al.* (1993) prepared nanogold-labeled DHFR, coupling 1.4 nm gold particles to a solitary cysteine in DHFR, unfolded the protein, and diluted it into an aqueous mixture with GroEL. Carrying out STEM analysis, gold particles were detected preferentially in end views in the central cavity of GroEL rings, i.e. along the seven-fold axis of symmetry. In side views, the gold particles were preferentially observed at terminal axial positions. This was consistent with binding of non-native polypeptide in the open central cavity at either (apical) end of the chaperonin cylinder. The conclusions from the STEM analysis were then supported by an analysis by Saibil and coworkers using cryoEM and image averaging (Chen *et al.*, 1994), comparing side and end views of malate dehydrogenase (MDH) bound to GroEL versus empty complex. In end projections, a density was observed in the central cavity. In side projections, a density was observed in the terminal position of the central cavity at the level of the apical domains (Fig. 2.7).

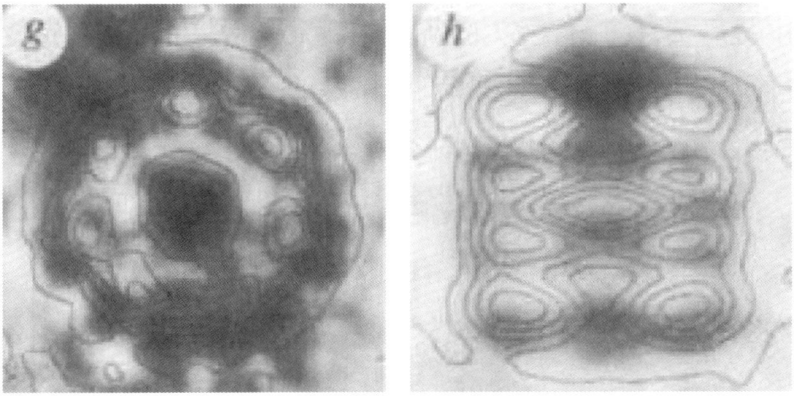

Fig. 2.7 Subtractive cryoEM analysis, GroEL–MDH binary complex versus GroEL standalone, showing the position of non-native MDH substrate protein while bound to GroEL as a dark mass. A dark mass over the central cavity in the end view *left* and present at the outer, apical level *right*. [From Chen *et al.*, 1994.]

Biochemical Observations on Polypeptide Localization and Folding in the *Cis* GroEL Ring underneath GroES

While the foregoing EM studies could not determine whether polypeptide can localize underneath GroES in an "encapsulated" *cis* GroEL ring, biochemical experiments in 1995 and 1996 supported that this occurs for several well-studied substrate proteins and that this is the location for productive folding. In particular, protease protection experiments and hit-and-run crosslinking experiments showed that a polypeptide could be localized underneath GroES inside the GroES-bound GroEL ring (Weissman *et al*., 1995). In a test of the folding-productivity of such a *cis* ternary complex, as compared with one where polypeptide was bound to an open ring opposite a GroES-bound ring (i.e., in *trans*), only the *cis* complex led to folding to the native state (Weissman *et al*., 1995). This indicated that productive folding commences within a *cis* cavity. To assess whether polypeptide can proceed all the way to the native form inside the *cis* cavity, additional experiments employed stable, long-lived, *cis* ternary complexes. In one case, a mutant single-ring version of GroEL (SR1) was used. Because it lacks an opposite ring from which the usual signal for evicting GroES arises *via* the binding of ATP, it forms a stable ternary complex after polypeptide and GroES are bound to it (Weissman *et al*., 1996). In such stable *cis* ternary complexes, polypeptide could proceed to efficiently reach the native state inside the *cis* cavity (Weissman *et al*., 1996; see also Mayhew *et al*., 1996).

Crystal Structure of GroEL

Structure Determination

Crystallization and data collection

In 1994, the crystal structure of GroEL in an unliganded form was reported at 2.8 Å resolution (Braig *et al*., 1994; PDB:1GRL).

Orthorhombic crystals were obtained from a *trc*-promoter-driven, massively overexpressed PCR-produced *E. coli* GroEL coding sequence containing, unknown at the time, two PCR misdirected amino acid substitutions (R13G, A126V) in what proved to be the coding sequence for the equatorial domain. These mutations do not interfere with overall folding function either *in vivo* or *in vitro*, although abrogation of normal negative cooperativity between rings for ATP binding and turnover was reported (Aharoni and Horovitz, 1996). The crystals were prepared in a 2 M ammonium sulfate mixture. Native data were collected initially at the F1 beamline at CHESS using propane-frozen crystals. In practice, 0.2° oscillations were employed, and ~500 frames were obtained from image plates mounted six at a time on a Nonius detector, then scanned using a Fuji scanner, requiring ~24 h of data collection. The crystals diffracted isotropically to ~2.7 Å and occupied a C2221 lattice. The large unit cell measured 178 × 204 × 278 Å. The volume of the asymmetric unit could accommodate one GroEL ring. A self-rotation function identified the seven-fold symmetry axis lying nearly parallel to c, and identified a dyad axis, indicating that the rings were back-to-back two-fold (crystallographic) symmetry–related.

Phasing and real-space non-crystallographic symmetry averaging

Phasing was accomplished by isomorphous replacement with a single ethyl mercury derivative that proved to fully occupy each of the three cysteines in a subunit, 21 in the a.u. The isomorphous difference Patterson map was thus complicated, with the number of peaks being 21^2–21. Z. Otwinowski designed a six-parameter search, inspecting for: position of the seven-fold symmetry axis in the *x–y* plane (two parameters), position of the heavy atom derivative in a reference subunit (three parameters), and angle of deviation of the seven-fold axis from the *c* axis, which was expected to be small. Once a first site (*i.e.* seven sites in the a.u.) was identified and fixed, the search was repeated for a second and then a third site. Initial SIR maps remained uninterpretable, with no resolvable boundary or envelope for the GroEL assembly. From the mercury sites, however, it was possible to accurately define non-crystallographic seven-fold matrices for NCS averaging. The first round of seven-fold averaging was carried out at 6 Å without using a solvent boundary. After this averaging, an envelope of the GroEL assembly became apparent. Successive rounds of NCS averaging were then carried out, commencing with amplitude data from 6 Å and proceeding to 2.7 Å through 400 cycles of phase improvement/extension, with a new envelope redefined every 20 cycles. This produced a map of sufficiently good quality that the backbone chain was traced through the equatorial and intermediate (lower) domains, but the apical domains remained poorly resolved. However, with the improved quality of new synchrotron data from EMBL outstation and re-refined NCS matrices using the program RAVE of Kleywegt and Jones (1999), better-resolved maps were obtained for the apical domains. In retrospect, NCS averaging succeeded with random starting phases, i.e. without even using initial experimental maps. Nevertheless, apical density for a portion of the main chain and for many of its side chains remained weak, which was consistent with the refined model where B factors for the apical domain were very high.

Second crystal form

The structure derived from the ammonium sulfate crystals was able to be matched with a model from a higher resolution PEG crystal of the same GroEL molecule in a monoclinic lattice (P21; 135.6 × 260.1 × 150.2, $\beta = 102.1$; PDB: 1DER; Boisvert *et al.*, 1996), which diffracted to better than 2.4 Å. In this second form, a tetradecamer was present in the a.u. In collaboration with Jimin Wang, isomorphous replacement was used to define a molecular envelope, and 14-fold NCS averaging was then carried out starting from a random phase set at 6 Å, gradually extending the data to 2.4 Å stepwise, and refining or redetermining the NCS matrices during the process (see Boisvert *et al.*, 1996). The derived model agreed well with the orthorhombic model, albeit this model also suffered from high B factors in the apical domains.

Initial refinement

Refinement was carried out in parallel for both the orthorhombic and monoclinic GroEL crystal forms using XPLOR (Brünger, 1992), including steps of positional refinement, B factor refinement, and simulated annealing. This brought R factors for the model to the low-to-mid 30s (Braig *et al.*, 1994). It was clear that the problems of refinement resulted from the poorly built apical domains, only poorly resolved in the maps. This could be due to (i) incorrectly placed apical domains; (ii) large isotropic motions of the constituent atoms of the domain; or (iii) a rigid body motion of the domain, which is hinged on the intermediate domain. The last such movement would abolish local seven-fold symmetry for the apical domains, diminishing the effectiveness of seven-fold NCS averaging. Any of the foregoing possibilities would increase B factors for the domain, leading to poor resolution. At that time, L. Rice in the lab of A. Brünger developed a torsional angle dynamics algorithm for simulated annealing, which increases the radius of convergence for refinement by allowing the system to be stably heated to much higher temperatures, with respect to the original Cartesian coordinate dynamics (Brünger *et al.*, 1997; Brünger *et al.*, 1998a,b). With the help of Brünger, torsional angle dynamics refinement was carried out by K. Braig and P. Adams, during which the seven-fold structural identity of the apical domain was relaxed (PDB: 1OEL; Braig *et al.*, 1995). The results were striking: in particular, a region at the top of the apical domain (aa 296–320) that had been very difficult to trace became resolvable, after it had earlier "flunked" the Bowie–Eisenberg 3D–1D score (Luthy *et al.*, 1992) for having a structure compatible with well-refined ones of ~2 Å in the PDB.

Further refinement

A further refinement approach was subsequently implemented by C. Chaudhry and P. Adams, employing the translation–libration–screw (TLS) method to describe possible anisotropic motions of groups of atoms, such as those of a domain, against a fixed axis or a fixed direction (Chaudhry *et al.*, 2004). TLS motions of rigid bodies are fitted to diffraction data such that correlated anisotropic displacements are incorporated into only nine parameters per body through the refinement. This further improved the R factors, suggesting that local anisotropic motions resulting from TLS types of movement occur. Motions of the intermediate and apical domains were detected about the hinge points at the top and bottom of the intermediate domains. These motions correspond to pathways of movement recognized to occur when nucleotide and GroES drive rigid body movements that produce the domed GroEL/GroES complex [TLS refined structures: GroEL, PDB: 1SS8; GroEL/GroES (ADP-$AlF_3)_7$, PDB: 1SVT].

Architecture of GroEL

The model of GroEL revealed a cylinder of 135 Å diameter and 145 Å height, with a cylindrical central cavity measuring ~45 Å in diameter (Plate 2.1). The rings are composed of seven subunits exhibiting seven-fold rotational symmetry, positioned back to back, forming an extensive interface across a flat equatorial plane. There are seven molecular dyads in this plane (perpendicular to the seven-fold symmetry axis), producing two-fold relationships between a subunit in the top ring and one in the bottom ring.

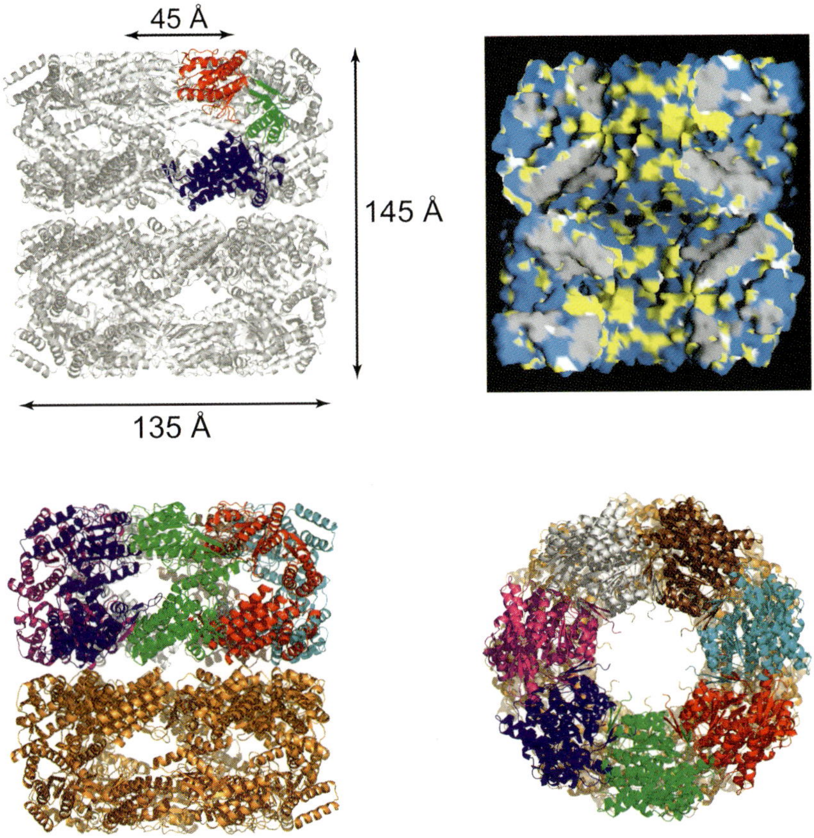

Plate 2.1. ***Architecture of GroEL***. The unliganded cylinder is composed of two back-to-back seven-member rings. Each subunit is two-fold-related to a subunit in the opposite ring (D7), with the two-fold axis positioned in the inter-ring interface between the "left" equatorial contacts (for example, a two-fold axis lies just below the left "foot" of the blue equatorial domain in the top ring in the top left panel). The central cavity of an open ring is ~45 Å in diameter. Note that the cavity is blocked at the equatorial level in each ring by the collective of seven subunit C-termini that are not crystallographically resolvable (see text and Fig. 2.6). In the top right panel, the cavity wall is lined with hydrophobic residues at the apical domain level, shown to be involved with binding non-native polypeptides (yellow; blue denotes polar surface). [Taken from PDB: 1OEL.]

GroEL subunits — three domains and a disordered C-terminus

The subunit polypeptide chain commences in density at residue 6 at the cavity wall of the equatorial domain, constitutes one part of the equatorial domain, ascends through the intermediate domain to form the apical domain, then descends through the intermediate domain and forms the remainder of the equatorial domain, ending at aa 523 near the N-terminus at the cavity wall as the end of a short β-strand pointing into the central cavity (Plate 2.2). The last 25 residues, the C-terminal tails of the GroEL subunits, are not crystallographically resolvable. The collective of this terminal 25-residue sequence, composed of a repeating GGM tripeptide through its last 13 residues, amounts to ~20 kDa

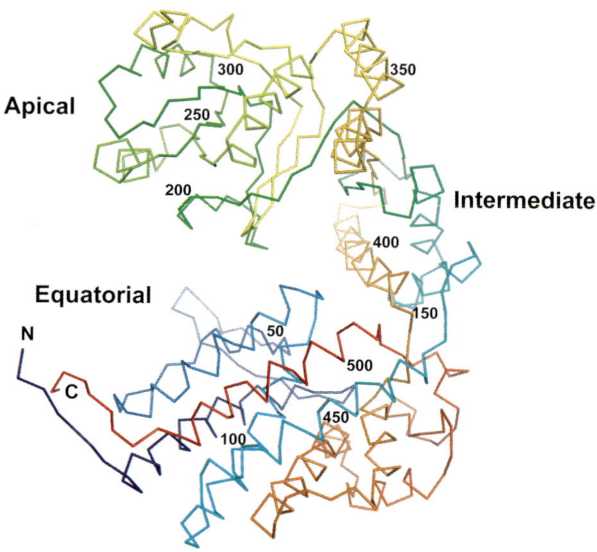

Plate 2.2. ***Chain trace of a GroEL subunit.*** Cα trace, showing the crystallographically resolvable chain, N corresponding to residue 4, proceeding to form part of the helical equatorial domain (blue), ascending through the intermediate domain to form the body and cavity surface of the apical domain (green), then descending back through the intermediate domain (yellow) to form an additional portion of the equatorial domain (orange/red), ending at C, residue 523. The C-terminal 25 aa of the subunit (aa 524–548) are not crystallographically resolvable, and are observed in EM as a collective mass in the central cavity (see text and Fig. 2.6). [Taken from PDB: 1OEL.]

of mass for the seven subunits of a ring and had been apparent as the electron density seen in EM images in the central cavity at the equatorial level of each ring (Saibil *et al.*, 1993 see Fig. 2.6). These residues are dispensable to the function of GroEL, because their coding sequence can be deleted from the chromosomal copy in *E. coli* and the cells maintain normal growth in a variety of media (Burnett *et al.*, 1993).

Equatorial domains

The equatorial domains at the waistline of the cylinder (aa 6–133 and aa 409–523, totaling 243 residues) are composed largely of long alpha helices and are well-ordered in all of the structure determinations (Plate 2.2). They provide the foundation for the GroEL structure and form most of the side-by-side contacts within a ring (982 $Å^2$ per subunit), as well as all of the contacts between the rings (418 $Å^2$ per subunit) across the equatorial plane (Plate 2.1). Concerning side-by-side contacts, each subunit projects an antiparallel β-loop (aa 36–50) from its body to the inner surface of its counterclockwise neighbor (looking at the top ring from outside), where it forms a parallel β-structure near its neighbor's C-terminal strand segment (aa 519–522) (Plate 2.3). This latter segment is supported by a short strand from the N-terminus of the same subunit (aa 6–8).

ATP binding site

Each equatorial domain houses a pocket in the top surface facing the inside of the cylinder that is the ATP-binding site, confirmed both by mutagenesis studies and by crystals of the monoclinic form of GroEL incubated with ATPγS (Plate 2.4; Boisvert *et al.*, 1996; PDB: 1DER). The ATP site differs somewhat from other ATPases in the geometry of its phosphate-binding (P) loop, insofar as it is composed of a helix–loop–helix arrangement (versus strand–loop–helix in many other NTP hydrolyzing components). A sequence conserved across the chaperonins is present in the loop, GDGTT. Mutation of the aspartate (D87) to lysine abolished ATP hydrolysis (Fenton *et al.*, 1994).

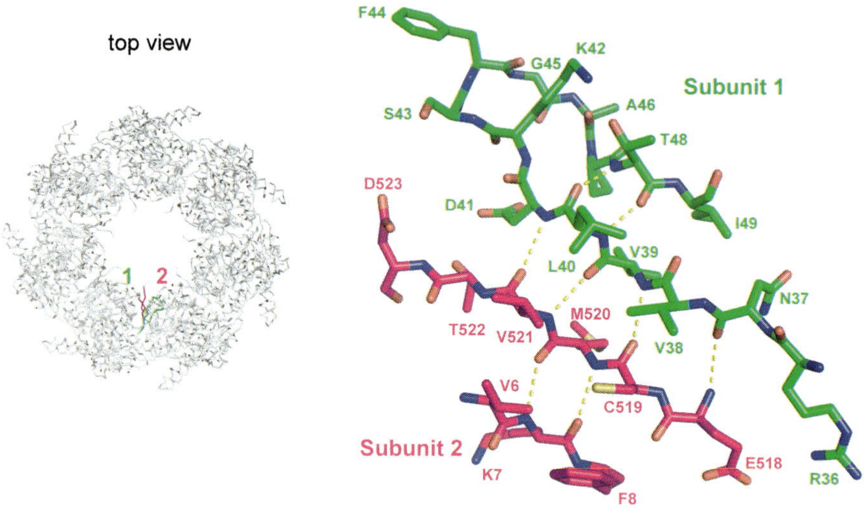

Plate 2.3. ***β-sheet intra–ring subunit–subunit interaction at the inside equatorial level.*** A *β*-hairpin (aa 39–49; also referred to as the stem–loop) reaches over from one subunit to form a *β*-sheet with N- and C-terminal strands of a neighbor. This interaction is present in all chaperonins and may play a role in the cooperativity of ATP binding and hydrolysis within a ring. [Taken from PDB: 1OEL. Adapted from Braig *et al.*, 1994.]

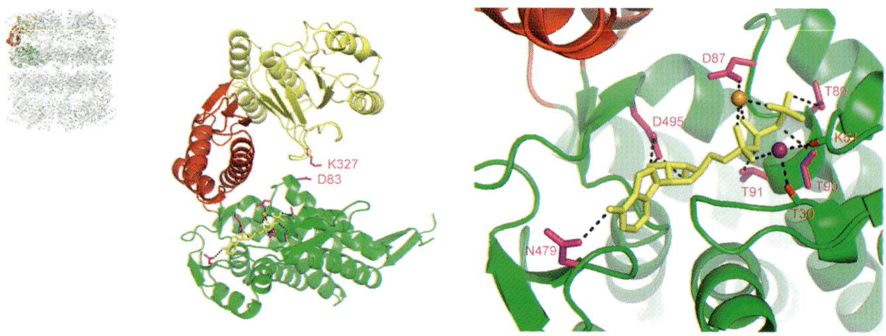

Plate 2.4. ***ATP binding pocket in the top equatorial surface of a subunit.*** An ATPγS-bound subunit, showing contacts of equatorial residues with the base and triphosphate moiety, including metal-mediated ones, Mg^{+2} in orange and K^{+} in purple (see text). [Taken from PDB: 1KP8.]

The equatorial domains, as a "base" of the machine, do not form contacts with other domains of GroEL, except for a salt bridge between D83 in the top surface of the equatorial domain with K327 lying at the inferior aspect of the apical domains. When these two amino acids are simultaneously altered to cysteine and oxidizing conditions are provided, they form a covalent connection between the two domains that locks the machine in the polypeptide-accepting (T) state (Murai *et al.*, 1996).

Apical domains

The apical domains (aa 191–376, totaling 186 aa) collectively form the terminal ends of the cylinder and make minimal contact with each other, potentially contributing to their high temperature factors (Plates 2.1 and 2.2). There are two salt bridge contacts (Plate 2.5), one an apical-to-apical bridge between neighboring subunits, K207–E255 (subunit to the leftward neighbor, if looking from the position of the cavity), and the other an apical-to-intermediate bridge between neighboring subunits, R197–E386 (subunit to the rightward neighbor from the vantage of the cavity). Mutational alteration of Arg197 to Ala leads to loss of positive cooperativity of ATP binding and hydrolysis within a GroEL ring, but also via allosteric action, to loss of negative cooperativity between rings (Yifrach and Horovitz, 1994, 1995). The apical domain (Plates 2.1 and 2.2) is composed of a central β-sheet whose strands are interconnected by loops and helices that form the cavity-facing peptide binding surface and a helix at the top aspect of the apical domain, followed after the sheet in the primary sequence by two long antiparallel helices forming a hairpin at the back aspect of the domain.

The apical polypeptide binding surface facing the cavity

The polypeptide binding surface is composed of a tier of structures: at the top, at the inlet to the cavity, is helix H (aa 233–242); beneath it is helix I (aa 258–266); and beneath helix I, innermost, is an underlying extended segment (aa 198–205) (Plate 2.6). These structures and the attachments from the central sheet that support them exhibit very high B factors, consistent both with the motion of the domain and with the potential local flexibility that accommodates to polypeptides binding at this surface (see below). Most striking on the tier of three

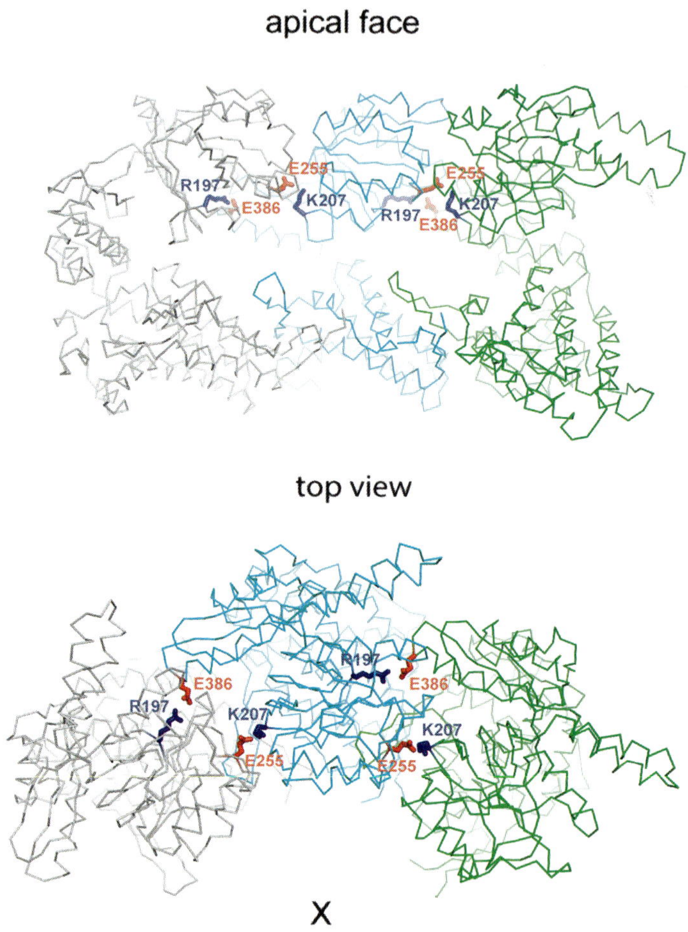

Plate 2.5. ***Electrostatic contacts between subunits of unliganded GroEL at the intermediate and apical domain levels.*** In addition to the numerous equatorial subunit–subunit contacts within a ring, there are two principal contacts in the upper two domains between neighboring subunits, salt bridges between apical R197 and intermediate E386; and apical K207 and apical E255. These are shown both in a view from the central cavity and in a view from the top. In the top view, X marks the seven-fold symmetry axis. [Taken from PDB: 1AON (*trans* ring).]

structures are the solvent-exposed hydrophobic side chains, aliphatics on the two α-helices and aromatics on the underlying segment. These constitute the polypeptide-binding surface. Mutational alteration of any single one of L234, L237, L259, V263, V264, Y199, Y203, and

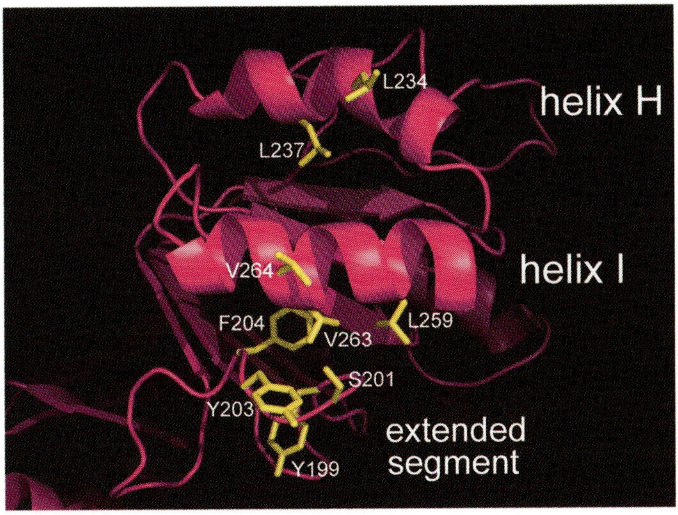

Plate 2.6. ***Apical polypeptide binding surface of a GroEL subunit.*** View from the central cavity of solvent-exposed hydrophobic side chains (yellow sticks) required for polypeptide binding, mapping to three structures — helix H at the inlet of the central cavity, helix I below it, and an underlying extended segment. Note that aliphatics are present in the top two structures and aromatics in the underlying segment. Polypeptide binding involves contact with the apical domain of at least three or four neighboring subunits (see text). [Taken from PDB: 1OEL.]

F204 to hydrophilic character (amounting to seven alterations of any given residue around a ring) abolished polypeptide binding *in vitro* and was incompatible with rescue of a GroEL-deficient *E. coli* strain (Fenton *et al.*, 1994). Thus, this supports the general idea that chaperone binding of non-native forms involves hydrophobic interactions between the apical surface and hydrophobic surfaces exposed in non-native substrate proteins. Such interaction prevents the non-native proteins from proceeding to form intermolecular interactions with each other that lead to aggregation. Direct support for hydrophobic interaction with non-native states comes from mutagenesis studies, from peptide binding studies, and from a co-crystal structure (the latter two described below).

In terms of substrate proteins that can be bound to the hydrophobic surface of an open ring, there seems to be no size limit, because an open ring can accommodate, in principle, a portion of any loosely structured polypeptide that exposes sufficient hydrophobic surface to become bound on the apical hydrophobic surface, the remaining portion lying outside the cavity in the bulk solution. This was supported by small angle neutron scattering studies showing a champagne cork topology for the bound substrate protein rhodanese, with one portion in the cavity and another portion outside in solution (Thiyagarajan *et al.*, 1996). Thus, a large substrate, such as one too large to be encapsulated by GroES, can in some cases be assisted by simple binding and release from an open GroEL ring. For example, the 80 kDa mitochondrial protein aconitase, which requires Hsp60 *in vivo* and GroEL *in vitro* for its productive folding, is too big to be GroES-encapsulated but nevertheless requires GroEL binding, followed by release, for its productive folding (Dubuquaié *et al.*, 1998; Chaudhuri *et al.*, 2001).

Intermediate domains

The intermediate domain (aa 134–190 and 377–408) is a slender antiparallel structure that covalently connects the equatorial domain and the apical domain of each subunit at the outer aspect of the cylinder (Plates 2.1 and 2.2). It exhibits hinge points at both the covalent connection to the equatorial domain (P137, G410) and the apical domain (G192, G375) (Plate 2.7). The ascending polypeptide chain departs from the equatorial domain neighboring the ATP-binding site, and at the top of the domain the chain is contiguous with an edge strand of the apical central β-sheet. In its descending limb, the intermediate domain polypeptide chain folds into a long α-helix, namely helix M (aa 388–408), that closes over the nucleotide pocket in the nucleotide-bound state of the ring (Plate 2.8) (PDB: 1AON, 1PCQ; EMDB 1998–2000), bringing a constituent residue, Asp 398, into contact with a water that can carry out an in-line attack on the γ-phosphate (Plate 2.15). An additional contact, as mentioned, occurs between E386, just preceding helix M and positioned at an "elbow" in the domain, and R197 in the neighboring apical domain at the C-terminal end of the

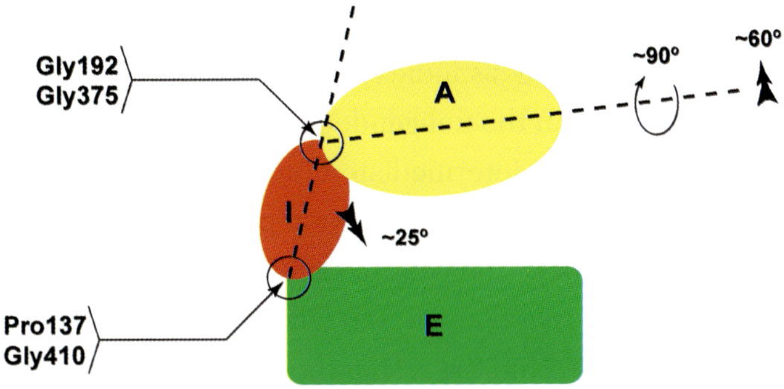

Plate 2.7. ***Schematic of intermediate domain "hinges" at the top and bottom of the domain, on which rigid body movements occur during ATP–GroES binding.*** The overall rotations that occur in the transition from unliganded GroEL to the ATP–GroES-bound domed folding-active state of GroEL/GroES are shown. [Adapted from Xu *et al.*, 1997.] Note that this transition is, as revealed by more recent EM studies, the sum of two sets of movement — the first directed by ATP binding, involving a sequence of downward rotation of the intermediate domains and elevation and counter-clockwise twist and separation of the apical domains, followed, upon docking GroES, by further elevation and a 120° clockwise twist of the apical domains (see Clare *et al.*, 2012; Plate 2.10; and Fig. 2.8).

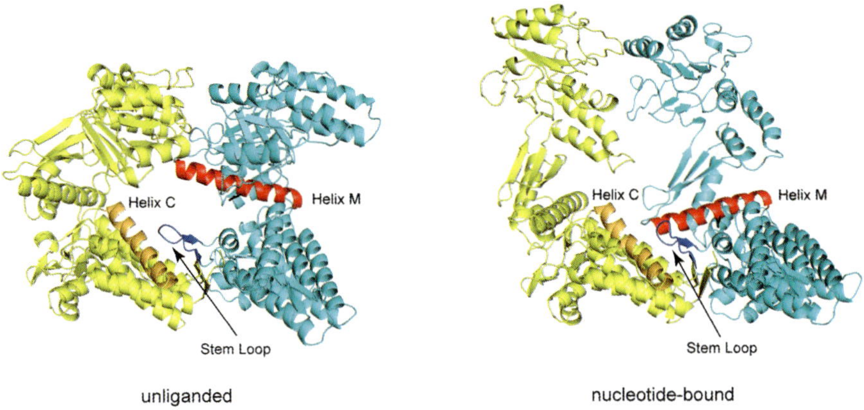

Plate 2.8. ***Movement of helix M upon ATP–GroES binding.*** This long intermediate domain α-helix (aa 388–408, red) rotates downward with the domain about the lower hinge (see Plate 2.7) in the presence of nucleotide (here ADP–GroES) to close over the nucleotide pocket, and forms contacts with the stem–loop of the same subunit (blue) and with helix C of the neighboring subunit. Helix M brings Asp398 (blue stick pointing downward from the middle of the helix) into the nucleotide pocket to activate a water for attack on the γ-phosphate (see Plate 15 for details). Note that, for simplicity, neither GroES nor nucleotide is shown in the right hand panel. View from inside the cavity of a ring. [Adapted from Xu *et al.*, 1997; PDB: 1OEL, 1AON.]

underlying segment (Plate 2.5). This contact is crucial to allosteric communication within the ring, as indicated by mutational analysis. Finally, it should be noted that, in the unliganded state of GroEL, there are side "holes" at the level of the intermediate domains (Plate 2.1), measuring 20 × 10 Å at the narrowest, enabling ready passage of nucleotide, ions, and solvent into and out of the central cavity. The floor of the portals is composed of an equatorial domain; the left wall and part of the ceiling, an intermediate domain; and the rest of the ceiling and the right wall, an apical domain.

Crystal Structure of GroES

Structure Determination

In 1996, Hunt, Deisenhofer, and coworkers reported a crystal structure of GroES at 2.8 Å resolution (Plate 2.9). This was determined from orthorhombic P212121 crystals containing one GroES heptamer in the asymmetric unit. Here, as with GroEL, phasing with (two) heavy atom derivatives did not produce an interpretable map but, upon optimization of the location of the seven-fold symmetry axis, lying nearly parallel to the crystallographic *c* axis, it was possible to employ seven-fold NCS averaging to produce a map from which a segmented polyalanine model could be built from most of the residues outside of the GroES mobile loop (aa 16–32). This map was improved to a point where a sequence assignment could be made by combination of the phases from the polyalanine model and experimentally derived phases, as well as the use of anomalous diffraction from a Se-Met 86 derivative. To verify the assignment, given the apparent imperfect seven-fold symmetry of the heptamer and considering the relatively high R factors, an imperfect seven-fold NCS averaging was performed, starting with the combined phases from the initial polyalanine model and 42 different symmetry operators to represent the relationship between a subunit and its six neighbors. This improved

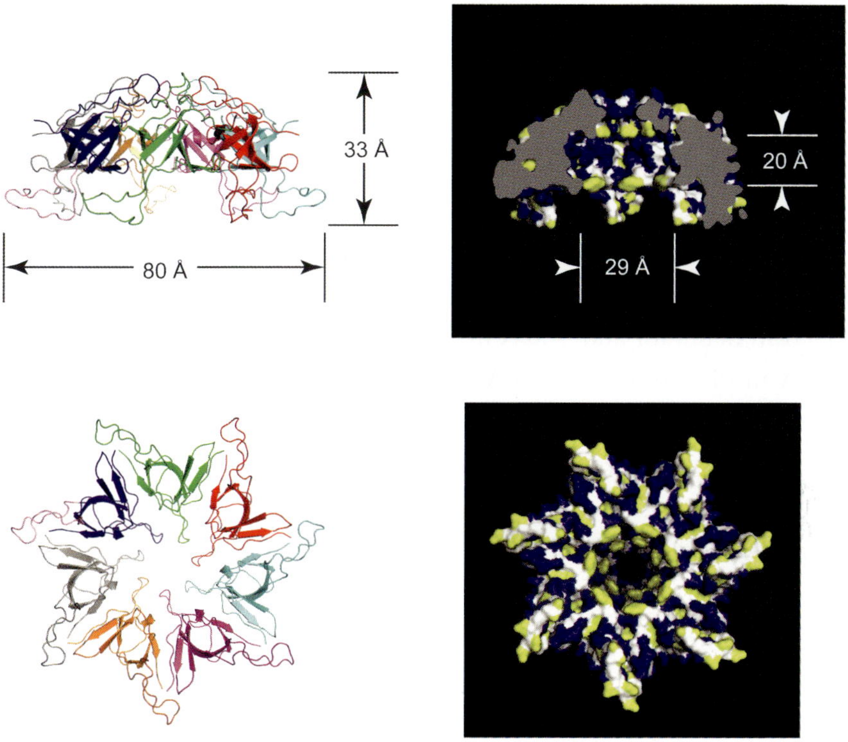

Plate 2.9. ***Architecture of GroES.*** GroES, the single-ring cochaperonin that complexes with an ATP-bound GroEL ring to form a folding chamber, shown in ribbons and space-filling views from the side (upper panels) and the underside (lower panels). In the left panels, the seven identical 10 kDa subunits are shown using a different color for each subunit. The body of a subunit is formed by a β-sheet; the roof of the complex is formed by the collective of loops emanating from the top aspect, and the contact points with GroEL are mediated by loops emanating downward and outward from the bottom aspect. The loops are disordered ("mobile") in standalone GroES (see text), but become ordered and crystallographically resolvable upon binding GroEL, as shown here in the GroES portion of the model of GroEL/GroES–ADP_7 from PDB: 1AON. The central cavity of GroES is revealed by the space-filling views, the upper showing a cutaway side view of the cavity and the lower looking up into it from the GroEL aspect. The GroES cavity while complexed with GroEL provides a smooth extension of the central cavity of GroEL (see text for further details). [From Hunt *et al.*, 1996.]

the R factors and produced well-defined side chain density for all except the mobile loop in six of seven subunits. In the seventh subunit, where the mobile loop makes a crystal packing contact, the entirety of the 97-residue chain could be observed.

Architecture

The model of the GroES heptamer revealed a dome-shaped molecule ~75 Å in diameter and ~30 Å in height, with an inside cavity measuring ~30 Å in diameter and ~20 Å in height (Plate 2.9). The body of each GroES subunit is composed of an antiparallel nine-stranded β-barrel that extends a β-hairpin from the top interior aspect, collectively forming a "roof" over the cavity. The single mobile loop that was trapped in the crystal contact was observed to project from the lower outer aspect of the body of its subunit, indicating that the mobile loops descend outward/downward to make contact with GroEL.

Mobile Loops of GroES as Putative Sites Contacting GroEL

The behavior of the mobile loops had already been divined through a prescient NMR study by Landry, Gierasch, and coworkers (1993). First, they observed in a one-dimensional experiment that, despite its ~70 kDa size, GroES surprisingly exhibited a number of discrete sharp signals, as opposed to the expected smear of broad resonances. Two-dimensional TOCSY and NOESY experiments were then carried out, allowing sequential assignment of a set of resonances to the mobile loops, composed of residues 17–32. When GroES was complexed to GroEL in ATP, the sharp signals became substantially broadened, suggesting that the loops had become immobilized in the complex, potentially by directly binding to GroEL. This latter conjecture was supported by observation of transferred NOEs upon incubation of a synthetic peptide including the mobile loop sequence with GroEL. A long range transferred NOE suggested that the bound mobile loop might be occupying a hairpin conformation. The interaction of the mobile loop with GroEL was further defined by GroEL/GroES crystal structures (see below).

GroEL–Nucleotide Complexes — X-Ray and cryoEM Structures

GroEL–ATPγS

The first GroEL–nucleotide X-ray structure involved the crystallization of GroEL (R13G, A126V) in the presence of ATPγS (PDB: 1DER; Boisvert *et al.*, 1996). Surprisingly, this produced the same monoclinic lattice as had been observed for earlier crystals of this standalone GroEL crystallized in PEG, indicating that no major domain adjustments had occurred. This contrasts with movements observed in both early negative stain EM studies (Saibil *et al.*, 1993) and later cryoEM studies, the latter observing rigid body movements of the intermediate and apical domains in the ATP-bound ring(s) (most recently, Clare *et al.*, 2012; EMDB 1998–2000). Thus, the crystal lattice appears to have constrained the domains from movements that are observed in solution. Nevertheless, nucleotide was readily detected in the equatorial binding pockets of the subunits, with full occupancy of both rings of GroEL (constituting the asymmetric unit). The nucleotide interacted only with residues within the equatorial domain. Two hydrogen bonds specify adenosine as the base (see Plate 2.4). The phosphate-binding loop recognized earlier (GDGTT) was observed to interact with nonbridging oxygens from all three phosphates.

Ion-Mediated Contacts Between Nucleotide and Equatorial Domain

Two ion-mediated interactions were also observed. A magnesium ion interacts with oxygens from all three of the phosphates (Plate 2.4, orange). Both the coordination of the ion and the anomalous signal observed following substitution of manganese for magnesium supported the identity as magnesium. A second ion (Plate 2.4, purple) links an α-phosphate oxygen to the main chain carbonyls of Thr 30 and Lys 51, the base of the stem–loop structure that reaches over to the neighboring equatorial domain; this may thus transduce a signal between subunits upon ATP binding to provide the positive cooperativity within a ring of ATP binding and hydrolysis (Gray and Fersht, 1991; Yifrach and Horovitz, 1995). This ion neighboring the α-phosphate oxygen was suspected to be potassium, considering earlier-observed effects of potassium to support ATP binding/hydrolysis (Todd *et al.*, 1993). The identity as potassium was strongly supported by a more recent study soaking the potassium mimetic, thallium, into GroEL crystals along with ATPγS (Kiser *et al.*, 2009; PDB: 3E76). A strong anomalous scattering signal was observed at the approximate position where the ion suggested to be potassium had been observed.

Reanalysis of GroEL–ATPγS

In 2003, the dataset for GroEL–ATPγS was reanalyzed by Boisvert and Wang, extending the data used from 2.4 Å to 2.0 Å (PDB: 1KP8; Plate 2.4). The stereochemistry at the nucleotide pocket was now more resolved. Magnesium was octahedrally coordinated, to one oxygen from each of the three phosphates, to the D87 side chain, and to two water molecules, with appropriate bond lengths. The other ion, thought to be potassium, was once again seen coordinating with an α-phosphate oxygen, carbonyls of T30 and K51, and with four water molecules, a total of seven ligands. But most surprising was that no sulfur atom could be identified on the γ-phosphate of the nucleotide, thus leaving one to conclude that ATP was the bound nucleotide. The investigators suggested that ATP could have been present as a contaminant in the ATPγS preparation at the 0.1–1.0% level, binding with selective affinity to the nucleotide pocket (at 2.5–25 μM final concentration); alternatively, hydrolysis of ATPγS followed by condensation of ADP molecules to form ATP and AMP could possibly have occurred in the solid phase of the crystals. The presence of ATP in the monoclinic crystals further reinforced the notion that the lattice had constrained what would be the normal ATP binding-directed movements of the intermediate and apical domains that were detectable in solution by the early EM studies. (In fact, the crystal lattice may have not only constrained movements but also allowed unscheduled ones: Wang and Boisvert observed that helix I in the apical domain became rotated along its axis.)

Early Solution cryoEM of GroEL–ATP

Further EM studies by Saibil's group in 1996 and 2001 demonstrated ATP-directed movements of the intermediate and apical domains of GroEL. In Roseman *et al.*, 1996, reconstructions of cryoEM images at 30 Å resolution showed that GroEL bound with any of ADP, AMP–PNP, or ATP (the last vitrified within 4 s of addition) exhibited vertical opening and twisting of the apical domain. The ATP structure exhibited the greatest degree of clockwise rotation of the apical domains, with the cavity-facing surface directed toward the intersubunit interface. This latter observation was noted to correlate with reduced affinity of GroEL for substrate proteins in the presence of ATP (Staniforth *et al.*, 1994), an apparent result of dislocating the hydrophobic polypeptide binding surface from facing the central cavity.

In Ranson *et al.*, 2001, a similar cryoEM study was carried out but with a larger dataset and using an electron microscope with a coherent source and correction for optical distortions, allowing ~10 Å resolution (EMDB 1047). In addition, for this study, a hydrolysis-defective mutant, D398A, was employed to prevent particles from hydrolyzing ATP prior to freezing. D398A complexes had been shown to bind ATP with normal affinity but hydrolyze it at a rate that is ~2% that of wild-type GroEL (Rye *et al.*, 1997). This mutant was studied by incubation with ATP for several seconds, followed by rapid freezing on the EM grid. The individual domains of GroEL, as obtained from the X-ray structure, were fitted into the electron density as rigid bodies. In the ATP-bound state, the GroEL particle was elongated and the rings were asymmetric with respect to each other. In one ring, there was a large downward rigid body rotation of the intermediate domains (~20°), forming contacts with the nucleotide pocket of the same subunit and with the top of the neighboring equatorial domain. The intersubunit salt bridge in unliganded GroEL (observed crystallographically) involving R197 and neighboring E386 was now apparently broken, freeing the apical domains to elevate and to undergo a small counterclockwise twist. These changes were interpreted to indicate that this ring now populated an R state, and the ring was proposed to be occupied by ATP. By contrast, the opposite ring remained similar to one of GroEL standalone, occupying a T state.

Consistent with one ring converting to an R state, which would exhibit reduced affinity for substrate polypeptide, the substrate protein Rubisco could be released from a single-ring version of GroEL (called SR1) in the presence of ATP (but not ADP) and captured by a "trap" version of the chaperonin. Interestingly, if both ATP and GroES were added, the Rubisco did not transfer to the "trap," indicating that GroES binds rapidly enough to prevent ATP-directed release of Rubisco.

The downward motion of the intermediate domains of one ring in ATP, particularly reflected in the movement of the long constituent helix M (aa 388–408), now contacting both the nucleotide pocket of the same subunit (including the stem–loop) and the top of the equatorial domain of the neighboring subunit (helix C; see Plate 2.8), provides a further avenue for subunit–subunit communication within a ring — an avenue for mediating cooperativity of ATP binding and hydrolysis within a ring [observed initially by Gray and Fersht (1992) and studied in detail by Yifrach and Horovitz (1995)]. This same intermediate domain movement is observed in the GroES-bound ring of X-ray structures of GroEL–GroES complexes (see below).

Recent cryoEM Study of GroEL–ATP — Structure Determination and a Trajectory of Movement

The nature of ATP-binding-directed movements has perhaps been best resolved through the recent cryoEM studies by Saibil and coworkers

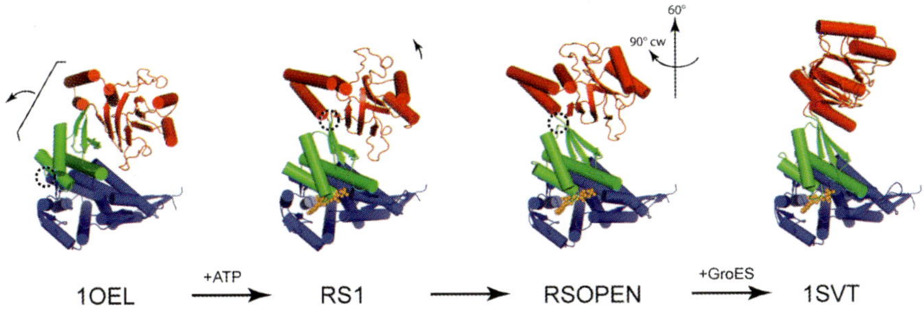

Plate 2.10. ***Movements of GroEL intermediate and apical domains upon binding ATP followed by GroES.*** ATP binding in the seven pockets of a ring produces a trajectory of movement as shown by cryoEM studies in Clare *et al.*, 2012. Here, the entire trajectory of a subunit from unliganded to ATP-bound to ATP (ADP–AlF_3)–GroES-bound is shown using crystal structures of unliganded start and GroES-bound end states, (GroES not shown) 1OEL and 1SVT, respectively, with cryoEM analyses of ATP-bound intermediate states, RS1 (EMDB 1998) and Rs-open (EMDB 2000). The rigid body movements along the trajectory are illustrated by arrows. Upon binding ATP, the apical and intermediate domains rotate sideways around the lower hinge of the intermediate domain (dashed circle; panel 1); the apical domain then elevates by rotation about the top hinge of the intermediate domain (dashed circle; panel 2), and the apical domains move outward from each other to produce Rs-open (Fig. 2.8), which is positioned for a direct "landing" of downward, extending GroES mobile loops to contact the apical domains; this is followed by further rigid body elevation of the apical domains along with a large 120° clockwise twist to produce the GroES-domed folding chamber into which polypeptide is released (see text).

(Plate 2.10; Clare *et al.*, 2012; EMDB 1998–2000). Once again, the D398A hydrolysis-defective mutant was employed with rapid freezing, thus trapping GroEL in an ATP-bound state. Here, an automated data collection system was used in collaboration with Clint Potter and Brigid Carragher (Lvumkis *et al.*, 2010), allowing large numbers of particles to be analyzed. Multivariate statistical analysis (MSA) was employed to subdivide the particles based on eigenimages, distinguishing conformational changes from orientational variation (for details of such analysis, see Cheng *et al.*, 2015). Six distinct classes were produced — three with ATP bound in one ring and three with ATP bound in both rings — and a resolution of 8–9 Å was reached. Once again, rigid body fitting of X-ray structure-derived domains was used, as well as energy minimization, to optimize side chain conformations. For GroEL with one ring ATP-occupied (states called Rs), three states were identified between the apo (T) state of a ring and the GroES-bound domed state, termed Rs1, Rs2, and Rs-open (see Plate 2.10; note that Rs2 is not shown). Three states were likewise observed for the doubly occupied complexes, termed Rd's. For both, a trajectory could be described, involving first a large, *en bloc* 35° sideways tilting rotation of the intermediate and apical domains about the lower "hinge" of the intermediate domain, forming Rs1 (Plate 2.10). This brings intermediate domain helix M and its constituent Asp398 required for ATP hydrolysis down onto the ATP binding pocket in the top of the equatorial domain (of the same subunit; see Plate 2.8). In the second step, an elevation of the apical domains occurs (about the upper "hinge" of the intermediate domain), forming Rs2. Finally, the apical domains move radially outward, separating further from each other and elevating 20°, producing the R-open state (Fig. 2.8 and Plate 2.10). In addition to analyzing these states with imposition of seven-fold symmetry, both the single- and double-ring-bound cases were analyzed with an absence of symmetry. This revealed significant asymmetry only in the case of Rs2, where one or two apical domains detach and elevate before the others.

Attitude of the Apical Domains in ATP and Putative GroES Collision State

In all three of these states, substrate polypeptide can remain bound to the apical domains of the ATP-bound GroEL ring, because helices H and I and the underlying segment remain facing the central cavity. Notably, in the R-open state, the mobile loops of GroES, if extended vertically downward from its body, would make direct contact with the hydrophobic cavity-facing helix H–I surface of the apical domains of GroEL (Fig. 2.8). This state is presumed to be the collision state of GroES with GroEL, albeit such an intermediate has not been stably captured and structurally analyzed. [See Miyazaki *et al.*, 2002 for a

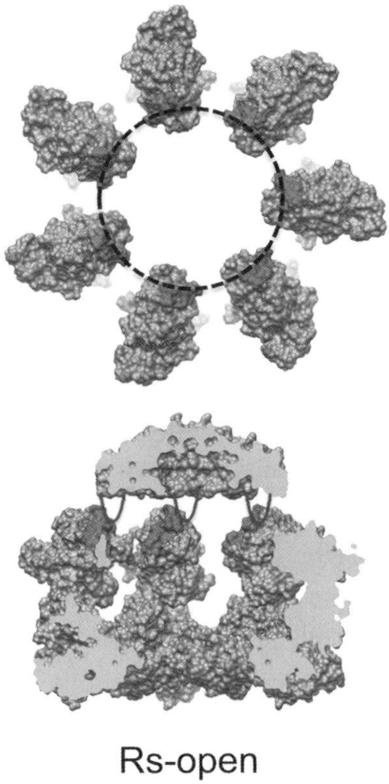

Fig. 2.8 The Rs-open state of apical domains in a ring of ATP-bound GroEL in end and cutaway side views, showing outward movement and separation of elevated apical domains, placing the hydrophobic apical binding surface (dark gray patches) in a position where mobile loops of GroES could putatively dock with the GroEL hydrophobic surface if directed downward (dashed circle in end view; GroES loops schematized in side view). This would constitute a collision state of GroES with GroEL. [From Clare *et al.*, 2012.]

cold-sensitive mutant of GroEL and Madan *et al.*, 2008 for a chemical modification of GroEL, both of which allow GroEL to bind GroES but prevent the release of a bound polypeptide into the encapsulated chamber. In both of these contexts, GroES is likely detained at the collision state or just beyond it. The collision state does not appear to be stably trapped, however, because these complexes proceed to reach a domed state.]

Putative Collision State

Presumably, the "landing" of GroES on GroEL in the R-open state would serve to encapsulate a protein while it is still bound to the apical hydrophobic binding surface, thus preventing polypeptide from a premature departure from the cavity (Fig. 2.8). At the same time, the hydrophobic binding site seems likely to be shared in some manner by both polypeptide and the GroES mobile loops which bind via a hydrophobic IVL "edge." [Notably, mutations of the hydrophobic residues in the apical cavity-facing surface of GroEL impaired both polypeptide binding and GroES binding, implying binding to the same surface (Fenton *et al.*, 1994).] It remains unknown whether arrival of the mobile loops of GroES on the apical surface of GroEL involves a displacement of polypeptide bound to the same apical domains to a point on the binding surface that is deeper into the cavity, for example onto the underlying segment, giving a model in which the same GroEL subunit can simultaneously share polypeptide and a mobile loop. Alternatively, the mobile loops could initially engage only apical domains not occupied by polypeptide, producing a model in which apical domains are not shared.

Notably, binding of larger substrates of GroEL requires three or more consecutive intact apical hydrophobic binding surfaces (Farr *et al.*, 2000). For its part, GroES requires five or more functionally intact (hydrophobic) mobile loops in order to be effectively recruited to a GroEL–polypeptide binary complex in ATP (Nojima *et al.*, 2008). If each of these five loops must become bound to a GroEL apical domain to recruit GroES, then this would imply that sharing of the same apical domains by polypeptide and GroES mobile loops must occur with at least several subunits.

Salt Bridges and the Trajectory

It is clear that the salt bridges R197–E386 (apical to neighboring intermediate) and E255–K207 (interapical) are broken during the ATP-directed movements observed by cryoEM, considering the large rigid body movements of the apical domains, with their complete separation produced by the point of reaching Rs-open (see Fig. 2.8). Saibil has proposed that, along the trajectory from T to Rs1 to Rs2 to Rs-open, a series of new interapical salt bridges are formed and progressively replaced ("click-stop"), until all are disrupted in the Rs-open state. This seems plausible, considering that it might transiently stabilize such states, allowing them to be captured on rapidly frozen grids as discrete states, albeit the collective of states may lie in what is a "shallow" energy landscape for the apical domain movement.

Effects of ATP Binding on the Inter-ring Interface — The Basis to Negative Cooperativity

Finally, the EM study revealed movements at the inter-ring interfaces that might explain the negative cooperativity for ATP binding and hydrolysis between rings reported early by Yifrach and Horovitz (1995) [fitting an overall model of nested cooperativity of GroEL in which there is concerted positive cooperativity between subunits within a ring (MWC) but sequential negative cooperativity between rings (KNF)]. Negative cooperativity must be mediated via the ring–ring interface, which involves two sites of contact of the equatorial domain of each subunit with, respectively, the same sites in two adjacent equatorial domains in the opposite ring, indicating a staggered arrangement of subunits (1:2) across the ring–ring interface (Plate 2.11). In the case of the "left" contact, this involves two-fold-symmetric van der Waals contact across the ring between residues A109 at the C-terminal end of helix D extending from the nucleotide pocket of the subunits in opposite rings. In the case of the "right" contact, this involves the pseudosymmetric electrostatic/van der Waals contacts involving residues R452/E461/V464. The cryoEM study allowed comparison of these contacts between unliganded and ATP-bound states. In GroEL with ATP bound in one ring, the equatorial domains of that ring were observed to downwardly tilt about the 452–461 contact, and they were also

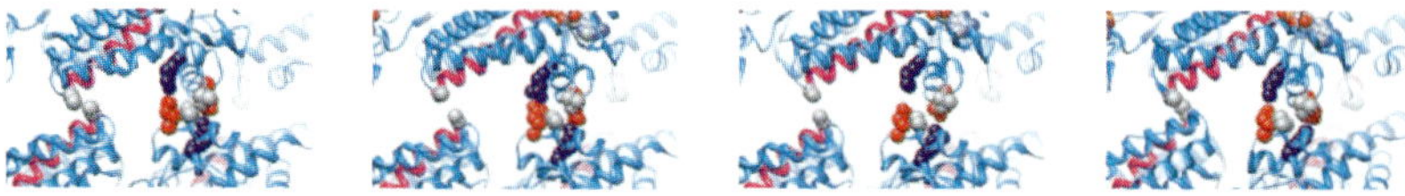

Plate 2.11. ***Ring–ring interface.*** Model of the staggered equatorial contacts across the GroEL ring–ring interface in several states: unbound GroEL (T) (PDB: 1OEL); ATP-bound (ATP in the top ring) (Rs1 and Rs-open from EMDB 1998 and 2000); and in the final domed GroES–$(ADP–AlF_3)_7$-bound top ring (PDB: 1SVT). In this diagram, a subunit from the top ring in the unliganded state makes "symmetric" contacts via its "left" (aa 109) and "right" (aa 452/461/464) sites with the respective left and right sites of two adjacent subunits in the bottom ring (*left hand panel*). Upon ATP binding (in the top ring), the equatorial domains downwardly tilt onto the 452–461 contact, altering the 109 (*left*) contact to a more vertical position (*panels 2 and 3*).

observed to undergo an anticlockwise rotation relative to the empty ring (Plate 2.11). These movements served to alter the position of the A109–A109 contact from tilted to vertical but also overall lengthened the contact by ~2 Å, thus altering the link between helices D and the respective nucleotide pockets in subunits of opposite rings. This potentially explains the source of the negative cooperativity.

Structures of GroEL-Polypeptide Complexes

Tag Peptide Extension Lodged on the Apical Surface

In 1997, Alan Fersht and coworkers presented a crystal structure of an isolated GroEL apical domain (aa 191–376) fused at its N-terminus with a 17-residue tag (Buckle *et al.*, 1997; PDB:1KID). With one apical domain in the asymmetric unit of the orthorhombic crystals, it was noticed that the C-terminal sequence of the tag lodged in the groove between helices H and I in the cavity surface of a neighboring molecule in the unit cell. The sequence, -7GLVPRGS-1, could be observed in density and adopted an extended conformation, stretching across the interhelical groove, with Leu-6 in the tag interacting with Leu237 on helix H and Val-5 in the tag interacting with Val264 on helix I (Plate 2.12, right panel). The next few residues run along helix I and then cross over

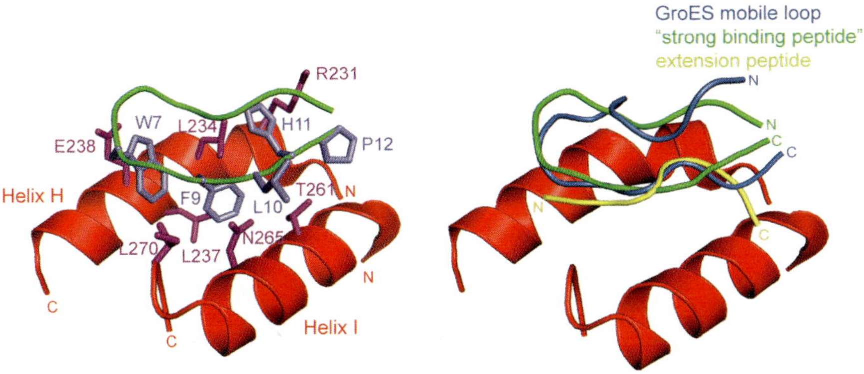

Plate 2.12. ***Crystallographic images of peptides bound to the GroEL apical domain.*** All three peptides — a segment of the GroES mobile loop (from PDB: 1AON), a synthetic "strong-binding peptide" (SBP, green, from PDB: 1DKD), and an extension peptide from a miniapical domain crystal lattice (yellow; from PDB: 1KID) — bind within a groove between apical helices H and I (*right panel*). The specific contacts of the side chains of the distal "leg" of the SBP hairpin (blue) in the groove are shown in the left panel — mostly hydrophobic contacts (W7, F9, L10; see text for discussion of physiological significance).

it. Additional hydrophobic interactions were reported for these residues, but also several hydrogen bonds were detected.

Selection of a Strong Binding Peptide and Cocrystallization

In 1999, Chen and Sigler presented a co-crystal structure of a 12mer peptide optimized for binding to GroEL by phage display (PDB: 1DKD). The strongest-binding peptide (SBP) was unrepresentative, because it contained two tryptophan residues within the short sequence, even though tryptophan is the least abundant amino acid in proteins, occurring at the ~1% level. Nevertheless, here two tryptophans allowed stronger binding to GroEL (in solution) than other less-favored peptides, exhibiting approximately micromolar affinity. For cocrystallization, the peptide was complexed with an isolated GroEL apical domain bearing an N-terminal tag nearly identical to that of Fersht and coworkers, but it crystallized in a different lattice, P21, with four peptide–apical-domain complexes in the asymmetric unit. The strong-binding peptide, SWMTTPWG-FLHP, was resolved in complex with each apical domain as a β-hairpin, with the distal sequence lodged in the cleft between helices H and I and the proximal sequence positioned away from the apical surface (Plate 2.12, left panel; note that the N-terminal tag did not resolve). Prominent hydrophobic interactions between the peptide and helices H and I were observed, for example between Trp7 and Leu 234 (on helix H), and between Phe9 and both Leu234 and Leu237 (Plate 2.12, left panel). As in the study by Fersht and coworkers, a number of hydrogen bond contacts were also observed. In a followup study of the peptide bound to all apical domains of the GroEL tetradecamer at 3.0 Å (PDB: 1MNF; Wang and Chen, 2003), it was noted that none of the hydrogen bonds involved side chains of the SBP peptide, all involving backbone interactions and thus not conferring any sequence specificity at the level of substrate. Interestingly, however, Ser1 of the peptide became ordered in this structure, forming a hydrogen bond through its backbone carbonyl with R268 of a neighboring apical domain, and at the turn in SBP, Pro6 forms a similar contact with R231 of the other neighboring subunit.

Local Flexibility of the Apical Substrate Binding Surface

Within the SBP-isolated apical domain asymmetric unit, the apical domain Cα rms deviation was greatest, when compared with that of the apical domain crystallized alone, in four distinct regions (notably, all of the molecules in the a.u. of the respective lattices exhibited no significant rms deviation with each other through these regions). The differences between peptide-associated and standalone mapped to: aa 207–212 contiguous with the underlying segment, helix H, helix I, and aa 301–310, the last a loop segment behind the inlet helix (H) that, notably, had been the most difficult region of GroEL to refine. These same regions also contributed the greatest rms difference when comparing the SBP–apical domain model with the earlier apical domain structure from the Fersht lab and several GroEL tetradecamer structures. This suggested that there is structural plasticity to the peptide backbone of the apical face. This presumably allows it to adjust to accommodate substrate protein binding, here involving peptide segments. Such adjustment likely contributes to favoring the overall hydrophobic interactions of a non-native protein with the apical face.

Biochemical Observations Concerning Optimized Hydrophobic Interactions with the GroEL Polypeptide Binding Surface

The SBP structural work complements biochemical observations concerning the binding to GroEL of α-helical peptides with hydrophobic amino acids variously positioned. These indicate that the strongest binding occurs in association with the maximum contiguous hydrophobic surface in the peptide (Wang *et al.*, 1999). In the context of a loosely structured polypeptide chain binding to the GroEL apical face, this might indicate that a certain degree of annealing could occur, probably mostly on the part of what is normally a weakly structured substrate protein, but also to some degree by the GroEL apical face, which optimizes the hydrophobic contact of substrate protein with GroEL. Binding has been shown to be multivalent, involving binding of substrate protein simultaneously to multiple consecutive surrounding GroEL apical domains, demonstrated both by mutational studies (Farr *et al.*, 2000) and by EM analyses discussed below. Such binding contributes a net affinity of GroEL for polypeptides like MDH (Ranson *et al.*, 1997) that is nanomolar as compared with the micromolar affinity of individual apical domains for peptides.

Topology of H–I Interhelical Peptide Binding — Resemblance to Mobile Loop Binding

The topology of peptide binding observed in the Chen/Sigler study was compared with binding of the GroES mobile loop to GroEL in the GroEL/GroES crystal structure (Xu *et al.*, 1997; see below) and as examined by transferred NOE NMR studies of Landry and coworkers (Shewmaker *et al.*, 2001). They observed from both the X-ray study and the transferred NOE studies that the β-hairpin formed by the GroES mobile loop upon binding GroEL closely resembled that formed by SBP upon binding to GroEL, thus raising the question whether SBP binding more mimics the behavior of binding of the GroES mobile loop to GroEL than the behavior of a bound substrate polypeptide. Because the peptide in the Fersht study lodges fairly similarly along the interhelical groove, here also a question is raised concerning whether the binding resembles that of an authentic non-native substrate protein versus that of the GroES mobile loop (see Plate 2.12, right panel). EM studies, as detailed below, indicate a broader topological occupancy of the apical binding surface of GroEL.

CryoEM Analysis of Polypeptide Substrate Binding — GroEL–MDH Complex

In 2007, Saibil and coworkers (Elad *et al.*, 2007) formed binary complexes of malate dehydrogenase (MDH) diluted from denaturant with GroEL and examined the topology of the bound substrate protein by cryoEM. In initial tests of such complexes, mostly end views were obtained that were not informative for the axial level of occupation of the apical binding surface. A series of manipulations were attempted to increase the number of side views, with the greatest increase (to 50–80% total complexes) occurring when the equatorial residue Asp473, facing outside the cylinder, was modified to cysteine and a 6-His tag covalently coupled with high occupancy (exerting no interference with MDH binding or ATP/GroES-mediated refolding). An initial 8000 particles were analyzed with seven-fold symmetrization of the map, followed by insertion of the GroEL domains as rigid bodies, revealing extra density in the central cavity at the equatorial levels for the subunit C-termini, as might be expected from earlier studies, but also density in the cavity of one ring lying at the level of helix I. The amount of density observed was considerably less than that of a 33 kDa MDH subunit, reflecting that this must be the most highly occupied site on the apical binding face where, even with seven-fold symmetry manipulation, the density is not smeared out.

Next, a dataset of 40,000 particles was processed without symmetrization. Multivariate statistical analysis was carried out on this dataset to detect significant variations likely to be structure-related. Eigenimages were classified with respect to substrate occupancy and distribution, dividing ultimately into smaller, more homogeneous subsets. The significant variations occurred in the central cavity. Projection matching of individual images to multiple models was carried out to refine the separation into five major classes (at a resolution of 10–11 Å, with FSC of 0.5). Three classes produced structures with defined extra density, revealing different levels of localization of non-native MDH within the central cavity in side views (Plate 2.13). Two classes showed strong interaction of non-native MDH with helix I and the underlying segment (aa 199–203), this portion of substrate thus lying relatively deep within the cavity. One class showed the substrate at the inlet to the cavity, contacting helices H and I. In end views, the densities of MDH in these three classes contacted three or four consecutive apical domains. This was consistent with an earlier mutagenesis study using covalent GroEL rings tandemizing seven GroEL subunits, produced by fusion at the coding sequence level, that allowed for combination and permutation of mutant (V263S) binding-deficient apical domain mutations, revealing that three or four contiguous binding proficient (wild-type) subunits are required for efficient binding of MDH or Rubisco (Farr *et al.*, 2000). The three classes occupied ~25, 40, and 60% of the volume of native MDH, respectively, indicating that only a portion of the molecule, in its ensemble average state, is present in density in these image classes. In the bound complexes, changes were observed in the architecture of the GroEL ring itself, with the bound apical domains bunched closer together.

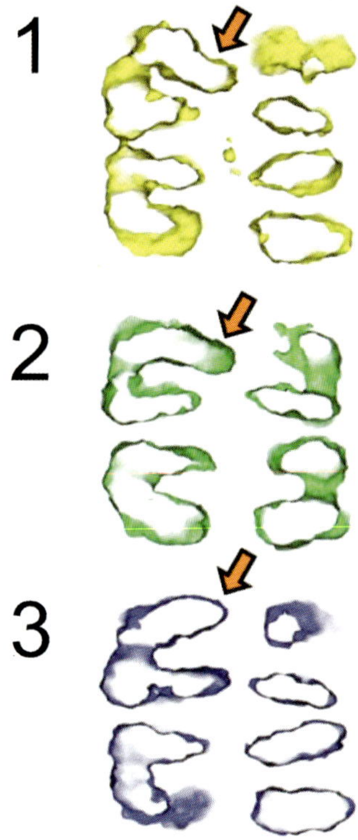

Plate 2.13. ***Non-native malate dehydrogenase (MDH) association with the GroEL apical domains observed by cryoEM.*** Binary complexes of MDH associated with GroEL were examined by cryoEM, revealing a variety of image classes. Side cutaway views of three such classes are shown here: one with the MDH density lodged against helix I and the underlying segment of a GroEL apical domain (1), one with the density lodged against helices H and I (2), and a third with the density lodged against helix H (3). Orange arrows indicate the density due to MDH. End views (not shown) show the MDH of these corresponding image classes lodged against three or four consecutive apical domains (see text for details). [Taken from Elad *et al.*, 2007.]

Surfaces Proximate to Bound Polypeptide

While density was present against the apical cavity-facing surface, where was the remainder of the polypeptide? This was assessed using a cysteine scan through a substrate protein, dihydrofolate reductase, and simultaneously across the surfaces of GroEL. After forming binary complexes between the various radiolabeled Cys-DHFRs and GroEL Cys surface variants, oxidative crosslinking was carried out and analyzed by gel analysis. Only GroELs bearing cysteine in the central cavity, including the equatorial stem–loop (e.g. F44) and flexible C-terminal tails (e.g. aa548), or at the very top aspect of GroEL at a loop behind the inlet, residue 315, became crosslinked. By contrast, no residues at the outside aspect could become crosslinked.

GroEL–T4 Capsid Complex

A second, similar cryoEM analysis was carried out by the Saibil group, examining the complex of non-native gp23, the major T4 viral capsid protein, with GroEL (Clare *et al.*, 2009; EMDB 1544). Here, five classes of particle were distinguished. Extra density was observed at the apical level of one ring in 40%, both rings in 40%, and neither in 20%. In the substrate-occupied rings, the substrate protein associated strongly with helix I and the underlying segment and appeared associated with at least five apical domains. The latter association may reflect the larger size of this substrate, 56 kDa, so large that evolution has provided a specialized phage-encoded GroES homologue called gp31 that enables a larger domed chamber to form, allowing for proper folding of gp23 (Hunt *et al.*, 1997, PDB:1G31; Clare *et al.*, 2006). In contrast with the MDH binary complex, in the gp23 complex no bunching of substrate-bound GroEL subunits was observed.

NMR Analysis of Polypeptide Complexed with GroEL

In 2005, Horst *et al.* reported on direct NMR observation of uniformly ^{15}N-labeled or ^{15}N-leucine-labeled human DHFR (80% deuterated) in complex with GroEL or SR1. In a number of different 2D TROSY experiments with GroEL, with the highest sensitivity from 2D [^{15}N-,^{1}H]-CRINEPT-HMQC-[^{1}H]-TROSY, only a small chemical shift dispersion was observed, the signals appearing in the random coil region of the spectrum, indicating that the bound substrate protein did not exhibit stable regular secondary or tertiary structure. Line broadening was also observed in both dimensions, potentially a function of different conformations of bound substrate protein (or of slow tumbling of the large complex). To analyze dynamics, ^{15}N-selected 1D ^{1}H NMR spectra of labeled hDHFR bound to SR1 were measured, using a CRIPT or INEPT magnetization transfer step. Instead of equal signal strengths with the two methods that would be predicted if the protein were rigidly bound to SR1, the INEPT transfer produced nearly three times higher signal intensity, suggesting internal mobility of the hDHFR relative to SR1. Model calculations of buildup curves over a range of different relaxation parameters supported the likelihood of slow internal motions. For both uniformly labeled DHFR and the leucine-labeled substrate protein, only a portion of the expected number of signals was observed, although the overlap of peaks prevented an exact tally. It seems clear, however, that only a portion of the polypeptide chain is present in an NMR-visible form. As evidence that the spectra of DHFR bound to GroEL were physiologically relevant, when ATP/GroES was added to the uniformly labeled or leucine-labeled DHFR complex, DHFR was released and the respective native, dispersed spectrum was observed.

Structures of GroEL/GroES–Nucleotide Complexes

CryoEM study

In 1996, Roseman, Saibil, and coworkers examined GroEL/GroES–ATP asymmetric complexes by cryoEM at ~30 Å resolution, comparing them with GroEL–ATP complexes. They observed that, in ATP alone, the apical domains appeared to exhibit a small elevation relative to the unliganded state, whereas, in the presence of GroES, they now underwent a large clockwise rotation (Fig. 2.9). Bridges of density from GroES to the rotated apical domains were observed — presumed to be the GroES mobile loops extending down to contact the GroEL apical domains. The idea was advanced that such rotation of the apical domains would remove the polypeptide binding surface from facing the cavity, rotating it into the intersubunit interface. This movement could potentially release polypeptide into the encapsulated chamber where folding could commence, as reflected in biochemical experiments (Weissman *et al.*, 1996; Mayhew *et al.*, 1996).

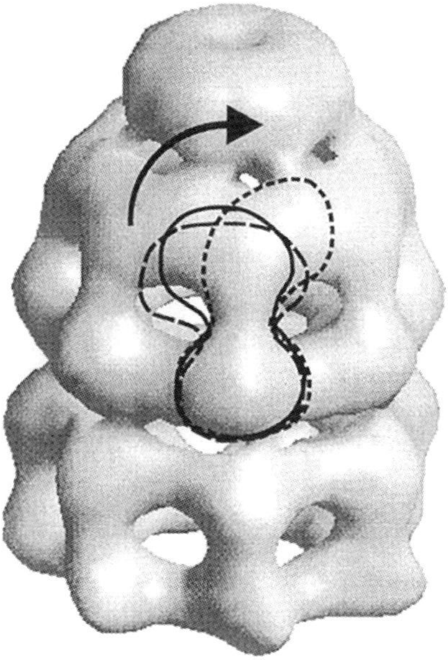

Fig. 2.9 GroEL subunit models superposed on the envelope of GroEL/GroES–ATP, from cryoEM studies of unbound, ATP-bound, and GroES–ATP-bound GroEL illustrating the elevation and twist of the apical domains of a GroES-bound ring. Long-dashed line, GroEL alone; solid line, GroEL–ATP; short-dashed line, GroEL/GroES–ATP. [from Roseman *et al.*, 1996.]

Crystal Structure of GroEL/GroES–ADP_7

Structure Determination

In 1997, Xu *et al.* presented a crystal structure of GroEL/GroES in ADP at ~3.0 Å resolution (Plate 2.14; PDB:1AON). Complexes of GroEL/GroES were formed in solution in MgADP, purified by gel filtration, and crystallized with PEG and sodium glutamate. Microseeding was used to generate single crystals. To successfully freeze these crystals, Zhaohui Xu carried out a myriad of trials, succeeding with "swishing" loop-captured crystals through an ethylene glycol solution and immediately freezing in a liquid nitrogen stream, then preserving the crystals in frozen propane. Crystals occupied a space group P21212, with a GroEL/GroES–ADP_7 asymmetric complex in the asymmetric unit. A molecular replacement approach was employed to identify GroEL using unliganded GroEL

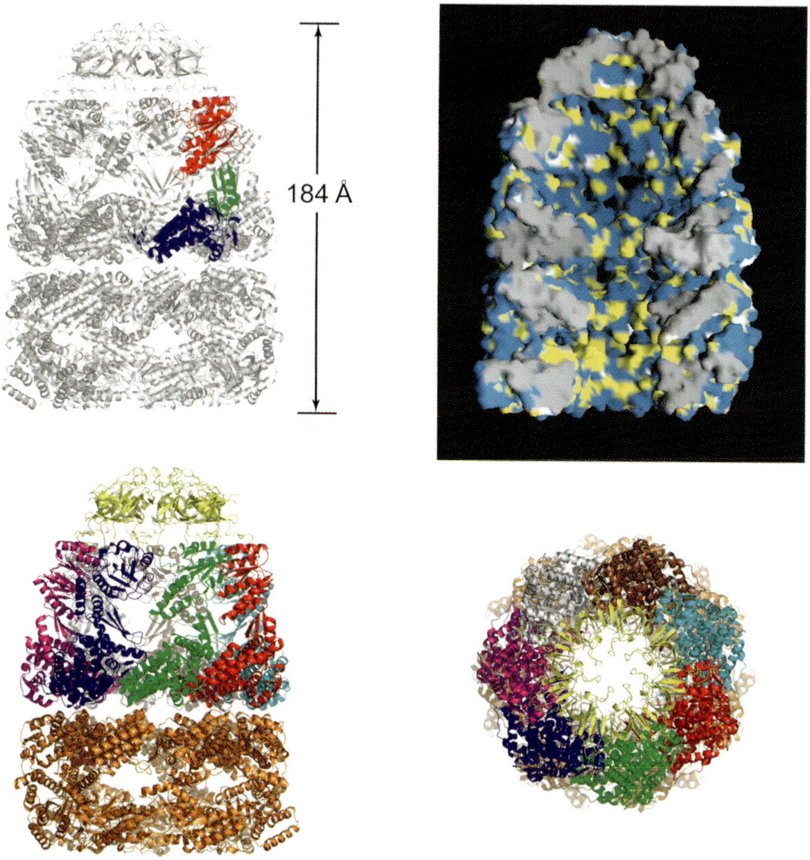

Plate 2.14. ***Architecture of a GroEL/GroES complex.*** Images of GroEL/GroES–ADP_7 from PDB: 1AON, showing GroES as a "lid" bound at one end of the GroEL double-ring via the descending mobile loops. The GroES-bound *cis* ring (multicolored in the lower panels) undergoes noticeable rigid body elevation and twisting of its apical domains (relative to the nonbound *trans* ring which resembles that of unliganded GroEL). The effect of GroES binding on the cavity lining is shown in the space-filling model in the top right panel: the open *trans* ring exposes the hydrophobic surface (yellow) that would enable capture of a non-native polypeptide. By contrast, the GroES-bound *cis* ring has undergone rigid body elevation and twisting of its apical domains such that the hydrophobic surface is removed from the cavity, contacting GroES mobile loops as well as forming an interface between apical domains (see Plate 2.16). The surface of the *cis* cavity is polar in character (rich in electrostatics), represented by the color blue (see Plate 2.17). (See text for discussion.)

to identify GroEL in the lattice. A cylindrical molecular boundary was placed around the complex in the asymmetric unit using GroEL's diameter and the height of the *c* axis, relative to which the seven-fold axis of the molecule was nearly parallel. Seven-fold NCS averaging with experimental phases from the molecular replacement model was then performed with low resolution data, revealing the bullet-shaped particle that had been seen in EM studies of GroEL/GroES–ADP complexes. Further phase extension with NCS averaging about the seven-fold axis produced a map with connected backbone density, and the subunits could be delineated. An envelope with one wedge consisting of a GroES subunit, the contiguous *cis* GroEL subunit and a contacting *trans* GroEL subunit was then used for subsequent averaging, placing spheres on the skeleton atoms. Subsequently, seven-fold NCS averaging was carried out, starting with random phases and a 12 Å map using only the intensity data at increasing resolution (with periodic envelope editing and upgrading of matrices). This last procedure removed any model bias.

Architecture and Rigid Body Movements in the *cis* GroEL Ring

In the 3.0 Å–derived model, a dome-shaped molecule is observed, with GroEL capped at the end of one ring by GroES, which closes off the cavity of the contiguous *cis* ring and anchors an elevated orientation of its apical domains (Plate 2.14). Accounting for the blockage of the cavity at the equatorial level by the mass of the C-terminal tails of the ring (unresolved here, as in the unliganded GroEL structure), the encapsulated *cis* cavity is ~80 Å in both height and diameter. This could accommodate a collapsed non-native, somewhat expanded polypeptide, in experimental contexts, of up to ~60 kDa (see Plate 2.17).

The apical *cis* ring subunit movements that contribute to *cis* cavity formation involve rigid body domain movements of the intermediate and apical domains (see Plate 2.10).

Intermediate Domain Downward Rotation

First, as in the EM structures in ATP alone, the intermediate domain rotates downward ~25° around the lower hinge (Pro137/Gly410), bringing its long helix M downward onto the nucleotide binding site on the top surface of the equatorial domain and forming contacts with the pocket and with a stem–loop in the same subunit and with the top of the equatorial domain (helix C) in the neighbor (Plate 2.8). The interactions sterically block the dissociation of ADP from the *cis* ring and imply a role of GroES binding in committing the bound physiological nucleotide, ATP, to *cis* hydrolysis. This was further supported by the subsequent, otherwise identical asymmetric structure of GroEL/GroES–$(ADP–AlF_3)_7$, where a trigonal AlF_3 metal complex was observed in the γ-phosphate position of the nucleotide pocket of the *cis* ring (Chaudhry *et al.*, 2003; see Plate 2.15). Here, Asp 398 on helix M from the intermediate domain has entered the nucleotide pocket of the same subunit to coordinate a water that is in-line with the aluminum, implying a mechanism of D398 activation of water for an attack on the γ-phosphate of ATP. Thus, the intermediate domain movement locks in and commits *cis* ATP to turnover.

In functional terms, as predicted particularly from the aluminum fluoride structure, when Asp 398 was mutated to Ala, ATP hydrolysis was reduced to ~2% wild-type level, despite normal ATP binding. This allowed dissection of the nucleotide cycle using D398A complexes, showing requirement for *cis* ring hydrolysis followed by *trans* ring ATP binding (and not hydrolysis) to release the *cis* ligands (GroES, polypeptide, and hydrolyzed ADP; Rye *et al.*, 1997).

Apical Domain Movements

The second striking movement in the GroEL/GroES–ADP_7 complex is the large rigid body apical domain movements in the *cis* ring which form the domed *cis* folding chamber. Directly compared with an unliganded GroEL ring, the apical domains are elevated by 60° and twisted around the long axis by ~90°, each apical domain associating with a

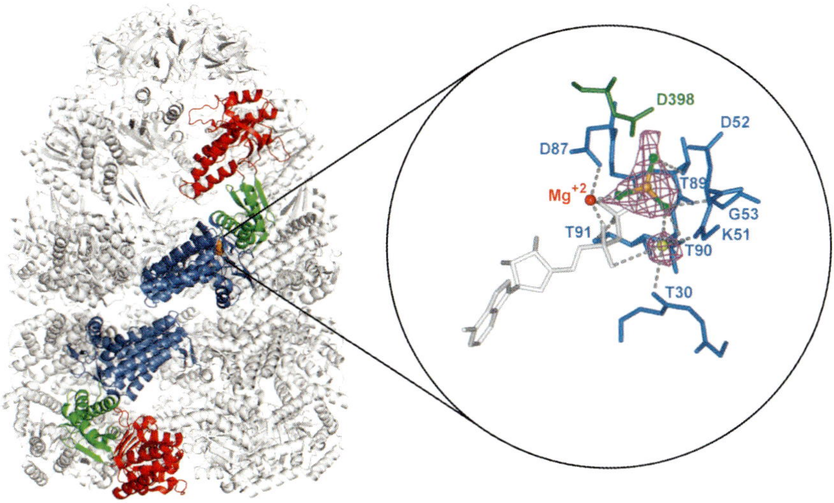

Plate 2.15. ***Nucleotide in cis ring of GroEL/GroES–(ADP–AlF$_3$)$_7$*** The architecture of a GroEL/GroES–(ADP–AlF$_3$)$_7$ complex is shown at the left and a blowup of the nucleotide pocket with contacts to the γ-phosphate and oxygens in a transition state mimicked by the trigonal AlF$_3$ complex is shown at the right. In particular, the side chain of D398 from the intermediate domain lines up with a water molecule (not shown) that is in-line for attack on the γ-phosphate. Phosphate oxygens are stabilized by the side chain of D52, D87-coordinated Mg+^{2}, and a potassium ion (yellow, in purple cage). (See text for additional details). [Taken from PDB: 1PCQ. Adapted from Chaudhry *et al.*, 2003.]

resolvable downwardly and outwardly positioned mobile loop of GroES (making contact with the apical domain 9 Å down and 20 Å out from the body of GroES). Given the recent cryoEM studies of GroEL–ATP described above (Clare *et al.*, 2012), the movements to arrive at the domed end state are clearly significantly more complicated, involving the initial movements directed by ATP binding (producing Rs1, Rs2, and Rs-open), a collision state that docks GroES to GroEL (potentially at the Rs-open state), and then a large clockwise twisting movement of the apical domains that produces the stable domed end state. If one takes Rs-open as a likely collision state for GroES, then the nature of the subsequent apical movements that produce the domed state observed in the X-ray structure can be deduced. In the putative collision state, the apical domains are counterclockwise-twisted by ~20°–30° (relative to unliganded GroEL) and already significantly elevated; thus, the subsequent production of the stable domed chamber must involve a large clockwise twist of ~120° and elevation of ~20°.

The Domed End-State *cis* Folding Chamber

In the final domed state, the GroEL hydrophobic polypeptide binding surface has been dislocated by the net elevation and clockwise twist away from facing the central cavity (Plate 2.16). This brings it to a position where contacts are formed between apical hydrophobic residues L234 and L237 of helix H and V264 of helix I of GroEL with an IVL (aa 25–27) in one edge of each of the GroES mobile loops (Plate 2.16). The other hydrophobic residues from helix I (L259,

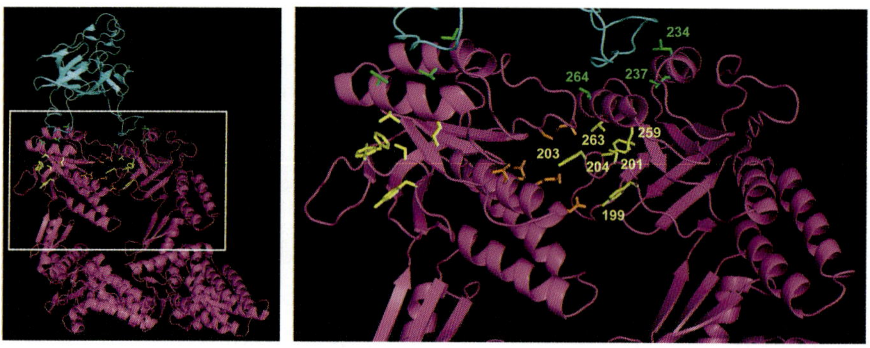

Plate 2.16. ***Removal of the hydrophobic polypeptide binding amino acids from facing the cavity upon ATP–GroES binding to GroEL.*** Left panel: Two subunits in the *cis* ring of GroEL in ADP–AlF3–GroES, including two respective contacted GroES subunits. The boxed area outlines the region containing the hydrophobic polypeptide binding amino acids, shown at higher resolution in the right panel. *Right panel*: Hydrophobic polypeptide binding residues of GroEL (L234, L237, V264) — in green — that contact the GroES mobile loop via its IVL side chains (not shown); hydrophobic residues of GroEL (L259, V263, Y199, S201, Y203, F204) — yellow — that move into the interapical region; side chains pointing into the interface from the neighboring apical domain, on helix J and following the loop, shown in orange, from bottom to top, D283, K286, L289, Q290 (on helix J), and S302 and I305 (on the loop following helix J), at a distance unlikely to form significant interactions with the yellow hydrophobics. [Taken from PDB: 1PCQ. See text for further details. Adapted from Xu *et al.*, 1997.]

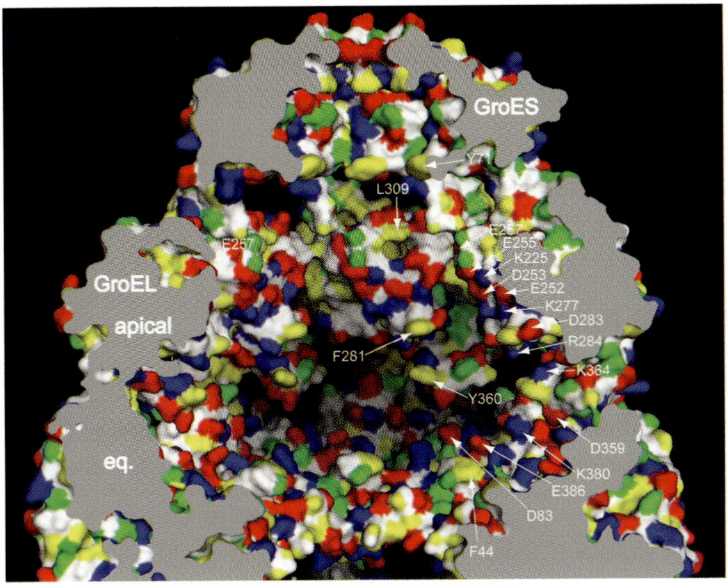

Plate 2.17. ***Electrostatic surface of the* cis *folding chamber.*** Cutaway view of the folding-active GroES-encapsulated *cis* chamber. Basic residues in blue; acidics in red; hydrophobics in yellow; polar residues in green; neutral residues in white. There is virtually no hydrophobic surface and a large amount of electrostatic surface, with a small excess of acidic residues. Individual residues are designated, but note that there are seven copies of each of these facing the chamber around the *cis* ring. [Taken from PDB: 1AON and from Farr *et al.*, 2007.]

V263) and the underlying segment (Y199, Y203, F204) are moved into the inter-apical interface, lying in proximity to helix J of the neighboring subunit and a following loop (see Plate 2.16 for polar residues that are situated on helix J and the loop — their character and distance make interaction with the hydrophobics unlikely).

As mentioned, the cavity of the GroEL/GroES *cis* chamber is ~75–80 Å in both height and diameter, and is now, in contrast with the hydrophobic lining of an open GroEL ring, polar in character, predominantly electrostatic (Plate 2.17). For example, cavity-facing residues include D224, K226, S228, E252, D253, E255, D283, R284, E304, K327, D328, D359, and E363. There is a net-negative charge (189 negatively charged side chains vs. 129 positively charged side chains per ring), which may act to repel incipiently folding proteins from the cavity wall, because the bulk of *E. coli* proteins have a pI below 7. There are only three hydrophobic side chains exposed to the *cis* cavity: L309, F281, and Y360. Mutation of these residues to hydrophilic/electrostatic character did not interfere with the folding function of the chaperonin system; nor did it affect the viability of *E. coli* (see Farr *et al.*, 2007).

The GroES Portion of the *cis* Chamber

The superior portion of the cavity is formed by the inside cavity of GroES. This is a smooth contiguous extension of the inner surface of GroEL, but above the position of the mobile loops it projects a hydrophobic "ridge," the side chains of GroES Tyr71, which jut out from the lowest aspect of the body of the GroES subunits. This residue was altered to hydrophilic (Arg or His) as an outcome of a screen that sought to optimize GFP folding at GroEL/GroES, but whether this alteration had a direct local effect on the folding protein or an indirect one mediated via an allosteric effect on the ATPase activity of GroEL was not clear (see Wang *et al.*, 2000; and comment in Horwich *et al.*, 2009). Thus, the nature of this residue and whether it has any effects on localization of substrate protein, or portions of it, within the cavity remain unclear.

Crystal Structure of *T. thermophilus* GroEL/GroES–ADP$_7$

In 2004, a crystal structure of an asymmetric GroEL/GroES complex from *T. thermophilus* was presented at 2.8 Å resolution (Shimamura *et al.*, 2004; PDB:4V4O). The overall architecture of this complex resembled that of *E. coli* GroEL with high conservation of cavity-facing residues, including the hydrophobic binding surface of helices H and I and the underlying segment, although the cavity wall of the *cis* chamber of this complex is much more acidic in character than that of GroEL/GroES. In a further experiment, crystals were dissolved and fractionated in SDS-PAGE, revealing a set of faint bands additional to GroEL and GroES, which when analyzed by MALDI MS corresponded to 24 different protein species that had apparently been trapped as substrates of the thermophilic chaperonin. Interestingly, their masses ranged from 13 kDa to 45 kDa, in agreement with the general mass of *cis* substrates of *E. coli* GroEL (~60 kDa being the limit). No common feature of these species could be identified. Notably, an asymmetry of the complex was observed in the apical region of the *cis* ring, attributed in part to its lack of intersubunit salt bridges, allowing more freedom of the apical domains. It was also speculated that perhaps substrate binding might produce asymmetry, although the diversity of substrates does not seem to lend itself to a specific asymmetry.

CryoEM Studies of Encapsulated Substrate Proteins

Two cryoEM studies have examined large substrates within folding-active GroEL/GroES complexes. The 50 kDa T4 phage capsid protein gp23 was examined after folding in the *cis* chamber of GroEL–gp31–ADP–aluminum fluoride complexes, a folding-active state (Clare *et al.*, 2009; EMDB 1548). Gp31 is a T4 phage-encoded version of GroES that contains a taller interior cavity than GroES as well as longer mobile loops (Hunt *et al.*, 1997; Clare *et al.*, 2006), enabling accommodation of the gp23 capsid protein. By contrast, GroES cannot encapsulate gp23. In the *cis* chamber, gp23 density occupies much of the chamber and could be fit with the major domain of the related gp24, for which there is a crystal structure. The density has an elongated shape that fits under the gp31 lid. The length of this density is too great in the axial direction to be accommodated by GroES, explaining the need for gp31 to promote T4 capsid folding and infection.

In a second study, a Rubisco subunit (51 kDa) was observed in the chamber of a complex formed by addition of ATP–GroES to D398A hydrolysis-defective GroEL, followed by rapid freezing (Chen *et al.*, 2013; EMDB 2327). Thus, Rubisco should likely have been released from the cavity wall and commenced *cis* folding in this experiment.

A density was observed in the *cis* chamber of nonsymmetrized complexes that could account for ~70% of the mass of native Rubisco. No secondary or tertiary structure could be delineated in this mass. The density was observed to be contiguous at the position of one asymmetric GroEL subunit with what appeared to be a resolvable C-terminal tail and the stem–loop that crosses from the neighboring subunit (aa 40–55). While such association, however transient, might be expected of a protein occupying the *cis* cavity, the investigators presented evidence to suggest that the encapsulation efficiency of GFP *in vitro* is reduced by the absence of the C-terminal tails. It has to be recalled, however, that deletion of the tails has no effect on the viability or growth rates of *E. coli in vivo* in a variety of media (Burnett *et al.*, 1993). Further, mutations in the stem–loop are without effect on polypeptide binding *in vitro* or ability to rescue GroEL-deficient *E. coli* (Fenton *et al.*, 1994). Thus, the significance of the observed physical contacts remains unclear.

Symmetric GroEL/GroES$_2$ Complexes

Crystal Structures

In 2014, Lorimer and coworkers presented crystal structures of GroEL/GroES$_2$–(ADP–BeF$_3$)$_{14}$ complexes at 3.7 Å — one devoid of substrate protein and another containing Rubisco (Fig. 2.10A; Fei *et al.*, 2014; PDB: 4PKO, 4PKN). Both crystallized in P212121 lattices, with slightly differing unit cell dimensions. Electron density for Rubisco, predicted to be able to reach native form in the respective complex, was not detected. This would not be expected considering the likely rotational freedom within the *cis* cavity.

In models of this structure derived without NCS operations, a degree of asymmetry was observed both in apical domain positioning and in positioning of the mobile loops of GroES. This was interpreted as reflecting "intrinsic flexibility" and "nonspecific interactions." Inspection of the electron density maps of either 4KPN or 4KPO shows that there is indeed subunit-to-subunit variation at these levels, but it also shows that

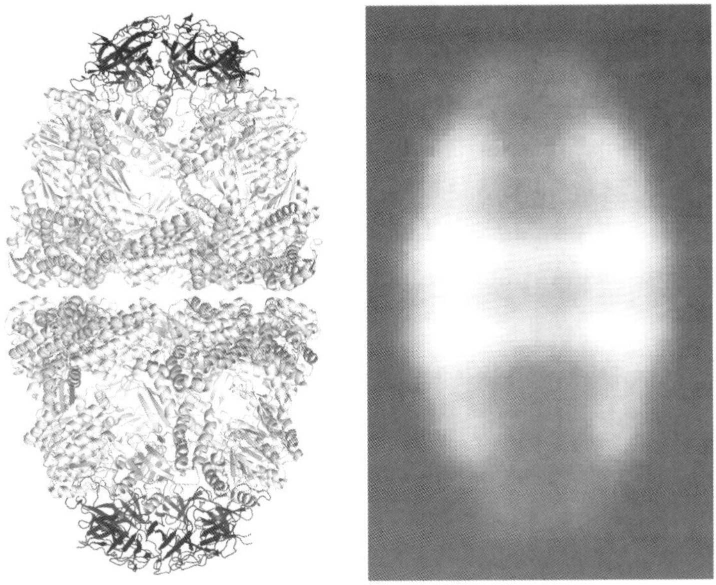

(A) (B)

Fig. 2.10 GroEL/GroES$_2$ symmetric structures by crystallographic and EM determination. [(A) From Fei *et al.*, 2014; (B) from Llorca *et al.*, 1994.]

the asymmetry may just as likely be a function of lattice contacts: symmetry mate equatorial domains enter at right angles and impose themselves directly upon the apical domains and the mobile loops. Further, in many cases, there is at best only partial density for the GroES mobile loops, obviating a call as to their positioning. It thus seems premature, with the local environment and electron density appearing as described, to draw any conclusions regarding apical and mobile loop flexibility as important for GroES binding.

EM Observations and Functional Considerations of GroEL–GroES$_2$ Complexes

EM studies

GroEL(GroES)$_2$ complexes have also been observed in EM studies as a variable percentage of particles when GroEL/GroES–ATP reactions are examined, with or without substrate protein (Fig. 2.10B; e.g. Llorca *et al.*, 1994; Sparrer *et al.*, 1997). In more recent biochemical studies *in vitro* using fluorescent reporters, a GroES/GroEL ratio ≥ 2 and a relatively high K^+ concentration (>100 mM) appear to favor such complexes (Yang *et al.*, 2013). The existence of $GroEL(GroES)_2$ particles with non-native protein substrate sequestered in both rings has been presented by several groups (e.g. Sparrer *et al.*, 1997; Takei *et al.*, 2012).

Proposed roles for $GroEL(GroES)_2$ complexes

How such substrate-containing $GroEL(GroES)_2$ particles would participate in the protein-folding cycle is not fully established, although model reaction cycles have been presented. In particular, Lorimer and his colleagues have proposed a "symmetric" cycle in which the binding of unfolded substrate protein to an open *trans* ring of an asymmetric "bullet" complex accelerates ADP/ATP exchange in the *trans* ring sufficiently to allow GroES binding there before ATP hydrolysis in the *cis* ring leads to GroES and protein substrate release (Ye *et al.*, 2013). Because of the substrate-protein-accelerated ADP exchange, symmetric complexes with non-native substrate proteins in both rings become a significantly populated intermediate state in the cycle. This proposal envisions the $GroEL(GroES)_2$ complexes as highly dynamic in the course of the folding cycle, with equally dynamic asymmetric complexes also participating.

This proposed cycle has several consequences. One is that it is highly dependent on the presence of sufficient non-native substrate protein to saturate a *trans* ring and change the ADP/ATP exchange rate [note that release of ADP has not been found to be limiting in another study (Tyagi *et al.*, 2010)]. At lower substrate concentrations, for example in an ongoing *in vitro* protein folding experiment with a limited starting substrate concentration, the cycle will revert to the asymmetric one previously described as the non-native substrate is folded. The symmetric cycle also implies that ATP hydrolysis could occur in either ring of the symmetric complex stochastically, changing the alternating ring model to more of a parallel processing one, although the first ring to bind ATP and GroES would seem to be favored temporally to be the first to complete ATP hydrolysis and be discharged of GroES and substrate (see Takei *et al.*, 2012). Finally, the "dwell time" of a non-native protein in the folding chamber would appear to be reduced in a fully symmetric reaction cycle, given the apparent change of rates. How this would affect the folding efficiency of different protein substrates is not clear. More binding and release cycles would be expected to occur, each capable of being productive as commented below, but effects on both rate and yield are difficult to predict, particularly in an *in vitro* reaction in which the non-native protein–chaperonin ratio is constantly changing.

Consideration of the reaction cycle in vivo

The relationship of the various models of the GroEL protein folding cycle to protein folding in the intact cell is unclear. Because there is one *GroEL* gene and one *GroES* gene in the *groE* operon, there is likely a 2:1 ratio of GroES heptamer rings to GroEL 14-mer double rings *in vivo*, fulfilling one requirement for participation of a symmetric complex. It has also been suggested that the chaperonin system is fully occupied by non-native substrates *in vivo*, but this estimate depends critically on what fraction of *E. coli* proteins occupy the chaperonin at any given time, a function of synthesis rates, relative concentrations, and association constants with GroEL, all parameters that are difficult to ascertain for the *in vivo* situation. The influence of ionic and metabolite conditions, particularly K^+ and ATP concentrations and the ATP/ADP ratio, will also play important roles in determining which model might apply. Perhaps the only data that would clarify this question would be cryoelectron tomography of chaperonin particles in flash-frozen *E. coli* cells grown under various conditions.

Modes for Kinetic Assistance During GroEL/GroES-Mediated Folding

Given the structural analyses presented here, there are several points in the GroEL/GroES reaction cycle, be it asymmetric or symmetric in behavior (or both), where assistance in polypeptide folding can be appreciated, with a variety of lines of experimental support.

Polypeptide Binding to an Open GroEL Ring

During the reaction cycle, open GroEL rings are accessible as *trans* binding-proficient rings opposite folding-active *cis* ones. Recent studies by both the Taguchi and Rye groups (Koike-Takashita, 2009; Lin *et al.*, 2013) indicate that *cis* ATP ternary complexes are able to accept polypeptide onto their *trans* rings during the time when the *cis* ring is folding-active (~10 s, the longest step of the asymmetric reaction cycle). Such a time might be significantly shorter in a symmetric reaction cycle as just discussed, but nonetheless binding to an open GroEL ring has been shown in fluorescent studies by Lin and Rye (2004), as well as Hartl and colleagues (Sharma *et al.*, 2008), to be associated with "stretching" of loosely structured non-native conformations, effectively reversing kinetically trapped topologies, presumably effected by a passive action of binding by multiple surrounding apical domains. This assisting action may be sufficient alone, without encapsulation, to promote the native state, as suggested by experiments in which *trans*-only GroEL/GroES complexes (with GroES loosely covalently tethered to one ring but preventing polypeptide access) can only bind and release non-native forms from an open ring without the step of encapsulation. These complexes were able to slowly refold stringent (GroEL/GroES-dependent) substrates like MDH and Rubisco at a rate that was 15–25% that of a normal GroEL/GroES system (where encapsulation occurs), and with recovery reduced to ~50% (Farr *et al.*, 2003). *In vivo*, a *trans*-only construct could rescue a GroEL-deficient strain to make colonies, but they were approximately one-tenth the normal size, further indicating the inefficiency of reaching the native state in the absence of ability to encapsulate (Farr *et al.*, 2003). Thus, the ability to bind substrate in an open ring is crucial to assisting production of the native state, but it is not very efficient alone, particularly as judged *in vivo*.

The Step of ATP–GroES Binding to a Polypeptide-Bound Ring

The *cis* ring changes reported by cryoEM upon ATP binding, involving particularly the separation of apical domains at the step of producing the Rs-open state (Clare, 2012), could in principle exert a pulling force on the bound polypeptide to further reverse wrong structures. This, as well as the further twisting apical movements following GroES collision, seems particularly appealing. Some evidence was found in an early hydrogen exchange study (Shtilerman *et al.*, 1999). The observations in that study, however, were not borne out by subsequent studies using the same hydrogen exchange protocol (Park *et al.*, 2005). On the other hand, a study examining ensemble FRET measurements of labeled Rubisco bound to the *trans* ring of an ADP–GroES asymmetric complex observed a rapid drop of FRET (consistent with an action of increasing the distance between fluorophores), followed by a slow rise, upon addition of either ATP–GroES or essentially ATP alone (in the presence of a GroES trap; Lin *et al.*, 2008). Similar observations were made with a double mutant of maltose binding protein (DM-MBP; Sharma *et al.*, 2008). In the latter study, however, an additional test was carried out to address whether the ATP-directed stretching (drop of FRET) was required for productive folding upon encapsulation, using the folding-productive transition state analogue, ADP–AlF_x. Addition of the analogue alone to DM-MBP–GroEL complexes produced 50-fold slower kinetics of stretching than ATP. Thus at short times after addition, little or no stretching had occurred, but addition of GroES led to rapid compaction of the substrate protein followed by fully productive *cis* folding. It therefore seems unclear that an ATP-mediated step of unfolding is mechanistically required for productive folding.

Cis Folding

The requirement of GroEL/GroES-dependent substrates to employ the *cis* chamber for their efficient folding seems well supported by a variety of reconstitution experiments. In one study, a GroEL/GroES-dependent ("stringent") substrate protein, rhodanese, could efficiently reach the native state inside of long-lived complexes of a single ring version of GroEL, SR1, complexed with GroES in the presence of ATP, with recovery of enzyme activity directly measured from the complexes (Weissman *et al.*, 1996). Notably, cycling GroEL/GroES reactions exhibit the same kinetics of rhodanese recovery as at SR1–GroES, despite the fact that polypeptide is released after every round of the ATP cycle (whether folded or not), implying that non-native states are rapidly rebound following release and spend virtually all of their time in the *cis* cavity attempting to reach native form. Importantly, if in a cycling reaction, access to the cavity of GroEL was blocked by sterically preventing the return of non-native substrate to the central cavity [by the rapid binding of tetrameric streptavidin to biotins placed at the inlet to the cavity (Brinker *et al.*, 2001)], production of the native state immediately halted, indicating that released non-native substrates cannot find their way to the native state in free solution. Of course, not only is encapsulation blocked in this experiment, but so is the binding to an open ring that is able to reverse kinetically trapped states. Yet, it seems clear that free solution is not conducive to productive folding, but rather is conducive to further misfolding and to aggregation, as was shown in the original studies of directly diluting stringent substrates into free solution (Goloubinoff *et al.*, 1989).

The importance of *cis* folding is further emphasized by evolutionary adaptation, for example in the case of T4 phage infection. Here, T4 has evolved a specialized "elongated" version of GroES, gp31, that functions with GroEL to mediate encapsulation and folding of the elongated T4 gp23 capsid protein (Clare *et al.*, 2009). Where GroES has been shown to be unable to encapsulate gp31 to assist its folding (Bakkes *et al.*, 2005), T4 brings in its own version of the cochaperonin to provide encapsulation and support *cis* folding.

Cis Folding Chamber — Active or Passive?

There is some discussion of whether the *cis* chamber might actively promote the native state through, for example, an action of confinement of substrate proteins or by the charged nature of its cavity walls (e.g. Tang *et al.*, 2006; Georgescauld *et al.*, 2014). Yet, a variety of data to date do not support such activity (Horst *et al.*, 2007; Farr *et al.*, 1997; Apetri and Horwich, 2008; Tyagi *et al.*, 2011; Motojima *et al.*, 2013; Ambrose *et al.*, 2015). For example, in Horst *et al.*, 2007, the trajectory of DHFR folding inside an SR1/GroES complex was compared with that in free solution by hydrogen–deuterium exchange/NMR and found to be identical. In Farr *et al.*, 1997, the effort to "stuff" the *cis* cavity to increase the confinement of substrate proteins by multiplying the GroEL C-terminal tails (Tang *et al.*, 2006) was found to be complicated by profound effects on ATPase activity, affecting the reaction cycle and rates of folding. In Apetri and Horwich, 2008, faster folding at GroEL/GroES than in free solution (under so-called "permissive" conditions — Tang *et al.*, 2006) could be accounted for by reversible aggregation in free solution that takes a substrate protein off-pathway, thus slowing its acquisition of native form. In Motojima *et al.*, 2012, the installation of a large number of positive charge substitutions in the GroEL cavity wall (as in Tang *et al.*, 2006), originally aimed at showing that reducing the net-negative charge of the *cis* cavity wall would result in slowing of the rate of folding, was shown instead to interfere with GroES binding, resulting in failure to form a *cis* cavity in the first place. Finally, in Ambrose *et al.*, 2015, aggregation in free solution was found to complicate an HX experiment (Georgescauld *et al.*, 2014) aimed at showing different routes of folding in a *cis* cavity versus free solution. Perhaps most striking, in Tyagi *et al.*, 2011, where GroEL/GroES produced the native state of the double mutant of maltose binding protein at a rate 5–10-fold greater than in free solution, relief of off-pathway multimolecular aggregation in free solution by an ionic adjustment increased the folding rate in free solution to equal that in the *cis* cavity. All of these results, particularly the last, argue for a passive chamber that supports the idea that the polypeptide chain in a *cis* chamber is using its primary structure to direct it to the native state, as Anfinsen originally proposed, but without the kinetic complication of aggregation.

Table 2.1 Crystal and CryoEM-derived Structures of GroEL, GroES, and their Complexes

PDB/EMD[1]	Chaperonin Structure		Mode[2]	Res. (Å)	Reference
1GRL	GroEL(R13G, A126V)		xtl	2.8	Braig *et al.*, 1994
1OEL	GroEL(R13G,A126V)		xtl	2.8	Braig *et al.*, 1995
1SSB, 1SX3, 1SVT	GroEL, GroEL/$(ATP\gamma S)_{14}$, GroEL/GroES/$(ADP\text{-}AlF3)_7$		xtl	2.8	Chaudhry *et al.*, 2004
—	GroES		xtl	2.8	Hunt *et al.*, 1996
1G31	gp31		xtl	2.3	Hunt *et al.*, 1997
1DER	GroEL(R13G,A126V)/$(ATP\gamma S)_{14}$		xtl	2.4	Boisvert *et al.*, 1996
1KP8	GroEL(R13G,A126V)/$(ATP\gamma S)_{14}$		xtl	2.0	Wang & Boisvert, 2003
3E76	GroEL/$(ATP\gamma S)_{14}$/Tl^{+}_{14}		xtl	3.94	Kiser *et al.*, 2009
4AAQ/1998	GroEL(D398A)/ATP_7	Rs1			Clare *et al.*, 2012
4AAR/1999		Rs2	EM	8.0–8.5	
4AAS/2000		Rs(open)			
2C7E/1047	GroEL(D398A)/ATP_7		EM	14.9	Ranson *et al.*, 2001
4KI8	GroEL(D83A-R197A)/ADP_{14}		xtl	2.7	Fei *et al.*, 2013
1KID	GroEL miniapical		xtl	1.7	Buckle *et al.*, 1997
1DKD	GroEL miniapical/SBP peptide		xtl	2.1	Chen & Sigler, 1999

(Continued)

Table 2.1 (*Continued*)

PDB/EMD[1]	Chaperonin Structure	Mode[2]	Res. (Å)	Reference
1MNF	GroEL$_{14}$/(SBP peptide)$_{14}$	xtl	3.0	Wang & Chen, 2000
—	GroEL/MDH	EM	11	Elad *et al.*, 2007
— 1544	GroEL/gp23	EM	10.7	Clare *et al.*, 2009
1AON	GroEL/GroES/ADP$_7$	xtl	3.0	Xu *et al.*, 1997
1PCQ	GroEL/GroES/(ADP-AlF$_3$)$_7$	xtl	2.8	Chaudhry *et al.*, 2003
4V4O	GroEL/GroES/ADP$_7$ *T.thermophilus*	xtl	2.8	Shimamura, *et al.*, 2004
3ZQ1/2327	GroEL(D398A)/GroES/ATP/ Rubisco	EM	15.9	Chen *et al.*, 2013
— 1548	GroEL/gp31/(ADP-AlF$_3$)$_7$/gp23	EM	10.1	Clare *et al.*, 2009
4PKO	GroEL/GroES$_2$/(ADP/BeF$_3$)$_{14}$	xtl	3.8	Fei *et al.*, 2014
4PKN	GroEL/GroES$_2$/(ADP/BeF$_3$)$_{14}$/ Rubisco	xtl	3.6	Fei *et al.*, 2014

[1]PDB: Protein Data Bank accession number; EMD: EM Databank accession number.
[2]xtl: crystal structure; EM: cryoEM structure.

References

Aharoni A, Horovitz A (1996). Inter-ring communication is disrupted in the GroEL mutant Arg13Gly; Ala126Val with known crystal structure. *J Mol Biol* **258:** 732–735.

Ambrose A, Fenton WA, Mason DJ, *et al.* (2015). Unfolded DapA forms aggregates when diluted into free solution, confounding comparison with folding by the GroEL/GroES chaperonin system. *FEBS Lett* **589:** 497–499.

Anfinsen CB (1973). Principles that govern the folding of protein chains. *Science* **181:** 223–230.

Apetri AC, Horwich AL (2008). Chaperonin chamber accelerates protein folding through passive action of preventing aggregation. *Proc Natl Acad Sci USA* **105:** 17351–17355.

Bakkes PJ, Faber BW, Heerikhuizen Hv, Van der Vies SM (2005). The T4-encoded cochaperonin, gp31, has unique properties that explain its requirement for the folding of the T4 major capsid protein. *Proc Natl Acad Sci USA* **102:** 8144–8146.

Barraclough R, Ellis RJ (1980). Protein synthesis in chloroplasts. IX. Assembly of newly-synthesized large subunit into ribulose bisphosphate carboxylase in isolated intact pea chloroplasts. *Biochim Biophys Acta* **608:** 19–31.

Boisvert DC, Wang J, Otwinowski Z, *et al.* (1996). The 2.4 Å crystal structure of the bacterial chaperonin GroEL complexed with ATPγS. *Nat Struct Biol* **3:** 170–177.

Braig K, Simon M, Furuya F, *et al.* (1993). A polypeptide bound by the chaperonin groEL is localized within a central cavity. *Proc Natl Acad Sci USA* **90:** 3978–3982.

Braig K, Otwinowski Z, Hegde R, *et al.* (1994). The crystal structure of the bacterial chaperonin GroEL at 2.8 Å. *Nature* **371:** 578–586.

Braig K, Adams PD, Brünger AT (1995). Conformational variability in the refined structure of the chaperonin GroEL at 2.8 Å resolution. *Nat Struct Biol* **2:** 1083–1094.

Brinker A, Pfeifer G, Kerner MJ, *et al.* (2001). Dual function of protein confinement in chaperonin-assisted protein folding. *Cell* **107:** 223–233.

Brünger AT (1992). *X-Plor: A System for X-Ray Crystallography and NMR.* (HHMI and Dept. MB&B, Yale University).

Brünger AT, Adams PD, Rice LM (1997). New applications of simulated annealing in X-ray crystallography and solution NMR. *Structure* **5:** 325–336.

Brünger AT, Adams PD, Rice LM (1998a). Recent developments for the efficient crystallographic refinement of macromolecular structures. *Curr Op Struct Biol* **8:** 606–611.

Brünger AT, Adams PD, Clore GM, *et al.* (1998b). Crystallography & NMR system: A new software suite for macromolecular structure determination. *Acta Cryst D* **54:** 905–921.

Buckle AM, Zahn R, Fersht AR (1997). A structural model for GroEL-polypeptide recognition. *Proc Natl Acad Sci USA* **94:** 3571–3575.

Burnett BP, Horwich AL, Low KB (1994). A carboxy-terminal deletion impairs the assembly of GroEL and confers a pleiotropic phenotype in *Escherichia coli* K12. *J Bact* **176:** 6980–6985.

Chandrasekhar GN, Tilly K, Woolford C, *et al.* (1986). Purification and properties of the groES morphogenetic protein of *Escherichia coli. J Biol Chem* **261:** 12414–12419.

Chapman E, Farr GW, Usaite R, *et al.* (2006). Global aggregation of newly translated proteins in an *Escherichia coli* strain deficient of the chaperonin GroEL. *Proc Natl Acad Sci USA* **103:** 15800–15805.

Chaudhry C, Farr GW, Todd MJ, *et al.* (2003). Role of the γ-phosphate of ATP in triggering protein folding by GroEL-GroES: function, structure and energetics. *EMBO J* **22:** 4877–4887.

Chaudhry C, Horwich AL. Brunger AT, Adams PD (2004). Exploring the structural dynamics of the *E. coli* chaperonin GroEL using translation–libration–screw crystallographic refinement of intermediate states. *J Mol Biol* **342:** 229–245.

Chaudhuri TK, Farr GW, Fenton WA, *et al.* (2001) GroEL–GroES-mediated folding of a protein too large to be encapsulated. *Cell* **107:** 235–246.

Chen D-H, Madan D, Weaver J, *et al.* (2013). Visualizing GroEL/ES in the act of encapsulating a folding protein. *Cell* **153:** 1354–1365.

Chen L. Sigler PB (1999). The crystal structure of a GroEL/peptide complex: plasticity as a basis for substrate specificity. *Cell* **99:** 757–768.

Chen S, Roseman AM, Hunter AS, *et al.* (1994). Location of a folding protein and shape changes in GroEL–GroES complexes imaged by cryo-electron microscopy. *Nature* **371:** 261–264.

Cheng MY, Hartl F-U, Martin J, *et al.* (1989). Mitochondrial heat shock protein HSP60 is essential for assembly of proteins imported into yeast mitochondria. *Nature* **337:** 620–625.

Cheng M-Y, Hartl F-U, Horwich AL (1990). Hsp60, the mitochondrial chaperonin, is required for its own assembly. *Nature* **348:** 455–458.

Cheng Y, Grigorieff N, Penczek PA, Walz T (2015). A primer to single-particle cryo-electron microscopy. *Cell* **161:** 438–449.

Clare DK, Bakkes PJ, van Heerikhuizen H, *et al.* (2006). An expanded protein folding cage in the GroEL–gp31 complex. *J Mol Biol* **358:** 905–911.

Clare DK, Bakkes PJ, van Heerikhuizen, H, *et al.* (2009). Chaperonin complex with a newly folded protein encapsulated in the folding chamber. *Nature* **457:** 107–110.

Clare DK, Vasishtan D, Stagg S, *et al.* (2012). ATP-triggered conformational changes delineate substrate-binding and -folding mechanics of the GroEL chaperonin. *Cell* **149:** 113–123.

Creighton TE. (1993). The hydrophobic interaction in *Proteins: Structures and Molecular Properties*. W.H. Freeman & Co, New York, pp.157–160.

Dubaquié Y, Looser R, Fünfschilling U, *et al.* (1998). Identification of *in vivo* substrates of the yeast mitochondrial chaperonins reveals overlapping but non-identical requirement for hsp60 and hsp10. *EMBO J* **17:** 5868–5876.

Elad N, Farr GW, Klare DK, *et al.* (2007). Topologies of a substrate protein bound to the chaperonin GroEL. *Mol Cell* **26:** 415–426.

Epstein CJ, Goldberger RF, Anfinsen CB (1963). The genetic control of tertiary protein structure: studies with model systems. *Cold Spring Harbor Symp Quant Biol* **28:** 439–449.

Ewalt KL, Hendrick JP, Houry WA, Hartl FU (1997). *In vivo* observation of polypeptide flux through the bacterial chaperonin system. *Cell* **90:** 491–500.

Farr GW, Furtak K, Rowland MB, *et al.* (2000). Multivalent binding of non-native substrate proteins by the chaperonin GroEL. *Cell* **100:** 561–573.

Farr GW, Fenton WA, Chaudhuri TK, *et al.* (2003). Folding with and without encapsulation by *cis*- and *trans*-only GroEL–GroES complexes. *EMBO J* **22:** 3220–3230.

Farr GW, Fenton WA, Horwich AL (2007). Perturbed ATPase activity and not "close confinement" of substrate in the *cis* cavity affects rates of folding by tail-multiplied GroEL. *Proc Natl Acad Sci USA* **104:** 5342–5347.

Fayet O, Ziegelhoffer T, Georgopoulos C (1989). The groES and groEL heat shock gene products of *Escherichia coli* are essential for bacterial growth at all temperatures. *J Bacteriol* **171:** 1379–1385.

Fei X, Yang D, LaRonde-LeBlanc N, *et al.* (2013). Crystal structure of a GroEL-ADP complex in the relaxed allosteric state at 2.7 Å resolution. *Proc Natl Acad Sci USA* **110:** E2958–2966.

Fei X, Ye X, LaRonde NA, Lorimer GH (2014). Formation and structures of GroEL:$GroES_2$ chaperonin footballs, the protein-folding functional form. *Proc Natl Acad Sci USA* **111:** 12775–12780.

Fenton WA, Kashi Y, Furtak K, Horwich AL (1994). Residues in chaperonin GroEL required for polypeptide binding and release. *Nature* **371:** 614–619.

Fujiwara K, Ishihama Y, Nakahigashi K, *et al.* (2010). A systematic survey of *in vivo* obligate chaperonin-dependent substrates. *EMBO J* **29:** 1552–1564.

Georgescauld F, Popova K, Gupta AJ, *et al.* (2013). GroEL/ES chaperonin modulates the mechanism and accelerates the rate of TIM-barrel domain folding. *Cell* **157:** 922–934.

Georgopoulos CP, Hendrix RW, Kaiser AD, Wood WB (1972). Role of the host cell in bacteriophage morphogenesis: effects of a bacterial mutation on T4 head assembly. *Nat New Biol* **239:** 38–41.

Georgopoulos CP, Hendrix RW, Casjens SR, Kaiser AD (1973). Host participation in bacteriophage lambda head assembly. *J Mol Biol* **76:** 45–60.

Glick BS, Brandt A, Cunningham K, *et al.* (1992). Cytochrome c1 and b2 are sorted to the intermembrane space of yeast mitochondria by a stop-transfer mechanism. *Cell* **69**: 809–822.

Goldberger RF, Epstein CJ, Anfinsen CB (1963). Acceleration of reactivation of reduced bovine pancreatic ribonuclease by a microsomal system from rat liver. *J Biol Chem* **238:** 628–635.

Goloubinoff P, Christeller JT, Gatenby AA, Lorimer GH (1989). Reconstitution of active dimeric ribulose bisphosphate carboxylase from an unfolded state depends on two chaperonin proteins and MgATP. *Nature* **342:** 884–889.

Gray TE, Fersht AR (1991). Cooperativity in ATP hydrolysis by GroEL is increased by GroES. *FEBS Lett* **292:** 254–258.

Hemmingsen SM, Woolford C, Van Der Vies SM, *et al.* (1988). Homologous plant and bacterial proteins chaperone oligomeric protein assembly. *Nature* **333:** 330–334.

Hendrix RW (1979). Purification and properties of groE, a host protein involved in bacteriophage assembly. *J Mol Biol* **129:** 375–392.

Hohn T, Hohn B, Engel A, Wurtz M (1979). Isolation and characterization of the host protein groE involved in bacteriophage lambda assembly. *J Mol Biol* **129:** 359–373.

Horst R, Bertelsen EB, Fiaux J, *et al.* (2005). Direct NMR observation of a substrate protein bound to the chaperonin GroEL. *Proc Natl Acad Sci USA* **102:** 12748–12753.

Horst R, Fenton WA, Englander SW, *et al.* (2007). Folding trajectories of human dihydrofolate reductase inside the GroEL–GroES chaperonin cavity and free in solution. *Proc Natl Acad Sci USA* **104:** 20788–20792.

Horwich AL, Apetri AC, Fenton WA (2009). The GroEL/GroES *cis* cavity as a passive anti-aggregation device. *FEBS Lett* **583:** 2654–2662.

Hunt JF, Weaver AJ, Landry SJ, *et al.* (1996). The crystal structure of the GroES co-chaperonin at 2.8 Å resolution. *Nature* **379:** 37–45.

Hunt JF, van der Vies SM, Henry L, Deisenhofer J (1997). Structural adaptations in the specialized bacteriophage T4 co-chaperonin Gp31 expand the size of the Anfinsen cage. *Cell* **90:** 361–371.

Hutchinson EG, Tichelaar W, Hofhaus G, *et al.* (1989). Identification and electron microscopic analysis of a chaperonin oligomer from *Neurospora crassa* mitochondria. *EMBO J* **8:** 1485–1490.

Ishihama A, Ikeuchi T, Matsumoto A, *et al.* (1976). A novel adenosine triphosphatase isolated from RNA polymerase preparations of *Escherichia coli. J. Biochem* **79:** 927–936.

Ishii N, Taguchi H, Sumi M, Yoshida M (1992). Structure of holo-chaperonin studied with electron microscopy. *FEBS* **299:** 169–174.

Kelley PM, Schlesinger MJ (1978). The effect of amino acid analogues and heat shock on gene expression in chicken embryo fibroblasts. *Cell* **15:** 1277–1286.

Kerner MJ, Naylor DJ, Ishihama Y, *et al.* (2005). Proteome-wide analysis of chaperonin-dependent protein folding in *Escherichia coli*. *Cell* **122:** 209–220.

Kiser PD, Lorimer GH, Palczewski K (2009). Use of thallium to identify monovalent cation binding sites in GroEL. *Acta Cryst* F **65:** 967–971.

Kleywegt GJ, Jones TA (1994). Halloween…masks and bones. In: S Bailey, R Hubbard, D Waller (eds.), *From First Map to Final Model*. SERC Daresbury Laboratory, pp. 59–66.

Koike-Takeshita A, Yoshida M, Taguchi H (2008). Revisiting the GroEL–GroES reaction cycle via the symmetric intermediate implied by novel aspects of the GroEL(D398A) mutant. *J Biol Chem* **283:** 23774–23781.

Landry SJ, Zeilstra-Ryalls J, Fayet O, *et al.* (1993). Characterization of a functionally important mobile domain of GroES. *Nature* **364:** 255–258.

Langer T, Pfeifer G, Martin J, *et al.* (1992). Chaperonin-mediated protein folding: GroES binds to one end of the GroEL cylinder, which accommodates the protein substrate within its central cavity. *EMBO J.* **11:** 4757–4765.

Lemaux PG, Herendeen SL, Bloch PL, Neidhardt FC (1978). Transient rates of synthesis of individual polypeptides in *E. coli* following temperature shifts. *Cell* **13:** 427–434.

Lin Z, Rye HS (2004). Expansion and compression of a protein folding intermediate by GroEL. *Mol Cell* **16:** 23–34.

Lin Z, Madan D, Rye HS (2008). GroEL stimulates protein folding through forced unfolding. *Nat Struct Biol* **15:** 303–311.

Lin Z, Puchalla J, Shoup D, Rye HS (2013). Repetitive protein unfolding by the *trans* ring of the GroEL–GroES chaperonin complex stimulates folding. *J Biol Chem* **288:** 30944–30955.

Llorca O, Marco S, Carrascosa JL, Valpuesta JM (1994). The formation of symmetrical GroEL–GroES complexes in the presence of ATP. *FEBS Lett* **345:** 181–186.

Lubin M (1969). Observations on the structure of RNA polymerase and its attachment to DNA. *J Mol Biol* **39:** 219–233.

Lüthy R, Bowie JU, Eisenberg D (1992). Assessment of protein models with three-dimensional profiles. *Nature* **356:** 83–85.

Lyumkis D, Moeller A, Cheng A, *et al.* (2010). Automation in single-particle electron microscopy connecting the pieces. *Methods Enzymol* **483:** 291–338.

Madan D, Lin Z, Rye HS (2008). Triggering protein folding within the GroEL–GroES complex. 2008. *J Biol Chem* **283:** 32003–32013.

Martin J, Langer T, Boteva R, *et al.* (1991). Chaperonin-mediated protein folding occurs at the surface of GroEL via a molten globule-like intermediate. *Nature* **352:** 36–42.

Mayhew M, da Silva AC, Martin J, *et al.* (1996). Protein folding in the central cavity of the GroEL–GroES chaperonin complex. *Nature* **379:** 420–426.

McMullen TW, Hallberg RL (1987). A normal mitochondrial protein is selectively synthesized and accumulated during heat shock in *Tetrahymena thermophila*. *Mol Cell Biol* **7:** 4414–4423.

McMullen TW, Hallberg RL (1988). A highly evolutionarily conserved mitochondrial protein is structurally related to the protein encoded by the *Escherichia coli* groEL gene. *Mol Cell Biol* **8:** 371–380.

Miyazaki T, Yoshimi T, Furutsu Y, *et al.* (2002). GroEL–substrate–GroES ternary complexes are an important transient intermediate of the chaperonin cycle. *J Biol Chem* **277:** 50621–50628.

Motojima F, Motojima-Miyazaki Y, Yoshida M (2012). Revisiting the contribution of negative charges on the chaperonin cage wall to the acceleration of protein folding. *Proc Natl Acad Sci USA* **109:** 15740–15745.

Murai N, Makino Y, Yoshida M (1996). GroEL locked in a closed conformation by an interdomain cross-link can bind ATP and polypeptide but cannot process further reaction steps. *J Biol Chem.* **271:** 28229–28234.

Neidhardt FC, VanBogelen RA, Vaughn V (1984). The genetics and regulation of heat-shock proteins. *Annu Rev Genet* **18:** 295–329.

Nojima T, Murayama S, Yoshida M, Motojima F (2008). Determination of the number of active GroES subunits in the fused heptamer GroES required for interactions with GroEL. *J Biol Chem* **283:** 18385–18392.

Ostermann J, Horwich AL, Neupert W, Hartl F-U (1989). Protein folding in mitochondria requires complex formation with HSP60 and ATP hydrolysis. *Nature* **341:** 125–130.

Park ES, Fenton WA, Horwich AL (2005). No evidence for a forced-unfolding mechanism during ATP/GroES binding to substrate-bound GroEL: no observable protection of metastable Rubisco intermediate or GroEL-bound Rubisco from tritium exchange. *FEBS Lett* **579:** 1183–1186.

Pelham HRB (1984). Hsp70 accelerates the recovery of nucleolar morphology after heat shock. *EMBO J* **3:** 3095–3100.

Pelham HRB (1986). Speculations on the functions of the major heat shock and glucose-regulated proteins. *Cell* **46:** 959–961.

Pushkin AV, Tsuprun VL, Solvjeva NA, *et al.* (1982). High molecular weight pea leaf protein similar to the groE protein of *Escherichia coli*. *Biochim Biophys Acta* **704:** 379–384.

Ranson NA, Burston SG, Clarke AR (1997). Binding, encapsulation and ejection: substrate dynamics during a chaperonin-assisted folding reaction. *J Mol Biol* **266:** 656–664.

Ranson NA, Farr GW, Roseman AM, *et al.* (2001). ATP-bound states of GroEL captured by cryo-electron microscopy. *Cell* **107:** 869–879.

Reading DS, Hallberg RL, Myers AM (1989). Characterization of the yeast HSP60 gene coding for a mitochondrial assembly factor. *Nature* **337:** 655–659.

Roseman AM, Chen S, White H, *et al.* (1996). The chaperonin ATPase cycle: Mechanism of allosteric switching and movements of substrate-binding domains in GroEL. *Cell* **87:** 241–251.

Rye HS, Burston SG, Fenton WA, *et al.* (1997). Distinct actions of *cis* and *trans* ATP within the double-ring of the chaperonin GroEL. *Nature* **388:** 792–798.

Saibil H, Dong Z, Wood S, auf der Mauer A (1991). Binding of chaperonins. *Nature* **353:** 25–26.

Saibil HR, Zheng D, Roseman AM, *et al.* (1993). ATP induces large quaternary rearrangements in a cage-like chaperonin structure. *Curr Biol* **3:** 265–273.

Schedl P, Artavanis-Tsakonis S, Steward R, *et al.* (1978). Two hybrid plasmids with *D. melanogaster* DNA sequences complementary to mRNA coding for the major heat shock protein. *Cell* **14:** 921–929.

Sharma S, Chakraborty K, Müller BK, *et al.* (2008). Monitoring protein conformation along the pathway of chaperonin-assisted folding. *Cell* **133:** 142–153.

Shewmaker F, Maskos K, Simmerling C, Landry SJ (2001). The disordered mobile loop of GroES folds into a defined β-hairpin upon binding GroEL. *J Biol Chem* **276:** 31257–31264.

Shimamura T, Koike-Takeshita A, Yokoyama K, *et al.* (2004). Crystal structure of the native chaperonin complex from *Thermus thermophilus* revealed unexpected asymmetry at the *cis*-cavity. *Structure* **12:** 1471–1480.

Shtilerman M, Lorimer GH, Englander SW (1999). Chaperonin function: folding by forced unfolding. *Science* **284:** 822–825.

Sparrer H, Rutkat K, Buchner J (1997). Catalysis of protein folding by symmetric chaperone complexes. *Proc Natl Acad Sci USA* **94:** 1096–1100.

Staniforth RA, Burston SG, Atkinson T, Clarke AR (1994). Affinity of chaperonin-60 for a protein substrate and its modulation by nucleotides and chaperonin-10. *Biochem J* **300:** 651–658.

Takano T, Kakefuda T (1972). Involvement of a bacterial factor in morphogenesis of bacteriophage capsid. *Nat New Biol.* **239:** 34–37.

Takei Y, Iizuka R, Ueno T, Funatsu T (2012). Single-molecule observation of protein folding in symmetric GroEL–$(GroES)_2$ complexes. *J Biol Chem.* **287:** 41118–41125.

Tang YC, Chang HC, Roeben A, *et al.* (2006). Structural features of the GroEL-GroES nan-cage required for rapid folding of encapsulated protein. *Cell* **125:** 903–914.

Thiyagarajan P, Henderson SJ, Joachimiak A (1996). Solution structures of GroEL and its complex with rhodanese from small-angle neutron scattering. *Structure* **4:** 79–88.

Tilly K, Murialdo H, Georgopoulos C (1981). Identification of a second *Escherichia coli groE* gene whose product is necessary for bacteriophage morphogenesis. *Proc Natl Acad Sci USA* **78:** 1629–1633.

Todd MJ, Viitanen PV, Lorimer GH (1993). Hydrolysis of adenosine 5'-triphosphate by *Escherichia coli* GroEL: effects of GroES and potassium ion. *Biochemistry* **32:** 8560–8567.

Tyagi NK, Fenton WA, Horwich AL (2010). ATP-triggered ADP release from the asymmetric chaperonin GroEL/GroES/ADP_7 is not the rate-limiting step of the GroEL/GroES reaction cycle. *FEBS Lett* **584:** 951–953.

Tyagi NK, Fenton WA, Deniz AA, Horwich AL (2011). Double mutant MBP refolds at same rate in free solution as inside the GroEL/GroES chaperonin chamber when aggregation in free solution is prevented. *FEBS Lett* **585:** 1969–1972.

Viitanen PV, Gatenby AA, Lorimer GH (1992). Purified chaperonin 60 (groEL) interacts with the non-native states of a multitude of *Escherichia coli* proteins. *Protein Sci* **1:** 363–369.

Wang JD, Herman C, Tipton KA, *et al.* (2002). Directed evolution of substrate-optimized GroEL/S chaperonins. *Cell* **111:** 1027–1039.

Wang J, Boisvert DC (2003). Structural basis for GroEL-assisted protein folding from the crystal structure of $(GroEL–KMgATP)_{14}$ at 2.0 Å resolution. *J Mol Biol* **327:** 843–855.

Wang J, Chen L (2003). Domain motions in GroEL upon binding of an oligopeptide. *J Mol Biol* **334:** 489–499.

Wang Z, Feng H-p, Landry SJ, *et al.* (1999). Basis of substrate binding by the chaperonin GroEL. *Biochemistry* **38:** 12537–12546.

Weissman JS, Kashi Y, Fenton WA, Horwich AL (1994). GroEL-mediated protein folding proceeds by multiple rounds of release and rebinding of non-native forms. *Cell* **78:** 693–702.

Weissman JS, Hohl CM, Kovalenko O, *et al.* (1995). Mechanism of GroEL action — productive release of polypeptide from a sequestered position under GroES. *Cell* **83:** 577–587.

Weissman JS, Rye HS, Fenton WA *et al.* (1996). Characterization of the active intermediate of a GroEL–GroES-mediated protein folding reaction. *Cell* **84:** 481–490.

Xu Z, Horwich AL, Sigler PB (1997). The crystal structure of the asymmetric GroEL–GroES–$(ADP)_7$ chaperonin complex. *Nature* **388:** 741–750.

Yang D, Ye X, Lorimer GH (2013). Symmetric GroEL:GroES2 complexes are the protein-folding functional form of the chaperonin nanomachine. *Proc Natl Acad Sci USA* **110:** E4298–E4305.

Ye X, Lorimer GH (2013). Substrate protein switches GroE chaperonins from asymmetric to symmetric cycling by catalyzing nucleotide exchange. *Proc Natl Acad Sci USA* **110:** E4289–E4297.

Yifrach O, Horovitz A (1994). Two lines of allosteric communication in the oligomeric chaperonin GroEL are revealed by the single mutation Arg196 → Ala. *J Mol Biol* **243:** 397–401.

Yifrach O, Horovitz A (1995). Nested cooperativity in the ATPase activity of the oligomeric chaperonin GroEL. *Biochemistry* **34:** 5303–5308.

Chapter 3

Structural Analysis of Type II Chaperonins in Archaebacteria and the Eukaryotic Cytosol

Arthur Horwich

Abstract

This chapter continues with the subject of chaperonin-mediated protein folding, dealing with a second family of chaperonins found in archaebacteria and the eukaryotic cytosol, termed type II chaperonins. The revelation in 1989 that organellar chaperonins are protein-folding machines prompted an immediate search for such machines in other compartments and kingdoms of life, resulting in the discovery in archaebacteria and, through homology, in the eukaryotic cytosol of additional double ring machines that mediate protein folding. Yet, this second family of chaperonins has some strikingly different properties: in contrast to the organellar (type I) chaperonins, these double ring assemblies have a built-in "lid" structure within their apical domains for encapsulating folding proteins within the cavities of their rings, as opposed to a detachable cochaperonin "lid." The polypeptide-binding properties of the apical domains also appear to differ. In addition, while type II chaperonins have nearly identical alpha-helical equatorial domains and a similar ATP-binding pocket and mechanism of hydrolysis, their inter-ring interface differs, allowing the use of steps of ATP binding and hydrolysis to produce different allosterically directed movements than the organellar chaperonins. The variety of differing features lead to machines and a reaction cycle that appear to particularly favor folding of the different set of protein substrates that co-evolved with the type II chaperonins.

*Yale School of Medicine, Howard Hughes Medical Institute, 145 Boyer Center for Molecular Medicine, 295 Congress Avenue, New Haven, CT 06510, USA.

Discovery of Type II Chaperonins in Archaea and the Eukaryotic Cytosol

Two studies in 1991 identified members of the type II chaperonin family. Phipps, Baumeister, and their colleagues purified and analyzed an abundant double ring assembly composed of two stacked eight-member rings from the cytoplasm of the extremely thermophilic archaebacterium *Pyrodictium*, grown at 100°C (Fig. 3.1A). Two distinct subunits of 56 and 59 kDa were observed, suggested to alternate within the rings, and subsequently demonstrated by observation in cryoEM of four fold symmetry (Nitsch *et al.*, 1997). The rings measured 16 nm in diameter and 15 nm in height, and exhibited a central cavity 5 nm in diameter. Purified complexes exhibited ATPase activity that was maximal at ~100°C. Remarkably, when *Pyrodictium occultum* cells were grown at 108°C, the complex became the single major species of the cell, constituting 73% of total soluble protein — a putative heat shock response of an already thermophilic organism, and a response reminiscent of induction of GroEL by heat shock. It was concluded that the complex, termed a thermosome, might function as a chaperonin, assisting protein folding under the thermophilic conditions of these organisms.

Shortly thereafter, Trent, Hartl, Horwich, and coworkers reported on a thermophilic archaebacterial ring assembly from *Sulfolobus shibatae*, composed of stacked nine-member rings containing subunits of ~55 kDa, called thermophilic factor 55 (TF55) (Fig. 3.1B). The *Sulfolobus* organism grows at 75°C, and the ATPase activity of the purified complex correspondingly peaked at that temperature. Consistent with the behavior of TF55 as a chaperone, it could bind *E. coli* proteins that become unfolded at 75°C, but not a native thermophilic protein. Most significantly, when the primary structure of the TF55 subunit was determined from a genomic clone, a 59 kDa polypeptide was predicted that bore a relationship — 40% identity along its entire length — to a eukaryotic cytosolic protein called t-complex polypeptide-1 (TCP1), found in all eukaryotes and known to be essential in *S. cerevisiae*, where a

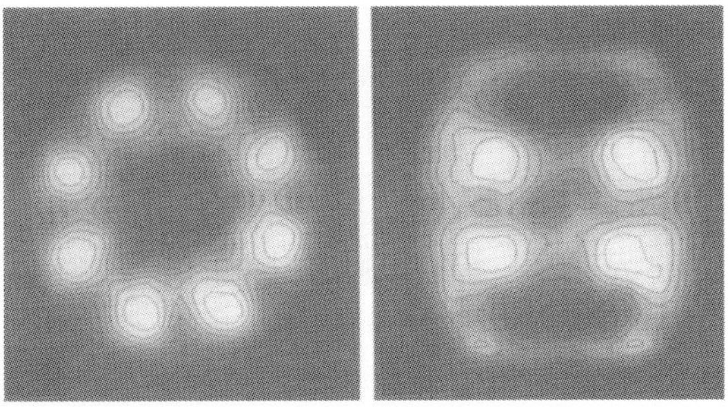

(A)

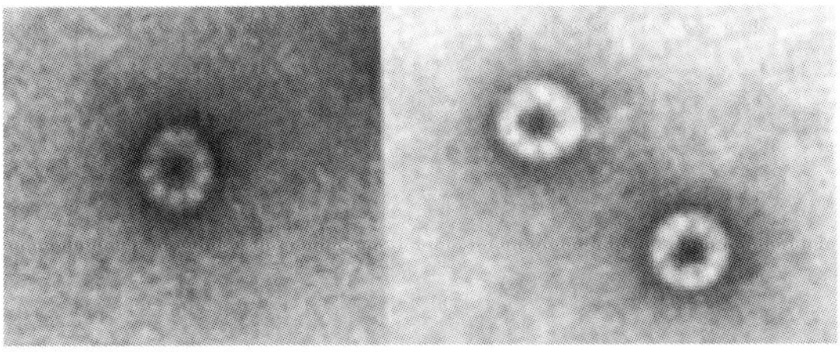

(B)

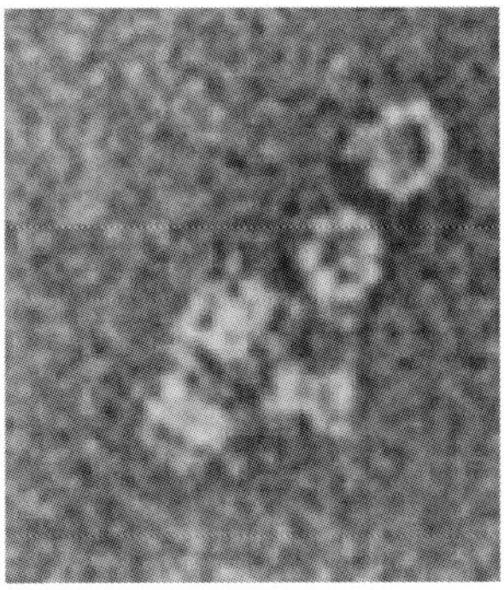

(C)

Fig. 3.1. Early negative stain EM images of type II chaperonin ring assemblies. **A**. Thermosome from thermophilic archaebacterium *Pyrodictium brockii*, with averaged end and side views, showing eight-member rings [from Phipps *et al*., 1991.] **B**. Chaperonin from thermophile *S. shibatae*, with nine-member rings [from Trent *et al*., 1991.] **C**. CCT chaperonin complex from eukaryotic cytosol (mouse testis germ cells), showing eight-member rings [from Lewis *et al*., 1992].

cold-sensitive mutant affected mitotic spindle formation (Ursic and Culbertson, 1991). Limited homology of TCP-1 with the bacterial GroEL chaperonin had already been suggested (Ahmad and Gupta, 1990; Ellis, 1990). Thus, it was strongly suggested that the archaebacterial cytosol and eukaryotic cytosol contained structurally related chaperonins.

TCP1 was shown to reside in a heterooligomeric eight-member double ring complex in the eukaryotic cytosol (Lewis *et al.*, 1992; Fig. 3.1C), and chaperone activity of the purified complex was soon demonstrated by an ability to bind and ATP-dependently refold nonnative actin, tubulin, and firefly luciferase (Gao *et al.*, 1992; Yaffe *et al.*, 1992; Frydman *et al.*, 1992). The ring assembly was called CCT (chaperonin-containing TCP1), or TRiC (TCP1 ring complex). It was not thermally inducible like its archaeal relatives, and each of its subunits was found to be essential for viability in yeast. Notably, GroEL could bind the substrate polypeptides of CCT in nonnative form but could not productively fold them in the presence of ATP and GroES (e.g. actin and tubulin; Tian *et al.*, 1995), indicating the unique ability of CCT to act on co-compartmentalized eukaryotic cytosolic substrate proteins.

Archaebacterial Chaperonins

Structure of the *T. acidophilum* Archaebacterial Thermosome in a Closed State

Structure determination and similarities to GroEL

In 1998, the crystal structure of the hexadecameric $(\alpha\beta)_4(\alpha\beta)_4$ thermosome from *T. acidophilum* was reported by Ditzel, Huber, Steinbacher, and their colleagues at 2.6 Å resolution (PDB: 1A6D; Plate 3.1). MIR phasing of I422 crystals with an $\alpha\beta$ dimer in the a.u. was employed, with chain tracing guided by the model for GroEL. As with GroEL, N- and C-terminal stretches of residues were not present in density, here the first 16(α) or 19(β) and last 26(α) or 22(β) residues. As with GroEL, the C-termini of these subunits have unusual character — α with GGM repeats and β with a run of serines followed by acidics. The fold of the equatorial and intermediate domains of the subunits was nearly identical to those domains of GroEL, with excellent superposition with the *cis* GroEL ring of GroEL/GroES. Notably, as in the *cis* ring, the intermediate domain of the thermosome including the long α-helix (here, helix L) is rotated downward onto the equatorial nucleotide pocket, and a correspondent acidic residue, D390, is swung into the ATP pocket (Plate 3.1).

Unique apical domain features

The apical domain was mostly superposable with the apical domain of GroEL, but it departs in three principal ways, as observed first in a standalone apical structure (Klump *et al.*, 1997). Most striking, an insertion is present at the cavity-facing aspect, a 28-residue apical domain insertion that forms a protrusion emanating from the cavity aspect at the top of each subunit, the collective forming a "lid" over the central cavity in the closed state (Plate 3.1). Each protrusion is formed from a triangle of two β-strands (S12 and S13) followed by an 18-residue α-helical extension at the N-terminus of what would be GroEL helix H (which lies at the inlet to the GroEL cavity). The eight S13 β-strands form a closed parallel β-octagon, an iris-like structure over the innermost aspect of the closed-off central cavity (Plate 3.1). Together with the packing of the legs of the triangle with legs of neighboring subunits, this seals off the central cavity. Both the sealing of the cavity and the orientation of the intermediate and apical domains as in a *cis* ring make clear that the state of the assembly in this crystal is a closed one, in which presumably a nonnative polypeptide would undergo folding.

A second item of difference with GroEL concerns the positions, on helices homologous to H and I in GroEL, where hydrophobic residues would be located for polypeptide binding to the cavity aspect of apical domains of an open ring. These residues are highly conserved in the organellar chaperonins, but in the thermosome subunits they are not present, being replaced with a variety of polar or even electrostatic residues. Thus, the nature of polypeptide binding seemed unclear (see below). A third item of difference is that the thermosome apical domain lacks the antiparallel helices that are present at the back of the apical domain of GroEL (GroEL helices K and L).

Chamber of a closed thermosome

Concerning the closed state, observed in this structure, the chamber surface of the closed thermosome is hydrophilic in character, resembling the GroEL/GroES cavity, with burial of nearly all of the surface-exposed

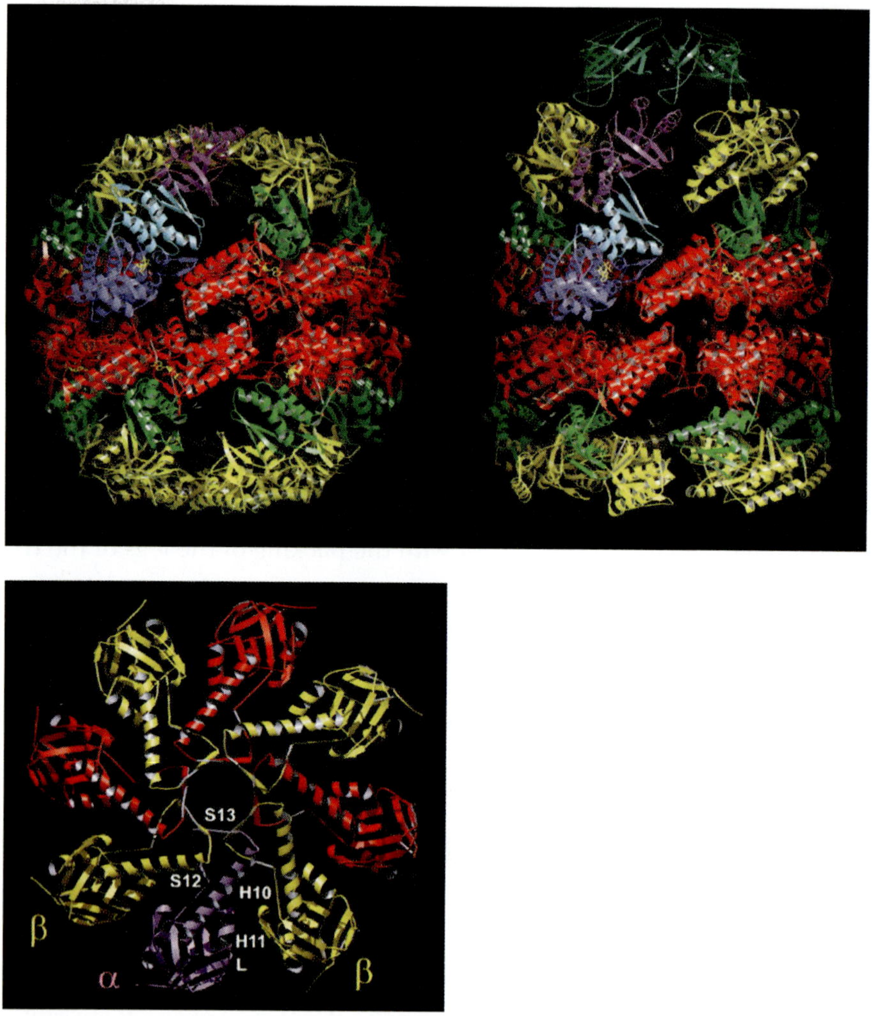

Plate 3.1. ***Model of the thermosome from T. acidophilum from crystallographic study (PDB: 1A6D).*** *Top image*: Comparison of the thermosome (left) with the asymmetric GroEL/GroES–ADP$_7$ complex (right). The equatorial domains of both chaperonins contain the same long conserved α-helices, shown in red — note the staggered relationship of the equatorial domains across the ring–ring interface in GroEL but the symmetric relationship in the thermosome. The intermediate domains of the two chaperonins also contain the same secondary structure and are shown in green, in orientation tilted down onto the equatorial domains in both rings of the "closed" thermosome and in the upper GroES-bound (*cis*) ring of GroEL (compare that with the untilted intermediate domains of the *trans* ring of GroEL). The apical domains, shown in yellow, differ between the two chaperonins (see text). *Lower image*: End view of a ring of the thermosome. α-subunits are colored red, or purple in the case of one α-subunit whose protrusion

hydrophobic residues of the subunits, including the protrusions, to subunit–subunit interfaces. This resembles the rigid body movement of GroEL's apical domains upon binding ATP and GroES, which removes the hydrophobic polypeptide binding surface from facing the *cis* cavity and directs it into subunit–subunit interfaces.

The surface of the closed thermosome chamber is at least as electrostatic in character as the *cis* GroEL/GroES chamber, containing eight positive and nine negative side chains per subunit. Yet, the volume of the thermosome chamber is significantly less than that of GroEL/GroES (see Plate 3.1). If the masses of the disordered C-termini are accounted for, the volume is estimated at ~75% that of the *cis* GroEL/GroES chamber, with a height of ~45 Å and a diameter of ~80 Å at the equatorial level, narrowing to ~55 Å at the apical level. The volume could accommodate proteins up to ~50 kDa, including, for example, an actin subunit (44 kDa).

Subunit arrangement across the ring–ring interface

Also differing from GroEL is the ring–ring interface (Plate 3.1). Instead of the staggered arrangement in GroEL where two contact points of a subunit in one ring meet contact points of two adjacent subunits in the opposite ring, here there is a 1:1 contact of subunits with each other across the interface, with twofold-symmetry-relatedness of each subunit to its identical neighbor across the equatorial plane. This has consequences for the relationship of left and right sites of the contact at the bottom aspect of each subunit. Instead of the homologous contacts formed in GroEL — for example, the helix D that emanates from the ATP pocket in the subunit of one ring interacting with the helix D from a subunit

◄

Plate 3.1. (*Continued*) secondary structure is labeled. The identically folded and 60% sequence-identical β-subunits are colored yellow. Note the fourfold relationship of the subunit types around the ring. The cavity of the ring is closed off by the collective of protrusions unique to the apical domains of type II chaperonins, composed of S12, S13, and the N-terminal portion of helix H10. An "iris" in the center is composed of eight copies of strand S13, one contributed from each of the eight subunits.

in the opposite ring — here the equivalent helix 5 of a left contact site interacts with a helix (H17) from the right contact site. Despite these heterologous contacts, there is nevertheless the same negative cooperativity for ATP binding and hydrolysis between rings as is present in the GroEL system, demonstrated by biochemical studies of Gutsche *et al.* (2000), Kusmierczyk and Martin (2003), and Bigotti and Clarke (2005).

Nucleotide binding resembles that at GroEL and informs about the hydrolysis mechanism by both machines

The closed state of the thermosome was observed without the presence of nucleotide in the crystallization mixture, and this was attributed to crystallization conditions that included 2M ammonium sulfate. This left unclear what allosteric movements are directed by ATP binding and hydrolysis, in particular whether the closed state is driven by ATP. Nevertheless, nucleotide binding in the equatorial pockets could be observed upon soaking the crystals with ADP, ATPγS, or ADP–AlF_x. Even with the intermediate domain rotated downward, nucleotide could enter the pocket, allowing assessment of the stereochemistry. In particular, the ADP–AlF_x soak (PDB: 1A6E) established the mechanism of ATP hydrolysis (Fig. 3.2). Mg–ADP–AlF_3 was bound in the ATP pocket with one fluorine contributing to the coordination sphere of Mg^{+2}, one forming a hydrogen bond to T96, and one to T97, the two threonines in the GDGTT Walker motif. A water at the apex of the trigonal bipyramid (not well resolved in GroEL/GroES structures) was fixed by hydrogen bonds to Asp 390 from the intermediate domain and Asp63 from the equatorial. The orientation of the water relative to the aluminum corresponds to being in-line to attack the γ-phosphate (Fig. 3.2).

Notably, a second archaeal thermosome crystal structure, of a recombinant all-α (mutant) subunit assembly of the *Thermococcus* KS-1 thermosome (natively an $\alpha\beta$ assembly), also in the absence of nucleotide but presence of high sulfate in the crystallization mix, was published in 2004, revealing an identical closed state and a similar subunit fold to that of the *T. acidophilum* structure (Shomura *et al.*, 2004; PDB: 1Q3Q, R, S and 1Q2V).

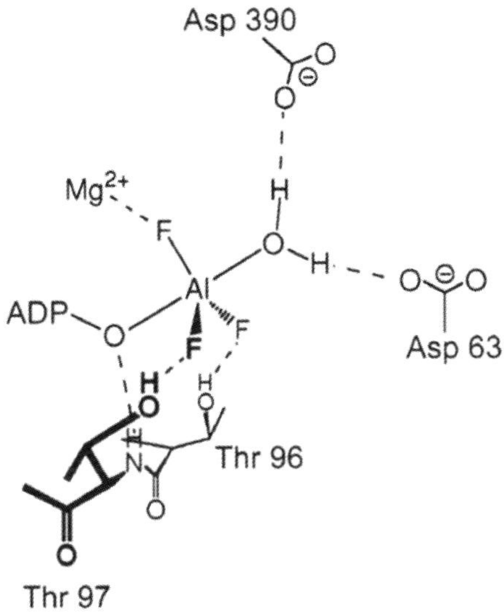

Fig. 3.2. Schematic of the AlF_3 moiety of the ADP–AlF_3 transition state mimic in the nucleotide pocket of the thermosome subunit, from PDB: 1A6E. Note that activation of water is mediated by Asp390, swung in from the down-tilted intermediate domain — similar to the behavior of Asp398 of the GroEL intermediate domain. Asp63 contributes to coordination of the water molecule. The fluorines are stabilized by Mg^{+2} and by T96, T97 from the conserved P-loop sequence GDGTT.

Open Conformation of Archaebacterial Type II Chaperonins by Electron Cryotomography and X-Ray Crystallography

EM Study

In 2010, a high resolution cryoEM study of both closed and open states of the hexadecameric homooligomeric chaperonin from the archaeon *M. maripaludis* was reported (Zhang *et al.*, 2010). An earlier study had shown that deletion of the apical protrusion segment (replacing it with a four residue linker segment) still allowed both substrate protein binding and hydrolysis of ATP (Reissmann *et al.*, 2007), and this "Δlid" form was used to avoid strictly end-on views obtained with the intact chaperonin. A nucleotide-free form of Δlid was employed for the open state (EMDB 5140), whereas the closed state was achieved by addition of ATP–AlF_x (EMDB 5138), owing to an earlier observation that such addition, associated with ATP hydrolysis (or the transition state thereof), produced a closed state (Meyer *et al.*, 2003; notably, in solution studies therein, none of AMP–PNP, ATPγS, or ADP–BeF_x could promote closure, whereas ADP–AlF_x could at least partially do so; and ATP, presumed to hydrolyze, could fully close the chaperonin). A 4.3 Å map of the closed state of the wild type (EMDB 5137) allowed chain tracing, with visualization of ~70% of side chains, revealing a structure akin to the crystallographic model from *T. acidophilum*. Notable were the extensive side-by-side contacts between the subunits at all levels.

The nucleotide-free open state maps of the wild type (EMDB 5139) and Δlid (EMDB 5140) were less resolved, at 10 Å and 8 Å, respectively, but these were notable for the complete absence of contacts above the equatorial domains, with the apical domains projecting up vertically in a standalone state devoid of any contacts (Plate 3.2; see also Clare *et al.*, 2008). The open-to-closed transition could be correlated at the equatorial level with several local changes at both intra-ring and inter-ring subunit–subunit interfaces. The changes at the ring–ring interface were most notable, driving/enabling a tilting ("rocking forward") of the entire subunit on the inter-equatorial interfaces such that the apical protrusions were moved forward over the central cavity in transitioning from the open to the closed state (see Plate 3.3).

Crystallization of Open Forms of Archaeal Chaperonin

In 2010, two crystal structures of archaeal chaperonins in an open state were reported. In Pereira *et al.*, the homooligomeric chaperonin from *M. maripaludis* lacking its lid domain (Δlid) was crystallized in an open state in ATPγS (PDB: 3KFK). The structure of an I222 crystal was resolved at 6.0 Å using molecular replacement. As in the cryoEM study, the open form showed apical domains rising up vertically, with no intersubunit contacts formed between intermediate or apical domains at any

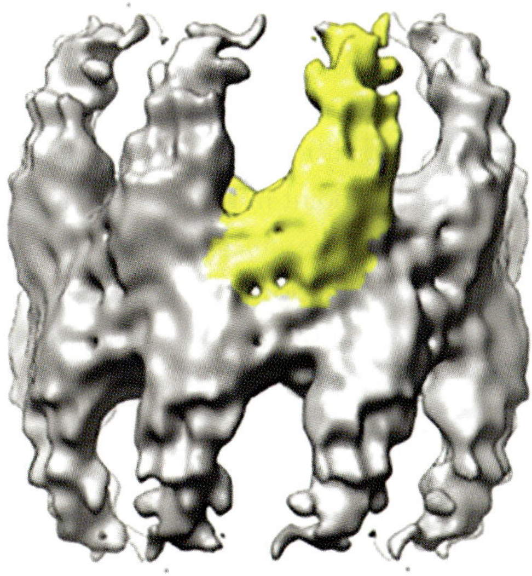

Plate 3.2. ***The open state of the wild-type M. maripaludis chaperonin* (*EMDB 5139*)** at 10 Å resolution, side view, showing lack of contacts above the equatorial domain [from Zhang *et al.*, 2010].

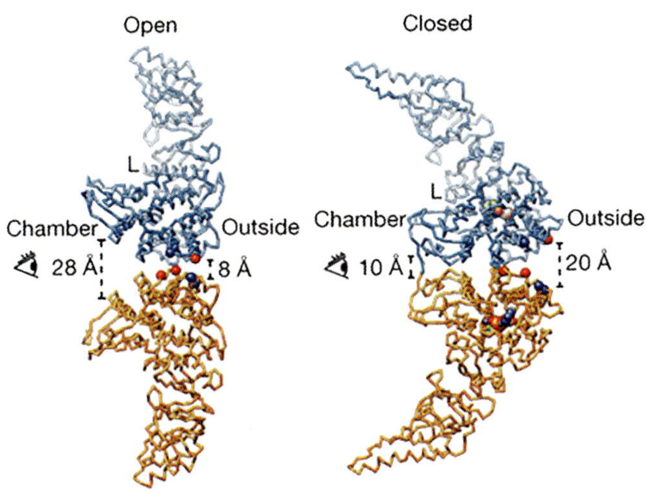

Plate 3.3. ***Side view comparison of pairs of subunits across the equatorial interface from open and closed states of the M. maripaludis chaperonin***, from the cryoEM study by Zhang *et al.*, 2010 (from EMDB 5139 and 5137, respectively). The subunits appear to rock forward on the equatorial interface to form the closed state. Notably, the equatorial β-sheets at the cavity aspect of the rings come close together upon closure of the chaperonin, whereas external contacts widen.

level. A closed structure of wild-type chaperonin was also resolved in an AMP–PNP-bound state (PDB: 3KFB), resembling that of *T. acidophilum*. Comparing the open and closed states of the *M. maripaludis* chaperonin, once again an inward-tilting movement of the entire subunit was appreciated, focused about a point in the equatorial domain at the ATP binding site (Pro41), moving the apical protrusions inward across the cavity. In addition, a 35° counterclockwise rotation of the entire subunit about the eightfold axis was observed. At a global level, these subunit movements direct an estimated change in height of the intact chaperonin from 204 Å in the open state to 162 Å in the closed, with the diameter maintained at ~160 Å. In the closed state, there are side-by-side subunit–subunit interactions at all levels, mostly hydrophilic in character.

Crystallographic Resolution of Adjustments of the Intra- and Inter-ring Subunit–Subunit Contacts in the Open-to-Closed Transition

Intra-ring subunit–subunit contacts in the open state (Plate 3.4, left panel) are confined to the equatorial level and are formed at the external aspect of the barrel by interactions between loops of adjacent subunits (H5/H6 loop and H16–H17 loop; not shown). At the internal aspect is the β-sheet contact of the first strand or last strand from one subunit with the stem–loop antiparallel β-strands (S2, S3) of the neighboring subunit (as in GroEL). In the closed state (Plate 3.4, right panel), this same stem–loop contact is preserved, but contacts now occur in the upper domains. The intermediate domain contacts the upper region of the equatorial domain of the neighboring subunit (loop S20/H19 contacting loop S7/S8). At the apical level there are also contacts (H11 and the S9/S10 loop of one subunit contact loop H12/S17 of the neighbor). Finally, there are the inter-subunit contacts of the built-in lid.

A nucleotide-sensing loop unique to type II chaperonins contributes to intra-ring subunit contacts. ATP turnover within a ring is positively cooperative, with a Hill coefficient of 2.0 (e.g. Reissmann *et al.*, 2007). Notably, the loss of the lid structure produces loss of

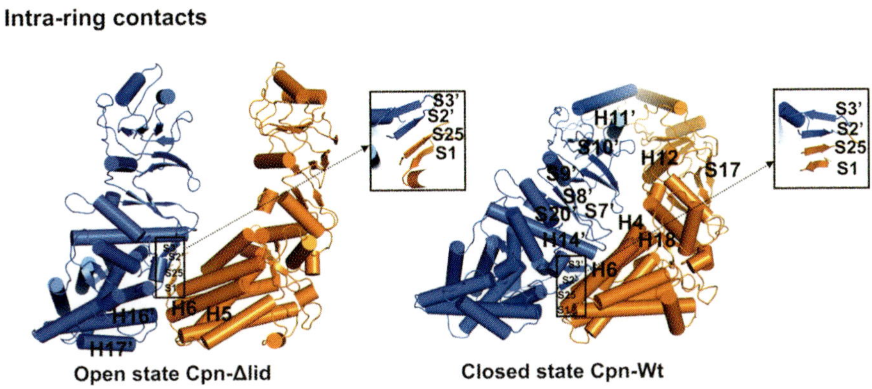

Plate 3.4. ***Intra-ring contacts in the open and closed states of the M. maripaludis chaperonin***. In the open state, contacts are restricted to the equatorial domain, whereas contacts occur at all three levels in the closed state. (See text for details.) [Taken from Pereira *et al*., 2010; PDB: 3KFK, 3KFB.]

positive cooperativity, indicating the participation of the lid in the allosteric actions of the machine (Reissmann, 2007). But cooperativity appears to be supported at an additional level, as indicated by the identification of a further structure that has been recognized from studies of closed *M. Maripaludis* crystal structures (Pereira *et al*., 2012). This is an intermediate domain loop (aa 160–169) containing a residue conserved in type II chaperonins, Lys161, that forms a contact with the γ-phosphate of ATP (Plate 3.5). This nucleotide-sensing loop (NSL) rearranges from a random coil in the presence of ATP forms [AMP–PNP (PDB: 3RUV) or ADP–AlF_x (PDB: 3RUW)] to alpha-helical in the presence of ADP (PDB: 3RUQ), simulating the effect of cleavage of the γ-phosphate. This structural adjustment affords an opening for product release from the nucleotide pocket. Such release was identified as the rate-limiting step of the nucleotide cycle (Bigotti *et al*., 2006). The neighbor of Lys 161 in the loop is Glu164, which contacts the stem–loop from the neighboring subunit in the ATP-like states. Such interaction is lost upon assuming the hydrolyzed ADP state. Both K161A and E164A mutants exhibited a slowed rate of ATP turnover, but lid closure could still proceed normally. It is presumed that K161 senses ATP and E164 couples the signal to the neighboring subunit.

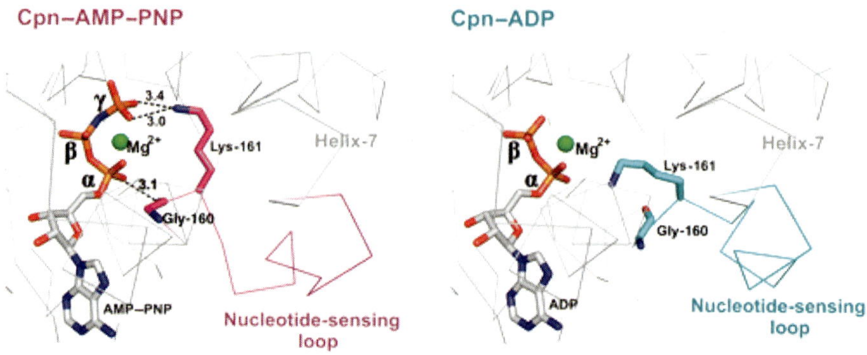

Plate 3.5. ***K161 in the nucleotide-sensing loop (intermediate domain loop 160–169)***, forms a contact with the γ-phosphate of ATP in the open state but not following hydrolysis. Its neighbor in the loop, E164 (not shown), contacts the stem–loop from the neighboring subunit in the ATP state, a contact lost after hydrolysis. This thus serves as a further line of inter-ring cooperativity. (See text for details.) [Taken from Pereira *et al.*, 2012.]

Inter-ring subunit–subunit contacts, noted here to be symmetric as in earlier studies, change from mostly hydrophobic contacts in the open state to several salt bridges in the closed state (Plate 3.6). In the open state, hydrophobic contacts are formed between side chains on H17 on one subunit contacting loops of H5/H6 and H15/H16 on the opposite subunit. In the closed state, electrostatic interactions occur between Arg18 and Arg29 on H1 in one subunit, with Glu33 and Asp112 of H1 and H5 in the opposite subunit. A constant salt bridge, Arg425–Asp457, is maintained in both states.

Apical Domain of the Open State and Putative Polypeptide Binding Site

At the apical level in the open state, two hydrophobic patches were observed — one on the protrusion structure and the other between helices 10 and 11 from the body of the apical domain (corresponding to helices H and I in GroEL). While the hydrophobics in the protrusion structure are conserved, the protrusion could be deleted without effect

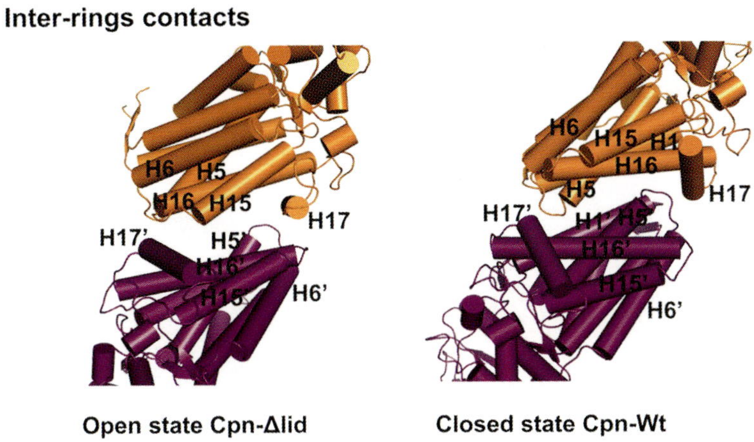

Plate 3.6. ***Inter-ring contacts in the open and closed states of the M. maripaludis chaperonin***. Hydrophobic contacts in the open state are replaced with electrostatic contacts in the closed state. (See text for details.) [Taken from Pereira *et al.*, 2010; PDB: 3KFK, 3KFB.]

on substrate binding (Reissmann, 2007), indicating the protrusion residues to be unlikely to play a role in polypeptide binding. The second set of hydrophobic residues in the helix 10/11 interhelical groove (L269, V273, A274, I278 on helix 10; L293, A294, Y297, L298, I383 on helix 11) were proposed to play a role in polypeptide binding (see modeled structures in Plate 3.7). Consistent with such a role, in the closed state of the *M. maripaludis* chaperonin, the interhelical site is rendered inaccessible because a loop including aa 326–331 of a neighboring subunit makes contact (see Fig. 3.5 of Douglas *et al.*, 2011). Indeed, functional studies altering loop residues T327, N328, D330, and D331 simultaneously to alanine blocked release of a bound substrate from the *M. maripaludis* chaperonin even in the setting of ATP hydrolysis and lid closure (Douglas *et al.*, 2011; and see below). Additional support for inter-helical substrate binding derives from the finding that helix 11 is involved in binding of substrates by the eukaryotic cytosolic chaperonin, CCT (see below, under CCT).

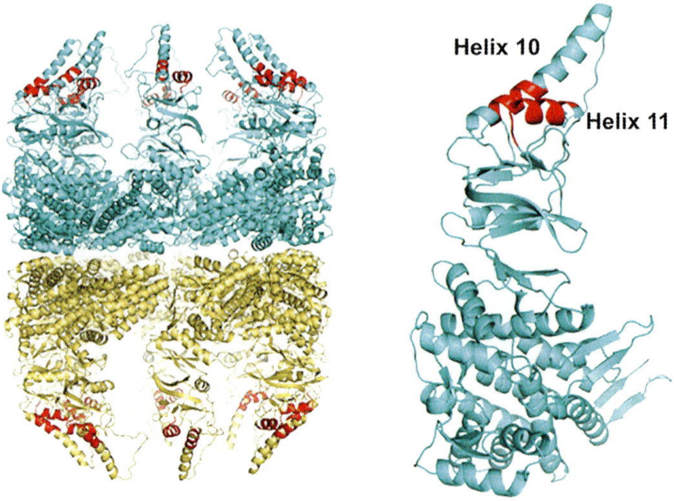

Plate 3.7. ***The modeled wild type open state derived from CpnΔlid***, showing a groove between helix 10 and helix 11, with hydrophobic side chains proposed to present as a binding site for nonnative protein. (See text for details.) [Taken from Pereira *et al*., 2010.]

Cavity of the Closed State of Archaeal *M. maripaludis* Chaperonin

In the closed state, the *M. maripaludis* cavity is, like that of *T. acidophilum*, electrostatic in character (Plate 3.8). This is a function of the significant subunit rotation in the open-to-closed transition. It brings to cavity exposure: Glu240, Glu243, Glu245, Asp247, and Glu249, provided by the overarching protrusions, and Lys216, Arg307, Arg308, and Lys310, provided by the apical faces.

In 2010, a second crystal structure of an open state of the chaperonin from the archaeon *Acidicanus tengchongensis*, an $\alpha\beta$ chaperonin that assembles into nine-member rings, was reported by Huo and coworkers (PDB: 3KO1). They crystallized an all-β subunit complex at 3.7 Å with a nine-member ring in the a.u. Here also an open-to-closed transition (closed provided by EM study) was accounted for by inward movement of the entire chaperonin subunit and by counterclockwise 30° rotation of the apical domain, including the lid protrusions. Rotation of intermediate or equatorial domains was not observed.

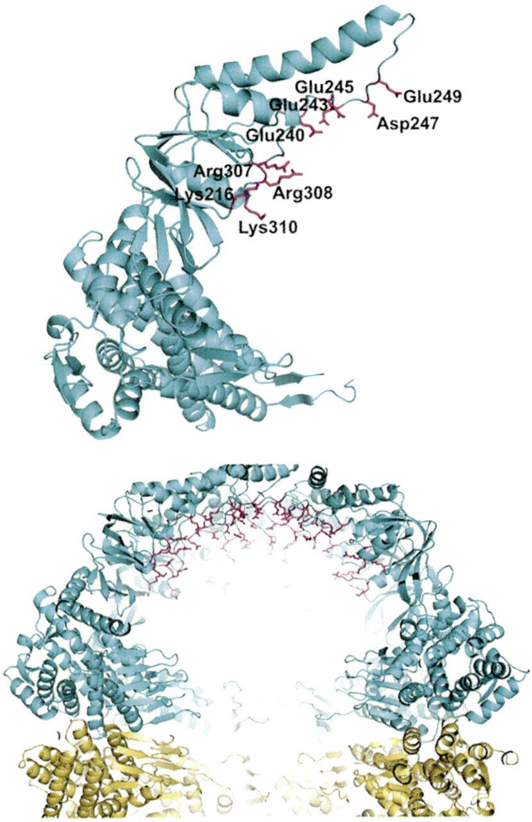

Plate 3.8. ***The closed state of the M. maripaludis chaperonin in AMP–PNP***, showing the electrostatic residues lining the cavity. [From Pereira *et al.*, 2010; PDB: 3KFB.]

ATP Binding Followed by Hydrolysis Produces, in Sequence, Counterclockwise Apical Domain Rotation Followed by Subunit Tilting That Closes the Cavity

In 2011, a cryoEM study of Zhang *et al.* at 8 Å resolution resolved movements produced by ATP binding as distinct from subsequent hydrolysis, using a D396A hydrolysis-defective lidless mutant of the *M. maripaludis* chaperonin (EMDB 5258). Analogous to D398A GroEL, this mutant, affecting the intermediate domain residue that activates water for ATP hydrolysis, binds ATP normally but cannot hydrolyze it (Reissmann, 2007). Here, the D396A Δlid chaperonin was observed in an

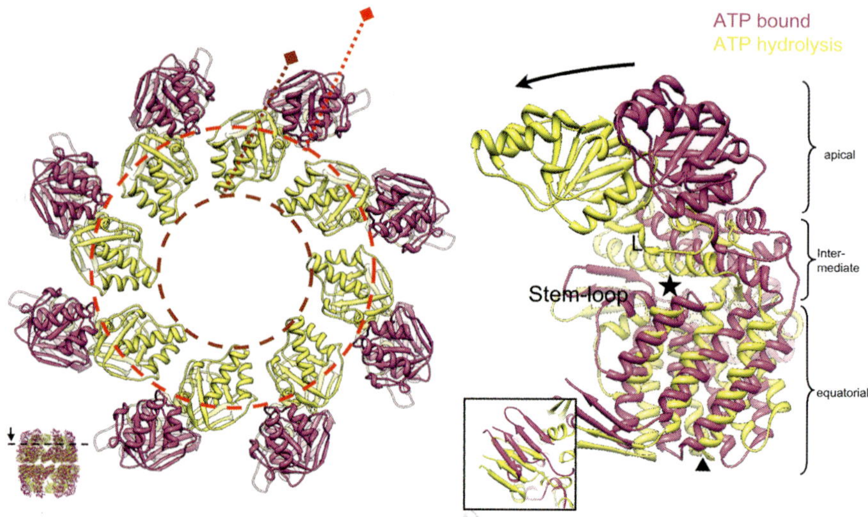

Plate 3.9. ***Closing of the M. maripaludis chaperonin*** occurs via subunit rotation, produced upon ATP binding, followed by subunit tilting to rock the subunit forward over the cavity. [From Zhang *et al.*, 2011.] The study of a D396A ATP-binding–proficient but hydrolysis-defective mutant allowed observation of the effect of ATP binding (EMDB 5258).

open state in the presence of ATP (Plate 3.9). This was compared in detail with previous EM studies of the nucleotide-free Δlid open state. The comparison indicates that, upon binding ATP, the apical domains move slightly inward toward the eightfold symmetry axis and make a 45° counterclockwise rotation as viewed from end-on (ATP-bound state in Plate 3.9). The apical domains, however, remain separated from each other, as in the nucleotide-free state. The intermediate domains, as might be expected based on GroEL studies of ATP binding, undergo a downward tilt around the bottom hinge, bringing helix L onto the nucleotide pocket.

Next, comparing the ATP-hydrolyzed state (ATP–AlF_x transition state and/or posthydrolysis) with the ATP-bound state (D396A), the apical domains move in closer to the eightfold axis (effectively to a position closing the lid if it were there) and form apical–apical interactions characteristic of the closed state (Plate 3.9, right panel). This hydrolysis-related transition is the product of the rocking/tilting forward of the entire subunit, serving to close the folding chamber. This latter movement seems uniquely allowed in the type II chaperonins by the 1:1 cross-equatorial subunit–subunit interactions. By contrast, the staggered relation of cross-equatorial subunits in group I chaperonins seems to prevent such rocking movement. Thus, even with the same ATP binding domain and hydrolysis mechanism, the differing ring–ring subunit contacts enable different actions of ATP binding and hydrolysis to translate into different rigid body movements of the respective machines to produce a binding-active open state and a folding-active closed-off cavity.

Biochemical Correlates Concerning Lid Closure and Productive Folding

In a parallel biochemical study, Douglas *et al.* (2011) demonstrated that ATP binding alone to the D396A hydrolysis-defective mutant of *M. maripaludis* (with intact lid) does not release a bound nonnative rhodanese, thus preventing productive folding from occurring in this state. Only when hydrolysis is allowed is rhodanese able to be released. In the case of Δlid, it is released into solution through the opening in the terminus of the chaperonin and fails to reach native form. In the case of the wild-type (lid-containing) chaperonin, the rhodanese is released into the closed cavity, where it can reach the native state. As with GroEL, at the end of a folding cycle the machine releases the substrate protein whether it has reached the native state or not. The nonnative molecules can return to the binding-proficient open state of the chaperonin for a further trial at folding. As with GroEL, such a return could be forestalled by capture of the released nonnative forms by a "trap" mutant version of GroEL, able to bind but not release rhodanese. Notably, as indicated above, the step of ATP hydrolysis that produces a closed hydrophilic folding chamber is associated with movement of a neighboring loop into the interhelical H–I groove of each subunit, putatively dislodging bound polypeptide in association with lid closure.

Chaperonin-Containing TCP-1 (CCT) in the Eukaryotic Cytosol

The type II chaperonin in the eukaryotic cytosol is referred to as CCT (chaperonin-containing TCP-1) or TRiC (TCP-1 ring complex). It exhibits an overall architecture that resembles that of the archaeal chaperonins; however, it differs from them in containing rings with eight related but distinct subunits (CCT1–8), arranged in a characteristic manner. The different apical domain surfaces of the eight individual subunits, each conserved across phylogeny, determine asymmetric polypeptide binding behavior. In addition, the subunits exhibit distinct equatorial ATP binding and hydrolysis ability, presumably affecting the process of polypeptide release during the transition from the open (binding-proficient) state to the closed (folding-active) state, potentially influencing the folding process. Thus, the distinct properties of both apical and equatorial domains likely determine the unique ability of CCT to mediate productive binding and folding of actin, tubulin, and other eukaryotic cytosolic substrate proteins.

Crystal Structure of a Closed Form of CCT

In 2011, Dekker, Pearl, Willison, and colleagues published a crystal structure of yeast CCT complexed with rabbit α-actin (formed in the presence of phosducin-like protein, a cofactor that favors actin binding) in the presence of ATP–BeF_x (PDB: 4V81). A 3.8 Å dataset was collected and the structure resolved using molecular replacement with the hexadecameric closed *Thermococcus* KS1 structure (Shomura *et al.*, 2004) as a search model. The CCT rings occupied a closed conformation (Plate 3.10), and the domains and apical protrusions of its subunits, as well as the β-strand iris of the closed apical domains, resembled those observed in the archaeal chaperonins.

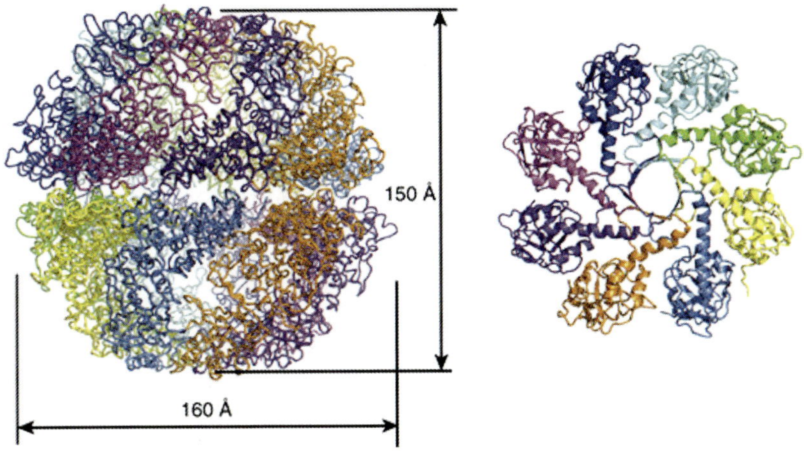

Plate 3.10. ***A crystallographic model of bovine CCT*** [from Dekker *et al.*, 2011; PDB: 4V81.] The closed structure formed in the presence of BeF_x closely resembles that of the thermosome.

Variable Nucleotide Occupancy of Equatorial Domains and an Unusual Track of the N-Terminus of One Subunit

Several observations at the equatorial level of the complex were novel. First, when the ATP binding pockets were examined, some were found to contain density that could be fitted with ADP–BeF, whereas others were devoid of density for base and sugar, indicating absence of nucleotide (Plate 3.11). This result differs markedly from the all-or-none occupancy of the archaeal chaperonin rings. A second observation involved the N-terminus of what later proved to be the CCT4 subunit (Plate 3.12A for subunit arrangement; Plate 3.12B schematizes the N-terminal trajectory of CCT4). Instead of being located at the cavity aspect of the chaperonin, the N-terminal-most sequence in density for CCT4 threads in from the outside of the complex via a channel between neighboring subunits. As a result, it cannot participate in the usual four-strand β-sheet interaction at the inside aspect of the subunit, potentially altering the strength of intra-ring subunit interaction of CCT4. Notably, CCT4 has more recently been shown to lie within the group of contiguous

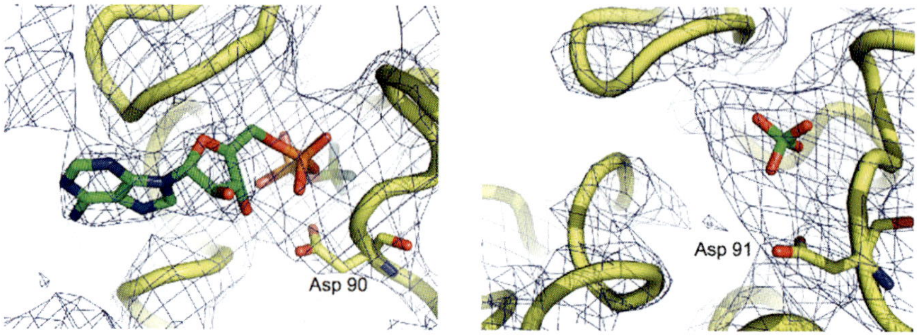

Plate 3.11. ***Variable nucleotide occupancy of CCT*** equatorial domains in structure determination of Dekker *et al.*, 2011.

subunits that have high ATP binding affinity (Reissmann *et al.*, 2012), and CCT4 itself also has high ATP hydrolysis activity, but the nature of these behaviors in relation to the novel track of the N-terminus remains to be determined.

Density in One Cavity of the Complex

Also of interest was a density present in one cavity of each of the two CCT complexes present in the asymmetric unit. A continuous density was resolved, with β-strand features, and considering the volume present, it was suggested that this might constitute the small domain of actin, which harbors a β-sheet. The density was observed to abut three contiguous subunits of CCT (the same in both complexes in the a.u.). Crosslinking experiments supported the view that actin can crosslink CCT subunits in the closed complex. Yet, the density remains to be confirmed as deriving from actin as opposed, for example, to representing a more ordered collective of the usually disordered C-terminal tails from several CCT subunits.

Crystal Structure of an Open State of CCT at Low Resolution

An open state of bovine CCT crystallized in the presence of ATPγS has also been determined at 5.5 Å resolution from orthorhombic crystals,

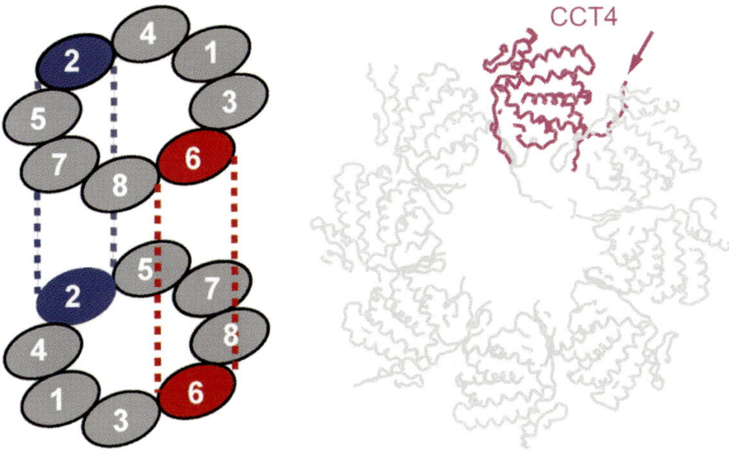

(A) (B)

Plate 3.12. ***Subunit arrangement in the CCT chaperonin***. *Panel A*: The subunit arrangement of CCT was determined using crosslinking and mass spectrometry (Leitner *et al*., 2012). Closed forms of either yeast or bovine CCT produced in ATP or ATP–AlF_x were crosslinked with DSS and digested with trypsin, crosslinked peptides were enriched, and examined by LC-MS-MS. Several hundred heterotypic crosslinks between different subunits were used to produce a topological order, employing the crystallographic model of the *M. maripaludis* archaeal chaperonin in the closed state as a reference (allowing 30 Å as a Cα–Cα distance between DSS-crosslinked lysines). This revealed that the rings are twofold symmetric with CCT2 and CCT6 engaged in homotypic ring–ring contacts, as shown in panel A. The topology was validated by a similar study of the open states of yeast and bovine CCT, comparing with the open state of the *M. maripaludis* chaperonin, and also by substituting pairs of cysteines for predicted near-neighbors at both intra-ring and inter-ring interfaces in yeast CCT, followed by inspection for disulfide crosslinking in the purified complexes after oxidation with $CuCl_2$. This, for example, confirmed homotypic interaction for CCT2 and CCT6 across the ring–ring interface. A further study by Kalisman *et al*. (2013) arrived at the same subunit arrangement through an independent approach of taking the crystallographic data of CCT in closed form (3.8 Å resolution; Dekker *et al*., 2011) and comparing this to constructions of CCTs with all possible subunit arrangements (several hundred thousand). The constructions employed the same backbone for all subunits and side chain rotamers adopted from a closed archaeal crystallographic model, and were searched for the best fit to the X-ray data. The single best fit model agreed with the previous crosslinking/MS-derived arrangement of Leitner *et al*. *Panel B*: Schematic of the top view of CCT equatorial domains, illustrating the track of the CCT4 N-terminus from the external aspect of the chaperonin.

with one CCT complex in the asymmetric unit (Muñoz *et al.*, 2011; PDB: 2XSM). Phasing employed a Ta_6Br_{12} derivative and was followed by solvent flattening, which allowed tracing of the equatorial and intermediate domains with the aid of a thermosome model. An averaged map was produced to allow resolution of the apical domains, which were traced employing the trace of the thermosome apical domains. A further NCSA-averaged map was produced, which identified the Cα positions of the equatorial and intermediate domains and 10 of 16 apical domains, in the case of the apical domains, 8 in one ring and 2 in the other. An additional large density was identified in the open cavity of both CCT rings — suggested, because the source of the CCT was testes, to be tubulin. Both mass spectrometry and EM experiments were marshalled to support this assignment, but it remains possible, in the absence of the ability to assign structure to the density, that this mass represents C-terminal tails of the CCT subunits. Notably, cryoEM analyses of open states of CCT at ~10 Å resolution consistently observe density in the central cavity of both rings, attributed to the subunit termini (Cong *et al.*, 2012). At the equatorial level, as with the closed complex, only a fraction of the nucleotide binding sites were occupied by nucleotide, supported by crystallizing with 2'-methylseleno-ATPγS and examining anomalous difference maps.

Transition Between Open and Closed States of CCT from CryoEM

An early symmetrized EM analysis (Booth *et al.*, 2008) and a more recent nonsymmetrized one (Cong *et al.*, 2012) have sought to resolve the mechanism of closure of CCT. These studies compared open CCT (nucleotide-free, ADP-bound, or AMP–PNP-bound) with closed (ATP–AlF_x). The recent nonsymmetrized analyses are at ~10 Å resolution and are probably the more informative. Open states were analyzed in the absence of nucleotide or the presence of AMP–PNP. Most striking was the pairing of apical domains in these states, apparent contacts forming between characteristic neighboring subunits at the

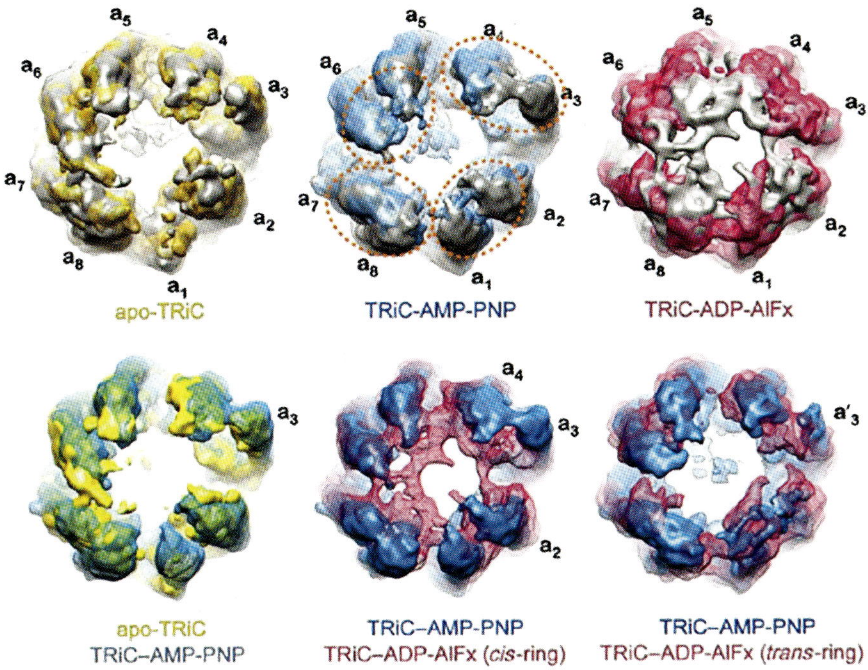

Plate 3.13. ***Pairing of apical domains in apo-CCT*** (EMDB 1960) and in AMP–PNP (EMDB 1961), but symmetrization in ADP–AlF_x (EMDB 1962). [From Cong *et al.*, 2012.]

apical level (Plate 3.13; subunits are arbitrarily labeled, as identification was not achievable at this resolution). This differs significantly from the archaeal chaperonins, where there is complete absence of contact between neighboring subunits above the equatorial level, in both cryoEM and X-ray studies. In ADP–AlF_x, an asymmetric complex with one ring open and the other closed was produced as a major species that was more symmetric around its rings, with neighboring apical domains all contacting each other (Plate 3.13). In ATP–AlF_x, both rings were closed and additional symmetry around the rings was observed. A movie of subunit closure comparing AMP–PNP (open) with ADP–AlF_x (closed) (Supplementary Movie 3 in the publication) shows what appears to be a rocking forward of the entire subunit, to propel the apical domain upward and its protrusion across the cavity opening.

This seems reminiscent of the movement of the archaeal subunits, although at greater resolution the movements may not be the same. In particular, it seems difficult to determine here whether there is a major counterclockwise twisting movement as reported in the earlier Booth *et al.* publication, where both the intermediate domain and the apical domain were interpreted to undergo a counterclockwise twist.

Asymmetry of Polypeptide Recognition by the Apical Domains

The study of polypeptide binding by CCT has been complicated by the presence of eight related but different apical binding domains and by the observation that the conserved hydrophobic residues present in the GroEL apical binding surface are not preserved in the corresponding secondary structures of CCT — there are, instead, polar or even charged residues at these positions. Thus, biochemical studies have focused on identifying the regions and residues in both substrate proteins and the apical domains that are involved with binding to CCT. Considering substrate studies, notable are detailed studies of the von Hippel Lindau tumor suppressor protein (VHL; Feldman *et al.*, 2003) and Gβ transducin (Kubota *et al.*, 2006), both of which identified hydrophobic amino acids along specific extended β-strands within these proteins as the points of recognition by CCT.

On the other side of the interaction, sites on CCT that bind the determinants in VHL were identified by crosslinking studies. First, the VHL determinants were found to be specifically recognized by two specific CCT subunits (Spiess *et al.*, 2006). Apical domains from these subunits in isolation could also be crosslinked with the VHL determinants. In subsequent mutagenesis experiments, helix 11 from the two apical domains (corresponding to helix 11 in the archaebacterial system and helix I in GroEL) was shown to contain residues crucial for binding VHL. When three surface-facing residues of this helix were swapped into a nonbinding apical domain, the latter became binding-competent. Notably, the helix contributes hydrophobic side chains to an interhelical groove reminiscent of that in the archaeal chaperonins. A flanking negative charge on helix 11 was proposed to potentially interact with a flanking positive charge in VHL, offering an explanation for subunit specificity for VHL. That is to say, at CCT, a combination of hydrophobic and polar interactions with its apical domains was proposed to dictate subunit-specific substrate recognition.

Binding of a Small Substrate to CCT Analyzed by NMR Chemical Shift Changes

A more recent study examining the binding of a small substrate, HIV peptide p6 (54 amino acids), to CCT is consistent with the idea of both hydrophobic and polar interactions (Joachimiak *et al.*, 2014). P6 associates with CCT3, and binding and chemical shift changes in uniformly ^{15}N-labeled CCT3 could be mapped (to the assigned CCT3 subunit) to helix 11 (aa 298–308) and its neighboring structures, an underlying "proximal loop" (aa 224–231) and a "distal" tyrosine (247), the latter lying at the hinge of the flexible loop of the protrusion (Plate 3.14). The overall region contained a mixture of hydrophobic and polar residues, respectively, in the helix, L299, M305, D298, Q301, H302, R306, and in the proximal loop, M229, H226, R228, R230, R231 (Plate 3.14). Single alanine mutations of these various residues were tested and some increased K_d for p6 by several fold (in SPR measurements), particularly in the helix, L299, Q301, H302, and in the proximal loop, R228, R230, R231. Thus, a mixture of hydrophobics and charged residues appears to contribute to binding this peptide. The elements in the p6 substrate itself contributing to its binding were also analyzed by chemical shift changes of the uniformly labeled peptide, identifying residues near its C-terminus, $_{41}$SLFGN$_{45}$. Mutational alteration to 5 ala blocked binding to both isolated CCT3 and intact CCT. A Rosetta dock analysis indicated that both polar and apolar interactions contribute to binding of p6.

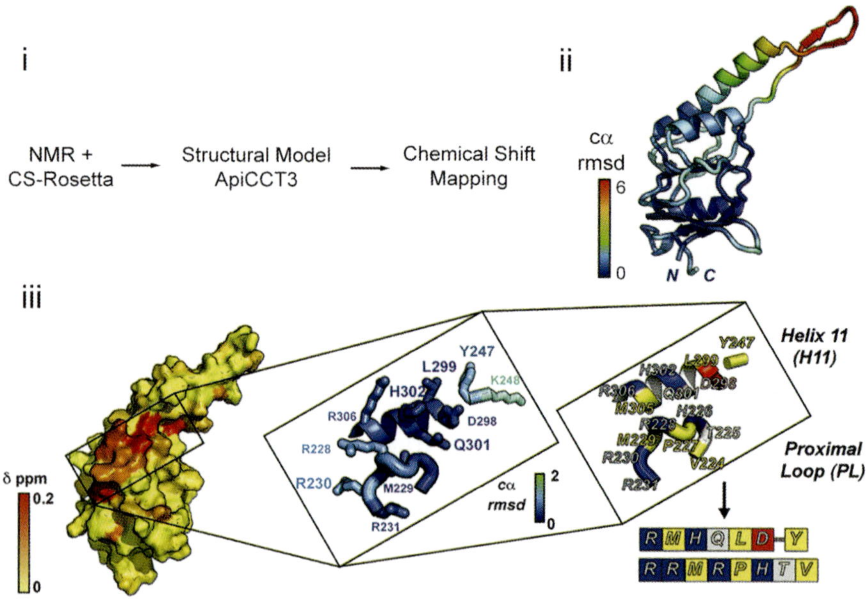

Plate 3.14. ***Mapping of p6 binding onto the isolated 15N-labeled CCT3 apical domain*** (**i**). **ii**. shows mobility of the protrusion relative to the body of the apical domain. **iii**. indicates a region of shift perturbation mapping, as shown to the immediate right as corresponding to a region of helix 11, a "proximal loop" underlying it, and Y247, lying distally. Further right are the residues constituting these structures. [Taken from Joachimiak *et al*., 2014].

Binding of Tubulin and Actin Substrates

An attempt was made to map the binding sites of full-length tubulin, actin, and Gag protein on CCT by covalent crosslinking of binary complexes with DSS, followed by trypsin digestion and mass spectrometry analysis. A very limited set of crosslinks was identified, perhaps out of the necessity of DSS to react with lysines, but there were, for example, two for tubulin, linking to CCT2 and CCT6, and one for actin, linking to CCT6 (Plate 3.15). The crosslinked lysines were indeed on the apical domain, in the case of actin at K222, directly within the underlying proximal loop. In the case of tubulin, however, the crosslinks mapped to helix 10 for both CCT2 (K263) and CCT6 (K278), equivalent to GroEL helix H. This implies that a wider range of apical binding surfaces may

	Position	CCT Subunit
Tubulin	K58	CCT2 K263
	K324	CCT6 K278
Actin	K330	CCT6 K222
Gag	K110	CCT6 K258
	K397	CCT2 K284
	K424	CCT7 K287

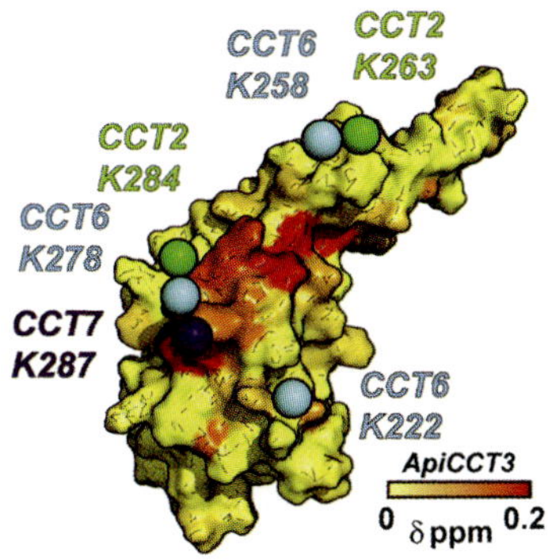

Plate 3.15. ***DSS Xlink mapping of tubulin, actin, and HIV Gag protein to CCT***, identifying the subunits and lysines crosslinked in a representative chaperonin subunit. (See text for discussion.) [Taken from Joachimiak *et al.*, 2014.]

be employed for binding tubulin. Alternatively, the immediate contacts with the apical domain may be localized within the helix 11 region, while the crosslinks occur at positions on the substrate that are slightly removed in space from the contacts. Perhaps most incisive on this point would be mutational analysis of the CCT2 and CCT6 apical domains of intact CCT.

Asymmetric ATP Binding and Hydrolysis by the Equatorial Domains

Once the arrangement of the CCT subunits within the chaperonin became established, it was possible to address the issues of partial occupancy of rings with nucleotide, as originally observed in the X-ray study by Dekker and coworkers, and of putative "sequential" ATP binding and hydrolysis within a ring, suggested by a study by Rivenzon-Segal and coworkers (2005). A biochemical analysis by Reissmann *et al*. (2012) was particularly informative. In an initial test with saturating ^{32}P-α-labeled ATP (and gel filtration), they observed substoichiometric binding with on average seven ATPs per CCT molecule. This was confirmed using ^{32}P-α-labeled 8-azido-ATP and photolysis, which covalently crosslinked 7–9 nucleotides per CCT 16-mer. In mass spectrometry studies, only four subunits were found to be crosslinked: CCT1, 2, 4, and 5. Thus, closure of both rings of a CCT complex was accomplished with substoichiometric binding and hydrolysis of ATP, as had been directly observed by Dekker *et al.*

To gauge the relative affinities of CCT subunits for ATP, a variety of azido-ATP concentrations were used. CCT5 was crosslinked, and to a lesser degree CCT4, at 0.2 mM; and CCT1, and to a lesser degree CCT2, at 0.5 mM (see Plate 3.15). Notably, 0.3 mM azido-ATP is sufficient to drive substrate folding, indicating that partial occupancy is sufficient for full function. The basis of varying affinity could be ascribed to varying sequences in the CCT subunit nucleotide pockets at the position surrounding the base and sugar, whereas there was high conservation between subunits around the triphosphate moiety.

In *in vivo* functional assessment, blocking ATP binding by altering the P-loop, GDGTT, to 5 Ala was lethal for yeast cells when the substitution was in any of the high affinity subunits but had no effect on growth when in the low affinity ones. This correlated directly with the ability of the isolated mutant complexes to close their chambers in the presence of ATP–AlF_x *in vitro*, with mutation of high affinity subunits blocking closure and mutation of low affinity subunits without effect. A single D → A substitution within the P-loop had a more confined effect, lethal only to cells bearing this substitution in CCT4. (Notably, this is the subunit with the unusual N-terminus that penetrates the chaperonin from outside.) ATP hydrolysis also was assessed *in vivo*, using a mutation altering the intermediate domain aspartate that attacks water to alanine, allowing ATP binding but not hydrolysis. This produced yeast inviability when present in CCT4 and severe growth defect when in CCT5, associated *in vitro* in the latter case with reduced lid closure.

Strikingly, the subunits with high affinity for ATP (CCT1, 2, 4, 5) map to a "hemisphere" of the chaperonin, apposed to a low affinity hemisphere (CCT3, 6, 7, 8) (Plate 3.16). The consequence of this arrangement is almost certainly, as suggested by Reissmann *et al.*, an asymmetric

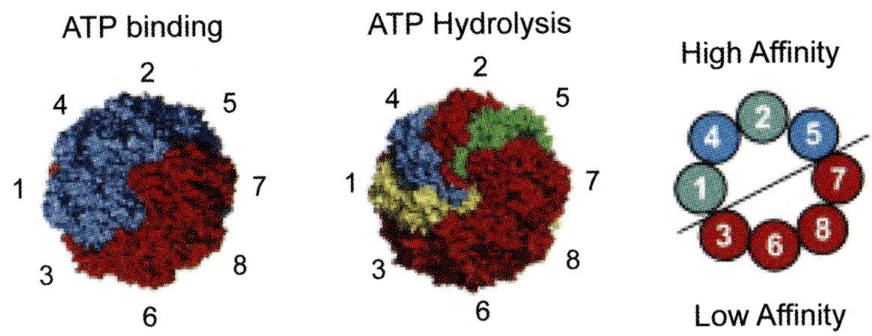

Plate 3.16. ***Differential affinity and turnover of ATP by CCT***, as shown in the studies by Reissmann *et al.*, 2012. A "hemisphere" of CCT has a strong affinity for ATP, and within that hemisphere CCT4 and CCT5 have the highest activity of ATP hydrolysis as revealed by mutagenesis studies (see text). [Taken from Reissmann *et al.*, 2012.]

power stroke, likely causing sequential substrate release. It would clearly be of interest to mutationally "exchange" high affinity and low affinity base–sugar binding regions to assess whether this would interfere with productive folding.

Concerning the specificity of CCT for directing productive folding of such substrates as actin and tubulin, one is left to conclude from the foregoing structural and mechanistic work that asymmetry, both of binding such substrates by specific apical domains through a putative subunit-specific arrangement of apolar and polar interactions and of release mediated by binding and hydrolysis of ATP in one hemisphere, contribute to its unique ability to mediate refolding of particular substrates in its encapsulated chamber. Presumably, the coevolution of the substrates, particularly such abundant structural ones as actin and tubulin, with this machine established the current variegated subunit architecture.

References

Ahmad S, Gupta RS (1990). Cloning of a Chinese hamster protein homologous to the mouse t-complex protein TCP-1: structural similarity to the ubiquitous "chaperonin" family of heat-shock proteins. *Biochi Biophys Acta* **1087:** 253–255.

Bigotti MG, Clarke AR (2005). Cooperativity in the thermosome. *J Mol Biol* **348:** 13–26.

Bigotti MG, Bellamy SRW, Clarke AR (2006). The asymmetric ATPase cycle of the thermosome: elucidation of the binding, hydrolysis and product-release steps. *J Mol Biol* **362:** 835–843.

Booth CR, Meyer AS, Cong Y, *et al.* (2008). Mechanism of lid closure in the eukaryotic chaperonin TRiC/CCT. *Nat Struct Mol Biol* **15:** 746–753.

Clare DK, Stagg S, Quispe J, *et al.* (2008). Multiple states of a nucleotide-bound group 2 chaperonin. *Structure* **16:** 528–534.

Cong Y, Schröder GF, Meyer AS, *et al.* (2012). Symmetry-free cryo-EM structures of the chaperonin TRiC along its ATPase-driven conformational cycle. *EMBO J* **31:** 720–730.

Dekker C, Roe SM, McCormack EA, *et al.* (2011). The crystal structure of yeast CCT reveals intrinsic asymmetry of eukaryotic cytosolic chaperonins. *EMBO J* **30:** 3078–3090.

Ditzel L, Löwe J, Stock D, *et al.* (1998). Crystal structure of the thermosome, the archaeal chaperonin and homolog of CCT. *Cell* **93:** 125–138.

Douglas NR, Reissmann S, Zhang J, *et al.* (2011). Dual action of ATP hydrolysis couples lid closure to substrate release into the group II chaperonin chamber. *Cell* **144:** 240–252.

Ellis RJ (1990). Molecular chaperones: the plant connection. *Science* **250:** 954–959.

Feldman, DE, Spiess, C, Howard, DE, Frydman, J (2003). Tumorigenic mutations in VHL disrupt folding *in vivo* by interfering with chaperonin binding. *Mol.Cell* **12:** 1213–1224.

Gao Y, Thomas JO, Chow RL, *et al.* (1992). A cytoplasmic chaperonin that catalyzes β-actin folding. *Cell* **69:** 1043–1050.

Gutsche I, Mihalache O, Baumeister W. (2000). ATPase cycle of an archaeal chaperonin. *J Mol Biol* **300:** 187–196.

Huo Y, Hu Z, Zhang K, *et al.* (2010). Crystal structure of group II chaperonin in the open state. *Structure* **18:** 1270–1279.

Joachimiak LA, Walzthoenl Liu CW, *et al.* (2014). The structural basis of substrate recognition by the eukaryotic chaperonin TRiC/CCT. *Cell* **159:** 1042–1055.

Kalisman N, Schröder GF, Levitt M (2013). The crystal structures of the eukaryotic chaperonin CCT reveal its functional partitioning. *Structure* **21:** 540–549.

Klumpp M, Baumeister W, Essen LO (1997). Structure of the substrate binding domain of the thermosome, an archaeal group II chaperonin. *Cell* **91:** 263–270.

Kubota S, Kubota H, Nagata K (2006). Cytosolic chaperonin protects folding intermediates of Gβ from aggregation by recognizing hydrophobic β-strands. *Proc Natl Acad Sci USA* **103:** 8360–8365.

Kusmierczyk AR, Martin J (2003). Nested cooperativity and salt dependence of the ATPase activity of the archaeal chaperonin Mm-cpn. *FEB Lett* **547:** 201–204.

Leitner A, Joachimiak LA, Bracher A, *et al.* (2012). The molecular architecture of the eukaryotic chaperonin TRiC/CCT. *Structure* **20:** 814–825.

Lewis VA, Hynes GM, Zheng D, *et al.* (1992). T-complex polypeptide-1 is a subunit of a heteromeric particle in the eukaryotic cytosol. *Nature* **358:** 249–252.

Meyer AS, Gillespie JR, Walther D, *et al.* (2003). Closing the folding chamber of the eukaryotic chaperonin requires the transition state of ATP hydrolysis. *Cell* **113:** 369–381.

Muñoz IG, Yébenes H, Zhou M, *et al.* (2011). Crystal structure of the open conformation of the mammalian chaperonin CCT in complex with tubulin. *Nat Struct Mol Biol* **18:** 14–19.

Nitsch M, Klumpp M, Lupas A, Baumeister W (1997). The thermosome: alternating α and β-subunits within the chaperonin of the archaeon *Thermoplasma acidophilum*. *J Mol Biol* **267:** 142–149.

Nitsch M, Walz J, Typke D, *et al.* (1998). Group II chaperonin in an open conformation examined by electron tomography. *Nat Struct Biol* **5:** 855–857.

Pereira JH, Ralston CY, Douglas NR, *et al.* (2010). Crystal structures of a group II chaperonin reveal the open and closed states associated with the protein folding cycle. *J Biol Chem* **285:** 27958–27966.

Pereira JH, Ralston CY, Douglas NR, *et al.* (2012). Mechanism of nucleotide sensing in group II chaperonins. *EMBO J* **31:** 731–740.

Phipps BM, Hoffmann A, Stetter KO, Baumeister W (1991). A novel ATPase complex selectively accumulated upon heat shock is a major cellular component of thermophilic archaebacterial. *EMBO J* **10:** 1711–1722.

Reissmann S, Parnot C, Booth CR, *et al.* (2007). Essential function of the built-in lid in the allosteric regulation of eukaryotic and archaeal chaperonins. *Nat Struct Biol* **14:** 432–440.

Reissmann S, Joachimiak LA, Chen B, *et al.* (2012). A gradient of ATP affinities generates an asymmetric power stroke driving the chaperonin TRIC/CCT folding cycle. *Cell Rep* **2:** 866–877.

Rivenzon-Segal D, Wolf SG, Shimon L, *et al.* (2005). Sequential ATP-induced allosteric transitions of the cytoplasmic chaperonin containing TCP-1 revealed by EM analysis. *Nat Struct Mol Biol* **12:** 233–237.

Shomura Y, Yoshida T, Iizuka R, *et al.* (2003). Crystal structures of the group II chaperonin from *Thermococcus* strain KS-1. *J Mol Biol* **335:** 1265–1278.

Spiess C, Miller EJ, McClellan AJ, Frydman J (2006). Identification of the TRiC/CCT substrate binding sites uncovers the function of subunit diversity in eukaryotic chaperonins. *Mol Cell* **24:** 25–37.

Tian G, Vainberg IE, Tap WD, *et al.* (1995). Specificity in chaperonin-mediated protein folding. *Nature* **375:** 250–253.

Trent JD, Nimmesgern E, Wall JS, *et al.* (1991). A molecular chaperone from a thermophilic archaebacterium is related to the eukaryotic protein t-complex polypeptide-1. *Nature* **354:** 490–493.

Ursic D, Culbertson MR (1991). The yeast homolog to mouse *Tcp-1* affects microtubule-mediated processes. *Mol Cell Biol* **11:** 2629–2640.

Yaffe MB, Farr GW, Miklos D, *et al.* (1992). TCP1 complex is a molecular chaperone in tubulin biogenesis. *Nature* **358:** 245–248.

Zhang J, Baker ML, Schröder GF, *et al.* (2010). Mechanism of folding chamber closure in a group II chaperonin. *Nature* **463:** 379–383.

Zhang J, Ma B, DiMaio F, *et al.* (2011). Cryo-EM structure of a group II chaperonin in the prehydrolysis ATP-bound state leading to lid closure. *Structure* **19:** 633–639.

Chapter 4

The Family of Small Heat Shock Proteins: Assembly and Binding Functions

C. Slingsby[1*] *and A. R. Clark*[2]

Abstract

Small heat shock proteins (sHsps) are a diverse family evolved from ancient stress proteins that are now associated with protecting microbial, plant, worm, insect and mammalian cells under stress, where prevention of aggregation of destabilized proteins is considered to be their general function. The sequences all contain a short α-crystallin domain (ACD) bracketed by variable sequence extensions that contribute to a dynamic continuum of oligomeric forms. A few large symmetric oligomeric assemblies from nonmetazoans have been resolved by crystallography, as well as a few metazoan subassembly forms. A structure-based sequence alignment of the ACDs from resolved structures with representative sequences of sHsps from species of high economic, model or medical value is presented. The ACDs are portrayed to emphasize their interaction interfaces. The construction of a range of nanoassemblies from a polyvalent building block by binding IXI/V sequence motifs in extensions into partner ACD pockets is illustrated. This interaction mechanism can also be used to connect sHsps to energy driven chaperone machines, as shown for the interaction with cochaperone Bag3 in myofibril Z-disks. The resolved complete assemblies are

[1] Department of Biological Sciences, Crystallography, Institute of Structural and Molecular Biology, Birkbeck College, Malet Street, London WC1E 7HX, UK.
[2] MRC Laboratory of Molecular Biology, Cambridge CB2 0QH, UK.
[*] Corresponding author; C.Slingsby@mail.cryst.bbk.ac.uk

built from a strand exchange dimer, whereas animal sHsps have an antiparallel (AP) dimer interface that forms a shared groove which can change shape. Spectroscopy has shown how the AP interface is regulated by pH. Animal sHsp full assemblies are undetermined, although some clues as to the structural role of the hydrophobic N-terminal extensions can be surmised from analysis of the resolved assemblies. A mechanism for prevention of protein aggregation is likely to involve a stress-regulated ensemble of interconverting destabilized sHsp assemblies when interfaces and extensions become more accessible to the proteome. Rapid developments in EM offer the possibility of resolution of an sHsp ensemble if the level of polydispersity can be lowered.

Introduction

Small heat shock proteins (sHsps) are part of the chaperone network that assists the folding, assembly or maintenance of the cytosolic proteome (Kim *et al.*, 2013). Unlike the Hsps of larger sequences (Hsp104, 90, 70, 60 and 40 KDa), they do not exert conformational change on their substrates, and are thus called "holdases" (Richter *et al.*, 2010). Although small in chain length, sHsps form large, dynamic assemblies which can readily reorganize in response to stress into forms that can soak up aggregating proteins ready for handing over to the ATP-driven "foldases" or degradation machines (Eyles and Gierasch, 2010; Haslbeck and Vierling, 2015). Closely related sHsp family members can coassemble (Basha *et al.*, 2012), with some family members under the control of the heat shock transcription factor, while others are constitutively active (de Thonel *et al.*, 2012; Waters, 2013).

sHsps are generally cytoprotective, providing thermotolerance, which is particularly important for plants, whereas in warm blooded animals they are often markers of disease, components of protein deposits, and have a disease phenotype when mutated (Boncoraglio *et al.*, 2012; Clark *et al.*, 2012). Understandably, they are of agricultural (Mickelbart *et al.*, 2015) and biomedical interest, particularly the human proteins (Arrigo *et al.*, 2007; Reddy and Reddy, 2015). sHsps are expressed in a range of human tissues and cell types, such as skeletal muscle, cardiac muscle, smooth muscle, fibroblasts, neurons and oligodendrocytes, where they are involved in a range of chronic and inherited diseases, such as myopathy and neuropathy (Houlden *et al.*, 2008; Selcen, 2011; Boncoraglio *et al.*, 2012), heart disease (Boelens, 2014), pulmonary congestion and fibrosis (Gunst and Zhang, 2008; Bellaye *et al.*, 2015) and multiple sclerosis (van Noort *et al.*, 2010), as well as cancer (Moyano *et al.*, 2006). A common function may be that of a holdase, binding stress-destabilized client proteins, but it is unclear as to the extent to which sHsp cytoprotective behavior can be attributed to lowering the levels of cellular protein aggregation (Mymrikov *et al.*, 2011). Although some members of the sHsp family protect against cell death induced by a variety of stimuli, these effects are generally assayed at functional end points of upstream apoptotic signals, making it difficult to pinpoint direct binding partners (Beere, 2005). Certain members of the sHsp family interact with universal adapter proteins such as 14-3-3 and Bag3, so they can indirectly affect many intracellular processes (Mymrikov *et al.*, 2011). As the structural basis of network biology progresses, the sHsp interactome grows. In one survey, the majority of proteins that interact with sHsps are linked to cell adhesion, tissue integrity, epithelial-to-mesenchymal transition and the cytoskeleton, whereas inhibition of protein aggregation or fibrillation is ranked second (Arrigo, 2013).

In 1982 it was discovered that sHsps from *Drosophila melanogaster* were sequence related to vertebrate eye lens α-crystallins (Ingolia and Craig, 1982). Ten years later Horwitz showed that stoichiometric amounts of α-crystallin prevented thermal aggregation of enzymes and that selective removal by ultracentrifugation of α-crystallins from total lens soluble cytoplasm rendered it susceptible to thermal aggregation at 60°C (Horwitz, 1992). This early study showed that highly destabilized α-crystallin prevented thermal aggregation of physiological substrates *in vitro*. sHsps are now generally considered to be the first responders to stress by capturing up to an equal weight of denaturing protein before it aggregates by forming a direct physical interaction, and keeping it accessible to other ATP-dependent components of the protein quality control network (Basha *et al*., 2012). There are two types of sHsp–substrate complex: one that maintains solubility of the aggregating protein that can handover substrate to the Hsp70 system, and the amorphous type, which also requires an ATP-driven disaggregase (Haslbeck and Vierling, 2015). sHsps may buffer aggregation of toxic soluble oligomeric intermediates which are less ordered than protein fibrils, the thermodynamic end state of aggregation (Kim *et al*., 2013) with NMR spectroscopy showing that αB-crystallin can prevent fibril formation from neutral solutions of a mutant beta2-microglobulin without forming a complex by rapid transient nonspecific interactions, indicating the absence of a preferred binding site (Esposito *et al*., 2013). A study using lifespan mutants of *C. elegans* showed that sHsps, but not other chaperones, become insoluble during aging, suggesting that they may be involved in a "protective aggregation response" whereby protein aggregation *in vivo* may occur not just as a consequence of proteostasis decline, but may also be induced to improve proteostasis by sequestering surplus, potentially harmful protein species (Walther *et al*., 2015).

The α-Crystallin Domain

sHsp sequences are classified in the Pfam database (Finn *et al*., 2014) as belonging to the clan Hsp20-like chaperone superfamily (CL0190), which contains four subfamilies. The different subfamilies fold into similar topologies, but are not considered to be evolutionarily related. sHsp sequences belong to the most highly populated subfamily, Hsp20/α-crystallin (PF00011), having around 10,000 sequences in the database, followed by the CS domain family (PF04969), with around 2600 sequences. Members of the Hsp20/α-crystallin subfamily all have an α-crystallin domain (ACD) of around 90 amino acids (de Jong *et al*.,

1998; Kriehuber *et al.*, 2010). Sequences have a fairly uniform architecture, in that the majority have just one ACD preceded by an N-terminal sequence of varying length that has no homology to other proteins, and followed by a C-terminal extension of variable length that generally contains a sequence motif similar to I-X-I/V (Poulain *et al.*, 2010; Hilton *et al.*, 2013).

sHsps are widespread among organisms but are more divergent than other Hsps, with rapid independent gene expansion occurring in the major higher organisms, resulting in large sHsp families in animals and plants (Franck *et al.*, 2004; Kriehuber *et al.*, 2010; Basha *et al.*, 2012; Waters, 2013). The human family comprises ten members labeled HspB1–B10 (Kappé *et al.*, 2003), with some poor at inhibiting protein aggregation *in vitro* and yet neuroprotective by unknown mechanisms (Carra *et al.*, 2013). HspB5 (αB-crystallin) is found throughout vertebrates and in many tissues, whereas the sister gene HspB4 (αA-crystallin) is the major lens crystallin and is specific to the lens, making it likely that it duplicated from αB around the time of the evolution of vertebrates, and hence the vertebrate eye. A likely orthologue of vertebrate HspB1 (Hsp27) is found in the urochordate, *Ciona intestinalis* (Q95P25). There are orthologues of human HspB8 in the the "living fossil" coelacanth, but not in cephalochordates. The sequence in *Drosophila melanogaster* labeled HspB8 (Q9VWG1) is considered a functional homologue.

The α-crystallin domain (ACD) has a β-sheet sandwich fold comprising strands β2–β9 (Slingsby *et al.*, 2013) (Fig. 4.1A). Sequences of the ACD from prokaryotes, fungi, protists, plants and several metazoan phyla are aligned and labeled with features from X-ray structures (Fig. 4.2). The distinction between metazoan and nonmetazoan sequences can be seen in the differing positions of a [redacted] between strands β3 and β4, and by the different lengths of the sequences between β5 and β7, which determine whether the β6 strand is present (Fig. 4.1A) or absent Fig. 4.3A). The β4 and β8 strands at the edges of the β-sandwich fold close to the C-terminal extension are highly conserved and create hydrophobic pockets (Fig. 4.1B), whereas the sheet edges close

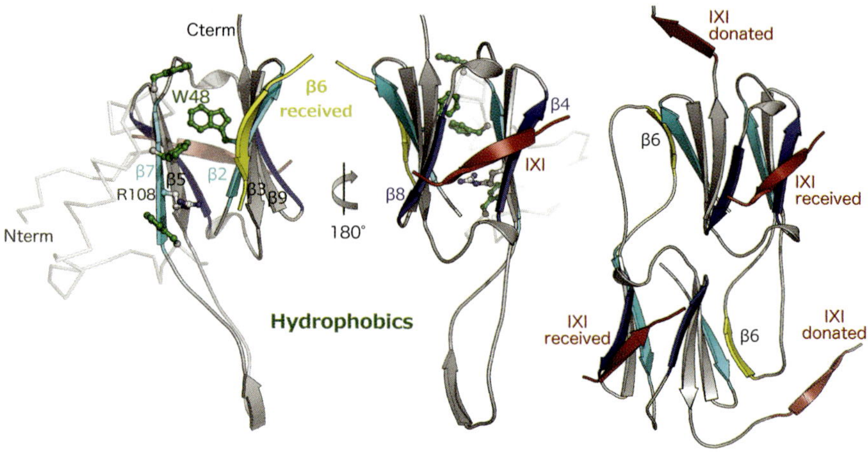

Fig. 4.1 *In nonmetazoans the α-crystallin domain (ACD) dimerizes by strand exchange and forms higher assembly through patching the sides of the fold by sequence extensions.* Panels A–C show structural features based on *Triticum aestivum* (wheat) Hsp16.9b (pdb 1gme). **A.** The ACD has a β-sandwich fold comprising strands $\beta 2$–$\beta 9$. The $\beta 6$ strand, which is exchanged between partner domains, is shown in yellow. The sides of the sheet close to the N-terminal extension have edges formed from strands $\beta 2$ and $\beta 7$ (cyan) that expose ACD core hydrophobic residues. The key hydrophobic residue Trp-48, positioned in the middle of the $\beta 2$ strand, faces Phe-110, Leu-113 (not shown) and superfamily conserved Arg-108 on the $\beta 7$ strand, which are being patched by Phe-10, Phe-13 and Phe-41 (all shown in green ball and stick) from the N-terminal extension (grey ribbon). **B.** The same molecule from A, rotated 180°, to show the $\beta 4/\beta 8$ strands (royal blue) that form hydrophobic pockets are being patched by the conserved IXI motif (plum ribbon) donated from the C-terminal extension of an adjacent dimer. **C.** The dimer is formed by reciprocal exchange of the $\beta 6$ strand (yellow) to form an antiparallel strand next to $\beta 7$ of the partner domain. The motifs in the C-terminal extensions that assemble the dimer into higher order oligomers are labeled as received or donated accordingly. Note the different orientations of the extensions in the dimer, allowed by flexibility in the linker between the IXI motif and the ACD. This facilitates flexible assembly, shown in Fig. 4.4. Complete assemblies observed in crystal structures are all built from dimers formed by the $\beta 6$ strand exchange mechanism.

to the N-terminal region are more variable with $\beta 7$ on one side, while the other edge can be formed by the $\beta 3$ (Fig. 4.3B) or $\beta 2$ (Figs. 4.1A and 4.3A) or $\beta 1$ strand (in *Methanococcus jannaschii*, Hsp16.5). The ACD core hydrophobics are exposed to some extent at the sides of the β-sandwich and act as magnets for sequence extensions from partner chains, a process termed "patching the ACD fold" (van Montfort *et al.*, 2001).

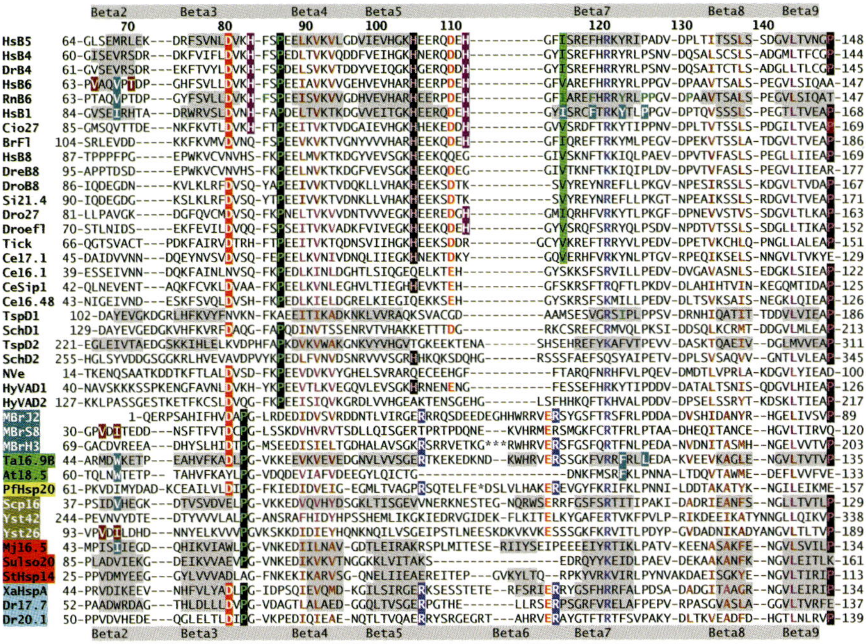

Fig. 4.2 *Structure-based sequence alignment of the α-crystallin domains.* Sequences of the ACD from bacteria, archaea, fungi, protists, plants, choanoflagellates and several metazoan phyla are aligned and labelled with features from X-ray structures. Uniprot codes are: **HsB5** (P02511), **HsB4** (P02489), **DrB4** (Q8UUZ6), **HsB6** (O14558), **RnB6** (P97541), **HsB1** (P04792), **Cio27** (Q95P25), **BrF1** (C3XVI4), **HsB8** (Q9UJY1), **DreB8** (Q6PFS1), **DroB8** (Q9VWG1), **Si21.4** (A0A0A7RI59), **Dro27** (P02518), **Droefl** (P82147), **Tick** (L7M4H7), **Ce17.1** (Q20660), **Ce16.1** (P34696), **CeSip1** (Q20363), **Ce16.48** (P02513), **Tsp** (Q7YZT0), **Sch** (P12812), **NVe** (A7RMQ7), **HyV** (T2MHG6), **MBrJ2** (A9UXJ2), **MBrS8** (A9URS8), **MBrH3** (A9VAH3), **Ta16.9b** (Q41560), **At18.5** (O64564), **PfHsp20** (Q8IB02), **Scp16** (O14368), **Yst42** (Q12329), **Yst26** (P15992), **Mj16.5** (Q57733), **Sulso20** (Q97W19), **StHsp14** (Q970D9), **XaHspA** (Q8PNC2), **Dr17.7** (Q9RTR5), **Dr20.1** (Q9RVB5). The beta strands are highlighted in gray in sequences where the structure was resolved by crystallography, and labeled above and below the alignment. The HsB5 numbering is shown above the sequence. In MBrH, *** represents an insert of 44 residues.

Assembly from a Strand Exchange Dimer in Prokaryotes, Yeast and Plants

It is instructive to scrutinize how sHsps assemble, as this likely is a clue to how they function as a holdase. Complete assemblies observed in crystal structures are typically built from dimers formed by a conserved $\beta 6$

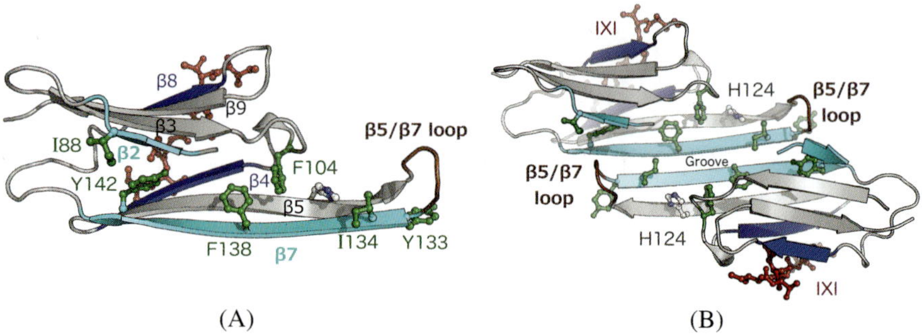

Fig. 4.3 *In metazoans the α-crystallin domain (ACD) dimerizes about an antiparallel (AP) interface.* Panels A and B show structural features based on human Hsp27/HspB1 (pdb 4mjh). **A.** In the monomer the segment between strands β5 and β7 is too short to form a β6 strand and is here labeled "β5/β7 loop" (brown). The β7 strand (cyan) is extended and faces the β2 strand (cyan) on one side of the β-sandwich fold, and on the other side of the fold strands β4 and β8 (royal blue) are patched by the C-terminal peptide shown in ball and stick (plum). The conserved hydrophobic residues (Ile-88, Phe-104, Phe-138, Ile-134, Tyr-133, Tyr142) that would be exposed in the monomeric species are depicted in green ball and stick. **B.** The assembled dimer, showing the two β7 strands (cyan) extending an AP β-sheet forming a shared groove bordered by the β5/β7 loops (brown) and lined with some of the conserved hydrophobic residues (green ball and stick). The conserved His-124 is shown in ball-and-stick representation, and colored gray and blue. The two sets of β4/β8 strands (royal blue) form pockets on either side of the dimer and are patched by IXI containing C-terminal peptides, shown in ball and stick (plum).

strand exchange mechanism (Fig. 4.1C); the higher assembly contacts, between dimers, are mediated by the sequence motif I-X-I/V in the C-terminal extension of one chain patching β4/β8 pockets in the ACD of another chain (Fig. 4.4). A conserved set of hydrophobic residues lines these pockets (highlighted in brown in Fig. 4.2), among them a highly conserved leucine. The first structure of an sHsp, from an archaeal thermophile (MjHsp16.5, pdb 1shs), revealed a highly organized 24-mer of ACDs assembled with the 432 symmetry of an octahedron (Kim *et al*., 1998), whereby each dimer lies along an edge of the octahedron and a single linker conformation between the ACD and the C-extension allows four dimers to assemble around each of the six vertices using an I-N-I into β4/β8 pocket mechanism (Fig. 4.4C, pdb 1shs). The spherical

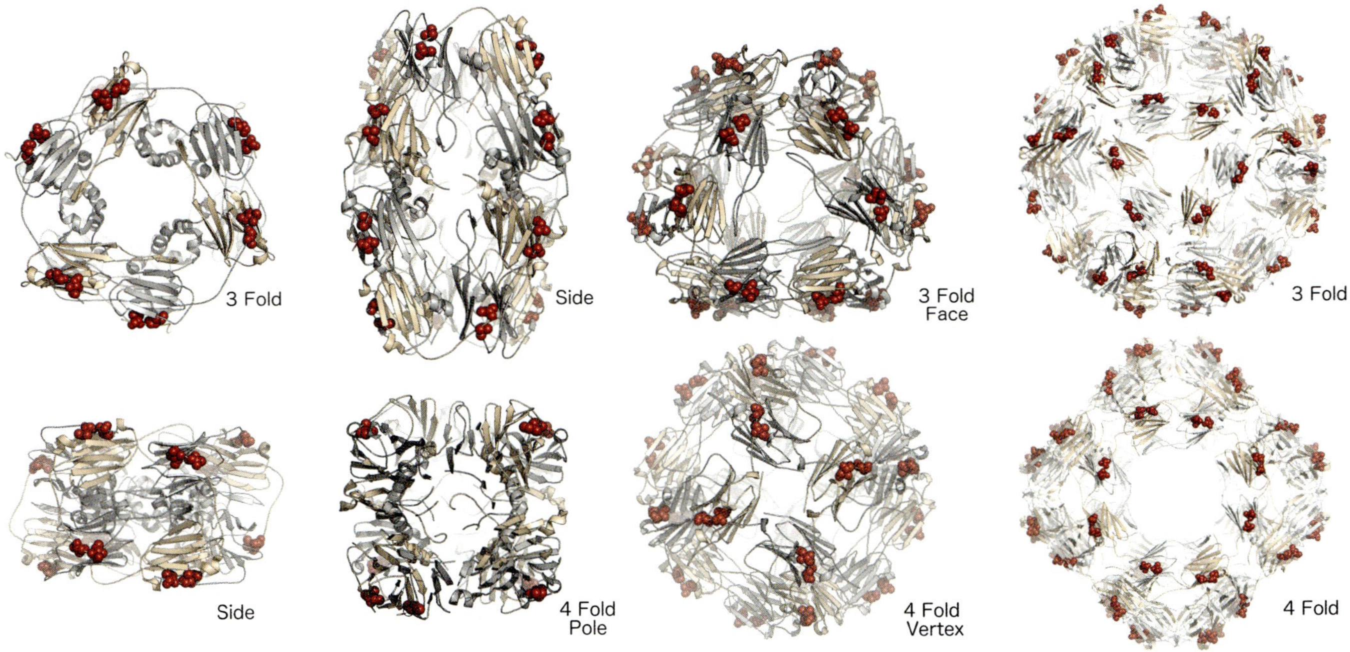

(A) (B) (C) (D)

Fig. 4.4 *Building sHsp nanoassemblies from strand exchange dimers.* All the structures were resolved by X-ray crystallography. Dimers, with the two chains colored gray and light pink, are assembled into higher oligomers by IXI motifs (depicted as red spheres) filling $\beta4/\beta8$ partner pockets and strapping the oligomers together with the pockets and straps arrayed across the exterior faces. **A.** Wheat Hsp16.9b (1gme) is a 12-mer whereby two linker conformations enable formation of a dihedral hexamer of dimers. The C-extension-into-pocket interaction directs a ring of dimers around the threefold axis using one linker conformation, while another links side-by-side dimers about a twofold axis, weaving

Fig. 4.4 (*Continued*) the two hexameric rings together. The helical N-terminal extensions of six chains are shown in gray in the center of the assembly. **B.** The yeast *Schizosaccharomyces pombe*, SpHsp16.0 (pdb 3w1z), is a 16-mer organized as an octomer of dimers in a spheroid assembly looking down the twofold from the side, and along the fourfold pole axis. Some N-terminal extensions in helical conformation are buried in the interior. **C.** The archaeal MjHsp16.5 (pdb 1shs) is a 24-mer of ACDs assembled with the 432 symmetry of an octahedron. All chains are identical and no N-terminal extensions were resolved. **D.** The MjHsp16.5 24-mer is converted to a 48-mer (pdb 4eld) by engineering an insertion into the N-terminal extension.

assembly is hollow in the crystal but has windows giving access to the 32-residue N-terminal sequences shown by electron microscopy to occupy the interior (Haley *et al.*, 2000).

Crystal structures of complete sHsp assemblies show how by relaxing dimer symmetry, different orientations of the C-terminal extension can direct assemblies of (lower) dihedral symmetry. For example, in *Schizosaccharomyces pombe* (SpHsp16.0), a 16-mer assembles with *422* symmetry (pdb 3w1z, Hanazono *et al.*, 2013), and in *Triticum aestivum* (wheat Hsp16.9b), a 12-mer assembles with *32* symmetry (pdb 1gme, van Montfort *et al.*, 2001). In wheat Hsp16.9b, two linker conformations enable formation of a dihedral hexamer of dimers enclosing six mainly helical N-terminal extensions (Fig. 4.4A). The yeast SpHsp16.0 is an octomer of dimers in a spheroid assembly where two dimers lie end to end along four lines of longitude (Fig. 4.4B). At each of the poles, the C-terminal I-A-I motifs bind into pockets about the fourfold axis, like they do around the vertices in the archaeal octahedron. The partner chains in the dimers use an alternative linker conformation to allow their C-terminal extensions to make a twofold interaction across the equator linking the two hemispheroids together (Fig. 4.4B). This makes a precariously flexible assembly, equivalent to removing four edges from around an octahedron, but it is strengthened by additional interactions between N-terminal extensions that make connections between staves and across the equatorial zone, involving an array of intrachain and interdimer phenylalanine and arginine interactions. The utility of the ACD dimer as a construction unit for a range of nanoassemblies was underscored by the conversion of the MjHsp16.5 24-mer to a 48-mer (Fig. 4.4D; pdb 4eld). An engineered insertion into the N-terminal extension close to the ACD caused the assembly to expand, and a new linker conformation between the ACD and the INI motif allowed the motif into pocket interactions to be conserved whilst maintaining the overall symmetry (Mchaourab *et al.*, 2012).

The building block of bacterial sHsps is also a strand exchange dimer. XaHspA, from *Xanthomonas axonopodis*, which causes citrus canker disease, forms a 36-mer in solution based on SEC, and the crystal structure (pdb 3guf) shows a strand exchange dimer that forms a trimer of dimers in the crystal lattice, similar to one hexameric disk in wheat Hsp16.9b (Fig. 4.4A) (Hilario *et al.*, 2011). Although there is an IQV motif in the C-terminal extension, the extension is disordered in the crystal, as is the N-terminal extension. DrHsp17.7 and DrHsp20.2 from *Deinococcus radiodurans*, an extremophilic, radiation-resistant bacterium, form dimers and higher assemblies, respectively (Bepperling *et al.*, 2012). The IDV motif in DrHsp17.7 has quite a short linker to the ACD, and in the crystal structure (pdb 4fei) the motif was bound in $\beta 4/\beta 8$ pockets of neighboring dimers in the crystal lattice, leading to the suggestion that this short linker contributes to its low assembly state (Bepperling *et al.*, 2012). It is noticeable that several hydrophobic-pocket-lining residues in Dr17.7 are not conserved (Fig. 4.2). The DrHsp17.7 dimer can prevent protein aggregation without forming a complex, presumably by weak and transient interactions. Negative stain EM showed that Hsp20.2 formed heterogeneous, mostly spherical oligomers with diameters between 90 and 180 Å, with the main population having a diameter of 130 Å. Hsp20.2 forms a 36-mer and 18-mer in solution by SEC and AUC, with no redistribution toward the smaller oligomer at 43°C. The 18-mer is considered to bind stably to client proteins. All three bacterial sequences have a conserved set of charged interactions buried in the interface close to the two-fold axis (D R R), in common with wheat Hsp16.9b, but which are not conserved in the archaeal and yeast sequences (Fig. 4.2). PfHsp20 (MAL8P1.78), which is involved in plasmodium mobility in the pathogenic protist *Plasmodium falciparum*, has also conserved this set of charges (Montagna *et al*., 2012) (Fig. 4.2).

Metazoan sHsps Have a Different Dimer Interface

A phylogenetic tree based on just the α-crystallin domains clearly shows a distinct clustering of the metazoan sequences distinguished chiefly by the shortness of the sequence segment between the β5 and β7 strands, precluding the formation of a β6 strand, and hence a strand exchange dimer (Kriehuber *et al*., 2010). Progress in 3D structure determination of animal sHsps has been hindered by the inability to identify a naturally occurring monodisperse assembly. However, removal of the N-terminal, or both sequence extensions from vertebrate sHsp sequences, creates monodisperse, smaller assemblies that can be resolved by crystallography or solution NMR (Basha *et al*., 2012; Delbecq and Klevit., 2013). Solution spectroscopy showed that the metazoan ACD formed an extended β7 strand (Mchaourab *et al*., 2009; Jehle *et al*., 2009) (Fig. 4.3A). The first vertebrate sHsp crystal structure, human αB ACD stripped of both extensions, showed a dimer interface distinct from the nonmetazoan structures whereby the elongated β7 strand hydrogen bonds with a partner domain to create an extended antiparallel (AP) β-sheet dimer interface (Bagnéris *et al*., 2009). The AP interface is maintained in constructs cocrystallized with I-X-I motif-containing peptides (Fig. 4.3B) or that included C-terminal extensions [*Danio rerio* αA-crystallin (HspB4), pdb 3n3e]. Three register shifts (I–III) have been observed (pdbs 3l1g; 3l1e; 3l1f; 3n3e) (Laganowsky *et al*., 2010; Laganowsky and Eisenberg, 2010), with solution NMR indicating that APII is more favorable in the human αB ACD dimer (Delbecq *et al*., 2012). The AP interface causes the geometric dispositions of the β4/β8 pockets to differ from the β6 strand exchange dimer in nonmetazoans, which will impact on the geometry of higher assembly.

Although these metazoan sHsp 3D structures are from vertebrates, the sequences indicate that a similar AP dimer would be formed in sHsps from urochordates (*Ciona intestinalis*, sea squirt Hsp27-like) and cephalochordates (*Branchiostoma floridae*, C3XVI4_BRAFL). Among the arthropods, sequences from the ACD of *Drosophila melanogaster* labeled Hsp27 and Droefl are very similar to those of the vertebrate B1/B4/B5/B6 group. DroB8 and the cold stress homologue SiHsp21.4 from the Chinese rice pest moth *Sesamia inferens* (Sun *et al.*, 2014) have most of the conserved features of the vertebrate ACD AP dimer interface (Fig. 4.2). Among the class Arachnida, the protein encoded by gene L7M4H7_9ACAR from *Rhipicephalus pulchellus*, a tick that is a vector of viral pathogens (Lutomiah *et al.*, 2014), appears to have the AP interface (Fig. 4.2).

In flatworms and the nematode worm, the interface varies. There is a crystal structure of Tsp36 (pdb 2bol) from the platyhelminth *Taenia saginata*, the beef tapeworm (Stamler *et al.*, 2005), where the sequence comprises two consecutive ACDs preceded by a long N-terminal helical extension (Fig. 4.5). In neither domain is the $\beta 5$–$\beta 7$ segment in a position to participate in dimerization by either a strand exchange or AP interface mechanism. The two domains form a four-domain assembly by the motif-into-pockets mechanism; the $\beta 4/\beta 8$ pocket of ACD2 is filled by the I-Q-P motif in the linker between the two domains, and the ACD1 pocket is filled by an I-F-P motif at the start of the N-terminal extension from a partner chain (Fig. 4.5). Similar two-domain sHsp sequences are often present in species that invade humans, such as blood, lung and liver flukes, black yeast fungi that colonize the lungs of cystic fibrosis patients and fungi that cause Darling's disease. However, in the nematode worm *C. elegans*, sHsp sequences such as those associated with lifespan modulation, Hsp16.48, Hsp 16.1 and Sip1 (Walther *et al.*, 2015) and protection from heat-induced necrosis, Hsp16.1 (Kourtis *et al.*, 2012), have only a single domain that aligns with the first domain of the tapeworm. These three sequences likely dimerize about an AP interface and engage in higher assembly using their IXI motifs in C-terminal extensions.

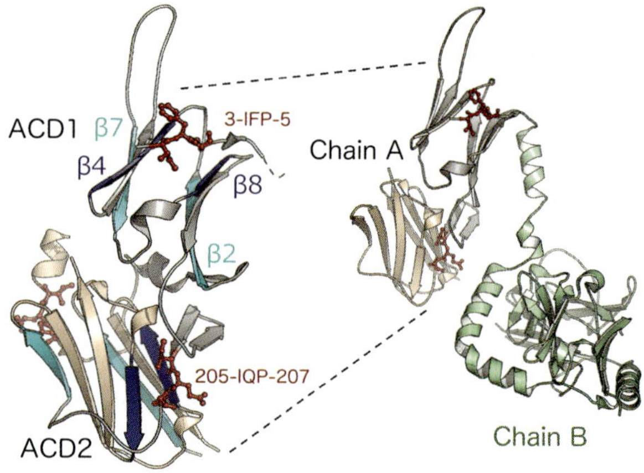

Fig. 4.5 *Parasitic worms have two ACDs that patch their pockets from linker and N-terminal IXI motifs to form a dimer wrapped in helical N-terminal extensions.* In Tsp36 from the platyhelminth *Taenia saginata*, the beef tapeworm (pdb 2bol), the polypeptide chain encodes two consecutive α-crystallin domains (ACD1, gray; ACD2, light pink) that form a four-domain assembly about a twofold axis. The IXI motif residues (shown as ball and stick, colored plum) from the linker between ACD1 (205-IQP-207) and ACD2 in each chain patch the $\beta 4/\beta 8$ pocket (shown in royal blue) of ACD2. The long N-terminal extensions are mainly helical and wrap around the four-domain assembly, with the figure showing the N-terminal extension from chain B colored in pale green. The N-terminal (3-IFP-5) motif/shown in plum ball and stick/from chain B patches the $\beta 4/\beta 8$ pocket of ACD1 from chain A (and vice versa for the partner chain).

There are examples of other ACDs with major changes to the segment length between the $\beta 5$ and $\beta 7$ strands. Among the three sHsp genes of the marine choanoflagellate *Monosiga brevicollis*, the closest unicellular relative of animals (King *et al.*, 2008), MBrJ2 has a typical ACD that clearly aligns with plants, yeasts and prokaryotes, whereas MBrH3 has an ACD with a considerably longer connection between strands $\beta 5$ and $\beta 7$ due to a 44-residue insertion which hints possibly at a transition structure toward the reengineered metazoan dimer interface. By contrast, there are ACDs where the connection between strands $\beta 5$ and $\beta 7$ is even shorter than in AP dimers. The crystal structure of such an ACD from archaeal *Sulfolobus solfataricus* Hsp20.1 (Q97W19_SULSO), which lives at low pH and 80°C, shows good superposition over the bulk of the β-sandwich fold (Liu *et al.*, 2015). The wild-type sequence has a long N-terminal sequence and forms a possible 24-mer, which disassembled when a truncation removed the V-D-I-K-V-E motif in the C-terminal extension. The crystal structure of the excised ACD did not form an AP interface, and although the asymmetric unit contained two monomers interacting via salt bridges, there was no firm evidence that they represented a dimer. *Arabidopsis thaliana* AtHsp18.5 and homologues are cytosolic plant sHsps that also lack a β6 strand, but the wild-type sequence does form a dimer (Basha *et al.*, 2013). At18.5 has lost all the conserved charges E R D R R characteristic of the strand exchange interface and Sulso20.1 has lost some of them (Fig. 4.2).

The Shared Groove

The X-ray structures of the vertebrate ACD dimers show a deep groove at the AP β-sheet dimer interface, with borders defined by strands β7 and β2 or β3 (Fig. 4.6A) (Clark *et al.*, 2011; Slingsby *et al.*, 2013). In full length αB (pdb 2klr), the region equivalent to the β2 strand is directed away from its own domain, allowing a new upstream region (β2a) to H-bond with the β3 strand from another dimer (Jehle *et al.*, 2010). Access to the groove is increased when the β2 strand is not present in the sheet, either because it is truncated in the construct (pdb 2y1y; pdb 3q9p) or it is elsewhere in the lattice (pdb 2wj5), and when the loop between strands β5 and β7 is flat (Clark *et al.*, 2011; Rajagopal *et al.*, 2015).

There is a kingdomwide, highly conserved arginine/lysine that in eukaryotes ion-pairs across the dimer interface with a conserved aspartate/glutamate in a sequence position that is characteristic of the type of dimer interface (Fig. 4.2). In the full assemblies the conserved arginine is facing toward the interior, and in the mammalian α-crystallin domain dimers the interface ion pair is found at either end of the deep groove (Fig. 4.6A). The acidic residue is not always conserved in the double

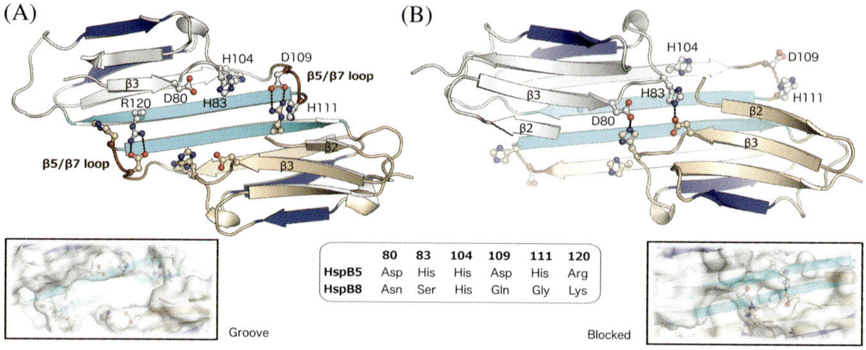

	80	83	104	109	111	120
HspB5	Asp	His	His	Asp	His	Arg
HspB8	Asn	Ser	His	Gln	Gly	Lys

Fig. 4.6 *Human HspB1 (Hsp27), HspB4 (aA-crystallin), HspB5 (αB-crystallin) and Hsp20 (HspB6) have a conserved network of charged interactions at the AP dimer groove which is blocked in the human R120G desmin-related myopathy mutant.* **A**. In the structure of the dimer interface in normal HspB5 (pdb 2wj7), the two β7 strands (cyan) at the interface of two ACD chains (colored gray and light pink) form a groove (see also box inset, "Groove"), surrounded by conserved residues, Arg-120, His-111, Asp-80, His-104, His-83, Asp-109. At each end of the groove, a conserved ion pair forms across the dimer interface between Arg-120 and Asp-109. **B.** In HspB5 R120G (pdb 2y1z), the rearrangement of the residues at the interface results in two new ion pairs across the interface that block the groove (also shown in panel inset, "Blocked"). The table shows the lack of conservation of this network in HspB8.

domain sHsps, which have neither a strand-exchange nor an AP dimer interface. In Sm-P40 from *Schistosoma mansoni*, both partners of the ion pair are replaced in ACD2. This protein is 15% of the endogenous soluble proteome of the mature egg when it passes through the human gut wall on its journey to water to hatch into the free swimming form which infects snails (Mathieson and Wilson, 2010).

3D structures of the core ACD dimer from several vertebrates, as well as human αB and Hsp27 (HspB1) cocrystallized with their respective I-X-V/I free peptide sequences (pdbs 4m5s; 4m5t; 4mjh), establish the generality of the AP interface in vertebrate sHsps, whilst further demonstrating the variability in the register at the twofold interface (Hochberg *et al.*, 2014). NMR, IM and nanoMS studies on oxidized and reduced αB E117C ACD dimers show that in solution this domain exists predominantly in the APII register, and that under reducing conditions there is a monomer–dimer equilibrium with fast subunit exchange dynamics (Hochberg *et al.*, 2014). The AP interface in αB-crystallin is destabilized at lower pH, and solution NMR (pdb 2n0k) has shown that stability is controlled by the conserved His-104 (Fig. 4.2), which is some distance from the interface (Fig. 4.6) (Rajagopal *et al.*, 2015).

The Closed Groove

Several inherited human diseases are caused by mutations to the conserved arginine (Boncoraglio *et al.*, 2012; Clark *et al.*, 2012). The mutation R120G in αB-crystallin causes myopathy, cardiomyopathy and cataract. The X-ray structure of the excised ACD dimer of human αB R120G close to physiological pH reveals a closed groove associated with rearrangement of an extensive array of charged interactions across the dimer interface (Clark *et al.*, 2011). His-83 and Asp-80 on movable arches on either side of the interface close the groove by forming two new salt bridges (Figure 4.6B). The set of charges at the dimer interface including Asp 109 and His-111 is characteristic of the vertebrate interface (HspB1–B6). The set is conserved in urochordate HspB1 (Cio Hsp27), and is largely conserved in amphioxus BF1 and insects, where the intervening sequence length between the strands β5 and β7 is exactly conserved (Figs.4. 2 and 4.6B). The set of charges is not conserved in vertebrate Hsp22 (HspB8), whereas the dimer-destabilizing histidine is conserved in both groups of sequences. Heterodimers are readily formed between HspB1 and B5, HspB1 and B6, and HspB5 and HspB6, but are less efficient with HspB8 (Mymrikov *et al.*, 2012).

Patching the B2/B7 Edge Strands

The exposure of core hydrophobics at the β2/β7 edge of the β-sandwich fold in the strand exchange dimer is different from that of the AP dimer interface. In wheat Hsp16.9b, the key hydrophobic residue Trp-48 positioned in the middle of the β2 strand facing Phe-110, Leu-113 and superfamily conserved Arg-108 on the β7 strand are patched by Pro-12,

Phe-10, Phe-13 and Phe-41 respectively from the N-terminal extension of the same chain (Fig. 4.1A). In yeast SpHsp16.0, ACD core equivalents are Val-41 Ile-104 and Pro-107, with Val-41 on the β2 strand patched by Tyr-23 from the N-terminal extension. In archaeal MjHsp16.5, Ile-47 is the equivalent residue on the β2 strand, and is patched by Ile-35 and Ile-37 of the β1 strand. In the tapeworm Tsp36 ACD1, the β2 strand is distorted and shortened, with the exposed β3 strand at the edge patched by Leu-57, Phe-58, Leu-60 and Phe-99 in the N-terminal extension.

The equivalent core hydrophobics in the structure of the N-terminally truncated human HspB1 are Ile-88, Tyr-142 and Pro-145, which line up with additional hydrophobics Ile-134 and Phe-138 on the β7 strand. The AP mode of dimerization places both sets of hydrophobics in proximity such that they line the shared groove (Fig. 4.3). In truncated rat Hsp20 (65–162), the β2 strand is not part of the sheet as the equivalent residue Val-67 is instead reaching into the β4/β8 pockets of another dimer chain in the lattice (Bagnéris *et al.*, 2009). This AP dimer is then left with an accessible shared groove in which lies a C-terminal extension from another dimer in the lattice (pdb 2wj5). Human N-terminally truncated Hsp20 (57–160) forms dimers in solution, whereas in a crystal lattice (pdb 4jus) it forms tetramers (Weeks *et al.*, 2014). The dimers form tetramers by reciprocal exchange of short flanking N-terminal sequence 61-ALPVAQVPTD-70 that fill partner β4/β8 pockets in either orientation of the strand. Consideration should be given to the role of the β2 strand when one is investigating the impact of truncation of sHsp N-terminal extensions.

The AP arrangement can thus be envisaged as patching exposed ACD core hydrophobics whilst forming a new groove that looks like a binding site. A candidate occupant is an N-terminal extension, by analogy with the way in which the N-terminal extensions in wheat, yeast and tapeworm patch the β2 and β7 edge hydrophobics. In fact, AP dimerization may be primarily patching ACD core hydrophobics, which would become exposed on dissociation and available for binding functions, such as with its own N-terminal extension. In this scenario, the monomer would be the chaperone (Fig. 4.3A), and other binding partners may have been appropriated for the shared groove, providing a structural basis for the diversification of functions in animal sHsps. The AP groove is proposed to be the binding site for cholesterol derivatives that appear to reverse mutant protein aggregation (Makley *et al.*, 2015).

Electron Microscopy Structures of sHsp Assemblies

As X-ray–diffracting crystals of sHsp oligomers are hard to produce, an alternative is to use single particle electron microscopy to image the assemblies. The domain layout of a dodecameric sHsp Acr1 from *Mycobacterium tuberculosis* was determined by electron microscopy. The MjHsp16.5 dimer coordinates (pdb 1shs) were fitted to a negative stain single particle reconstruction EM map calculated with 23 tetrahedral symmetry (pdb 2byu), whereby a trimer of dimers was placed along the dodecamer threefold axis (similar to the threefold views in Fig. 4.4), with the dimer twofold coincident with the dodecamer twofold (Kennaway *et al.*, 2005).

Insight into the structural basis of conformational variability in sHsps was gained by cryoEM of yeast Hsp26 (White *et al.*, 2006). Following multistatistical variance analysis and classification, 3D reconstruction revealed two discrete structures, both 24-mers with tetrahedral 23 symmetry, but of different dimensions. The structures, described as compact and expanded, are each interpreted as having 12 dumbbell-shaped units constituting the bulk of a porous shell. An ACD dimer and adjacent N-terminal structures were fitted into each unit, with the termini assembling the dimers into tetrahedral shells by trimeric contacts. Twelve C-terminals remain in the outer shell, where they are close enough to form a contact with the ACD of an adjacent dimer, but the other 12 C-terminals are too far away to form the motif-into-pockets interaction and instead form trimeric insertions. The compact and expanded forms differ significantly in their internal density with the compact assembly having a single, central mass of density inside the shell, whereas in the expanded assembly there are four densities inserted below the shell. Deletion of the complete N-terminal extension and part of the $\beta 2$ strand (1–95; Fig. 4.2) resulted in dimers. The study shows that domains can move with respect to each other, implying that they have flexible linkers which allow assemblies to take up different arrangements, yet under the overall constraints of a symmetric organization, as is the case in allosteric systems. Earlier cryoEM studies of Hsp26 complexed with denatured substrate showed much bigger particles that have structural features indicative of radical reorganization of the native assembly (Haslbeck *et al.*, 1999).

Caenorhabditis elegans CeHsp17.1 when expressed in *Escherichia coli* enables the cells to grow at a lethal temperature of 50°C, and lowers the level of insoluble protein (Zhang *et al.*, 2015). The CeHsp17.1 protein, His-tagged at the C-terminal, was largely recovered from the insoluble fraction and then purified under denaturing conditions. The structure, resolved through single particle 3D reconstruction by cryoEM at around 14.5 Å resolution and based on tetrahedral symmetry, showed the oligomers to be spherical with a diameter of ~130 Å, and composed of 12 rodlike "building blocks." The map was fitted with the coordinates of one dimer of wheat HSP16.9b, and tetrahedral symmetry applied. As the sequence of CeHsp17.1 in the ACD is more similar to that of vertebrates (Fig. 4.2), a better interpretation would have been to fit an AP dimer. Unlike the EM tetrahedral structure of Acr1 12-mer from *Mycobacterium tuberculosis*, here the dimer is the asymmetric unit. The resulting spherical oligomer is interpreted as having 24 subunits, consistent with its apparent molecular mass of 460 kDa as determined by size exclusion chromatography with the model having the N-terminal extensions buried in the center (Zhang *et al.*, 2015). There is no evaluation of the level of polydispersity of the sHsp particle images.

A large dataset collected by cryoEM from human αB-crystallin revealed size polydispersity. The particles were classified into a series of assemblies spanning the range of sizes measured by nanoMS, but with a significant population of particles assigned as 24-mers with tetrahedral 23 symmetry. This population was used to reconstruct a 3D map to which ACD dimers (2klr) were fitted as a rigid body, and the extension sequences modeled into the remaining density (Braun *et al.*, 2011). The resulting model (pdb 2ygd) is a trimer of dimers modeled as a large AP β-barrel with the threefold axis passing through the middle of the barrel and through density modeled as the self-associating N-terminal extensions (Braun *et al.*, 2011). The N-terminal region was modeled as helical and fitted to density on the surface of the 24-mer, contrary to the widely held assumption that it would be more buried.

The accuracy of nanoMS technology for determination of the size distribution of oligomers reveals that sHsp assemblies constitute a continuum of subunits, indicating a lack of symmetry. However, in yeast Hsp26, where heterogeneity is low, structural analyses have shown that even within a symmetric assembly of fixed oligomer number, the domain quaternary organization can vary, suggesting models for activation of sHsp assemblies. These limited studies show the importance of looking for and accounting for heterogeneity in sHsps. However, if the sHsp sample constitutes a continuum of particle masses, differing only by one subunit (20 kDa), then a large proportion of species are unlikely to have symmetry. A limiting factor of high resolution structures of sHSPs by 3D EM is the correct classification of each of the particles in a dataset into the different homogenous populations and the generation of many correct starting models (one for each species in the sample). Generation of a starting model involves identifying with confidence three different 2D views from the same oligomeric species, then reconstructing the 3D volume. The challenge of correctly selecting (from the many views of many particles) three distinct views, all from the same particle, has a statistically low chance of success and validation of success is challenging. A further difficulty for sHsps is that visualization of secondary structural elements is often used as additional evidence of a correct cryoEM structure, but sHSP ACDs are mainly composed of β-sheets, which require around 4.5 Å resolution to resolve in a map.

The N-Terminal Extensions

These are the least-defined structural feature of sHsps, being highly variable in sequence but usually hydrophobic. Where they have been resolved in crystal structures they are helical and make mainly intra-chain contacts, patching the $\beta2/\beta7$ edge of the ACD. In metazoans, several sHsp family members have IXI motifs close to the N-terminal, and the X-ray structure of the double ACD tapeworm Tsp36 showed how these N-terminal motifs could also bind into $\beta4/\beta8$ pockets and support assembly by wrapping around the domains (Fig. 4.5). Single domain Hsp20 (HspB6) does not have a C-terminal IXI, but does have the motif I-P-V-P-V close to the N-terminal containing potentially overlapping pocket fillers. Biophysical evidence indicates that the N-terminal domain of Hsp20 is not a random ensemble and is partially protected from enzyme digestion (Weeks *et al*., 2014). When this motif was mutated to G-P-G-P-V, the SAXS profile did not change, suggesting that this motif does not stabilize the dimer by binding partner ACD pockets (Weeks *et al*., 2014).

Specific family members of various vertebrate, insect, worm, spider and choanoflagellate sHsps have a region of sequence conservation in the N-terminal region (van Montfort *et al*., 2001; Heirbaut *et al*., 2014) (Fig. 4.7). This would suggest that although multiple rounds of gene

```
HsB5            1-MDIAIHHPWIRRPFFPFHSPSRLFDQFFGEHLLESDLFPTSTSLSP-46
HsB4            1-MDVTIQHPWFKRTLGPFY-PSRLFDQFFGEGLFEYDLLPFLSSTIS-45
HsB1       1-MTERRVPFSLLRGPSWDPFRDWYPHSRLFDQAFGLPRLPEEWSQWLGGSSW-51
HsB6       1-MEIPVPVQPSWLRRASAPLPGLSAPGRLFDQRFGEGLLEAELAALCPTTLA-51
HsB8    1-MADGQMPFSCHYPSRLRRDPFRDSPLSSRLLDDGFGMDPFPDDLTASWPDWAL-53
Dro27           1-MSIIPLLHLARELDHDYRTDWGHLLEDDFGFGVHAHDLFHPRRLLLP-47
Droefl          1-MSVVPLMFRDWWDELDFPMRTSRLLDQHFGQGLKRDDLMSSVWNSRP-47
```

Fig. 4.7 *A conserved region in the N-terminal sequence of some animal sHsps.* A sequence alignment of the N-terminal extensions from human HspB1, B4, B5, B6 and B8 along with Dro27 (P02518) and Droefl (P82147 lethal(2), essential for life) from *Drosophila melanogaster*. During normal development Droefl is expressed in larval body wall muscles accumulating around Z-disks (Wojtowicz *et al*., 2015). The conserved regions are highlighted in gray, green and red. Residues in bold RLPEE were shown in HspB1 to destabilize assembly.

duplication in the different lineages have repeatedly resulted in multigene families in which the N-terminal extensions have rapidly diverged, some sequences have maintained a conserved trait, which likely entails conservation of some structural or functional role.

The role of the N-terminal extension in eukaryote sHsp assembly and polydispersity is largely unknown. In the case of vertebrate sHsps, the oligomers are extremely large, polydisperse and undergo rapid subunit exchange (Mchaourab *et al.*, 2009; Hochberg and Benesch, 2014). Solid state NMR spectroscopy of αB-crystallin assembly (HspB5) resolved the ACD dimer and a dimer–dimer exchanged β2a strand, but the N-terminal extensions were invisible (Jehle *et al.*, 2010). Radically different models of the αB-crystallin full assembly that "aim to provide a structural rationale for known properties of αB-crystallin" have been evaluated by Delbecq and Klevit (2013). We have speculated that the N-terminal extensions might bind into the AP groove in a dynamic way and that the groove is a candidate binding site for the unfolded polypeptide chain (Clark *et al.*, 2011; Slingsby *et al.*, 2013).

sHsp Polydispersity and Holdase Mechanism

Low temperature crystallography and solid state NMR spectroscopy (Jehle *et al.*, 2010) give a uniform picture of the docking of IXI/V sequence motifs into partner ACD pockets in full sHsp assemblies and subassemblies. Solution state NMR at 22°C has shown that peptides mimicking the IXI/V motif can bind into ACD dimer pockets (Delbecq *et al.*, 2012), although other solution NMR studies have indicated that in full length αB assemblies the motif interchanges between highly populated flexible states and a sparsely populated bound state (Baldwin *et al.*, 2012). The interaction, which can take place with the motif strand running in either direction through the pockets, coupled with a flexible hinge, is a common theme for construction of sHsp nanoassemblies (Fig. 4.4). Furthermore, IXI/V motifs from other regions of the sequence can also dock into ACD pockets (Fig. 4.5), highlighting how these nanoassemblies have the potential for flexible polyvalency. It has been suggested that by having multivalent binding ability for aberrant proteins, the sHsps may act to seed and concentrate aggregate material (Walther *et al.*, 2015). The idea that this interface region can act as a substrate binding site was strengthened when $\beta4/\beta8$ pockets in αB-crystallin ACD were shown by NMR spectroscopy to bind amyloid $A\beta_{(1-40)}$ and to prevent fibril formation (Mainz *et al.*, 2015). We have previously speculated that the motif-into-pockets interaction might enable α-crystallin to transform over its long lifespan in the eye lens into the water-insoluble form that is characteristic of high refractive index regions of mature lenses (Slingsby *et al.*, 2013; Slingsby and Wistow, 2014).

The most generalized model for a small heat shock assembly is of a vessel formed from ACDs that encapsulates to some degree more disordered hydrophobic N-terminal extensions (Haslbeck *et al.*, 2005). An example of how a static sHsp oligomer can function as a holdase was shown using the engineered expanded assembly of the hyperthermophile Mj16.5 (Fig. 4.4D) which can bind destabilized T4/L lysozyme without becoming heterogeneous, as happens in most other cases of substrate-bound sHsps, allowing single particle reconstruction of bound complexes using cryoEM (Shi *et al.*, 2013). Comparison of the expanded assembly with and without bound substrate indicates that T4/L gains access to the interior and binds both the N-terminal region and the inner surface of the ACDs accompanied by some rearrangement of the N-terminal domain. In the symmetric assembly all the $\beta4/\beta8$ pockets are fulfilling assembly duties, being occupied by partner INI motifs. It is possible that a hydrophobic N-terminal sequence could bind a denatured client substrate, assisted possibly by the $\beta2/\beta7$ ACD edge, if the $\beta1$ strand peeled away.

A crystal structure of R107G of MjHsp16.5 (pdb 4i88) shows that it forms a 24-mer isostructural with the wild type octahedral assembly (Quinlan *et al.*, 2013), even though the superfamily conserved Arg-107 can no longer form an intradimer stabilizing ion pair with partner Glu-98 (Quinlan *et al.*, 2013). CryoEM of the wild type at both room temperature and 60°C indicated two populations of particles with the same symmetry and the same 124 Å diameter, but they differed in the presence or absence of central density (Quinlan *et al*., 2013). CryoEM studies of R107G MjHSP16.5 at room temperature indicated two types of 24-mers of 124 Å diameter, whereas at 60°C there was an empty 124 Å diameter assembly and an expanded 24-mer of diameter 136 Å with significant changes in the fourfold symmetry axis and dimer–dimer interface (Quinlan *et al*., 2013). The authors argue that the conserved arginine contributes to quaternary assembly, and that at elevated temperature the resulting perturbed assembly functions as a better holdase.

Mesophilic sHsp assemblies can be studied in the laboratory under stress conditions similar to those encountered during their lifespan, such as mild heat or events leading to phosphorylation. Under equilibrium conditions sHSPs can act as sensors of protein stability with binding affinities correlated with the level of substrate destabilization and coupled to the activation state of the sHsp oligomer (Mchaourab *et al*., 2009). The activity of sHsps and their mutants is generally evaluated by inhibition of stress-induced aggregation of model substrates, and there is a large literature indicating that a range of sHsp assembly forms can perform this role. The yeast *Schizosaccharomyces pombe* (SpHsp16.0) X-ray structure (Fig. 4.4B) showed the bulk of the hydrophobic N-terminal extensions encapsulated within a vessel with much of their hydrophobic character unsatisfied. There is no evidence of marked disassembly at heat shock temperatures, based on SAXS, although it is suggested that the lack of perfect symmetry weakens the assembly, lowering the barrier for exposure of the N-terminal extensions (Hanazono *et al*., 2013). In the case of monodisperse pea Hsp18.1, at heat shock temperatures the bulk of the dodecamers redistribute toward a polydisperse distribution of higher order 13–20-mers (Stengel *et al*., 2010). In plants AtHsp18.5 functions as a dynamic dimer (of unknown interface) with sequence extensions exposed to solvent, yet forms large stable complexes with denatured substrate. The sequence shows that the ACD has conserved the $\beta 4/\beta 8$ binding pockets, and that it has IXI motifs in both extensions, with the N-terminal extension required for chaperone activity (Basha *et al.*, 2013). The native dimer thus has four IXI motifs and four potential docking points for substrate (Trp-48, shown in Fig. 4.1, is conserved in the sequence), giving it sufficient valency to build a complex.

It has recently been pointed out that size techniques based on mass transport, such as SEC, AUC and native gels, will give potentially misleading results if the rate of oligomer dissociation is comparable to or faster than the timescale of the measurement (Hochberg and Benesch, 2014). In αB-crystallin the subunit exchange rate can vary by four orders of magnitude over a pH range of 5–9 and at temperatures between 24°C and 50°C, whilst the mean number of subunits in the oligomers is confined to around 24–28, although there is significant variation in the ratio of even- to odd-numbered oligomers in the distributions (Hochberg and Benesch, 2014). As these numbers do not fit symmetric assembly, it was proposed that αB is a distribution of interconverting polyhedra via the motif-into-pockets interaction (Baldwin *et al.*, 2011b).

sHsp solution dynamics is considered to play a role in their chaperone function by creating an equilibrium between low and high affinity substrate-binding forms (Mchaourab *et al.*, 2009; Basha *et al.*, 2012; Hochberg and Benesch, 2014; Stengel *et al.*, 2010; Haslbeck and Vierling, 2015). Although the rate-limiting step in subunit exchange is considered to be the dissociation of subunits from the oligomers, there is no consensus that subunit exchange rates correlate with chaperone activity (Mchaourab *et al.*, 2009; Baldwin *et al.*, 2011a). Nevertheless, the motif-into-pockets interaction provides a range of options for rapid reassembly into less organized assemblies that presumably favors a holdase mechanism.

A dynamic AP interface in metazoan sHsps has been linked to holdase function. A KD for the aB ACD dimer measured by ITC and NMR was around 2 μM at pH 7.5, 25°C, with the interface decreasing in strength around 15-fold by dropping to pH 6.5 or increasing the temperature to 37°C (Rajagopal *et al.*, 2015). There is conformational variability in the loop between the $\beta5$ and $\beta7$ strands: it can be flat or "sitting-up" (Clark *et al.*, 2011) (Fig. 4.6). Based on solution NMR, it is proposed that protonation of His-104 at the base of the loop may perturb intersubunit contacts of the sitting-up $\beta5/\beta7$ loop made by His-111, leading to dimer destabilization at pH 6.5 (Rajagopal *et al.*, 2015). These authors propose that His-104 perturbs a dynamic network of electrostatic and H-bonding interactions which is responsible for modulating the stability of the AP dimer interface (Fig. 4.6). The R120G human disease mutation site, also in this region, perturbs the charge network, resulting in closure of the AP groove (Fig. 4.6). Thus, a small drop in pH perturbs a weak dimer interface which correlates with increased αB-crystallin holdase activity, the effect mimicked by mutation of His-104 (Fig. 4.6) that shifts the ACD from dimer to folded monomer (Rajagopal *et al.*, 2015). NanoMS and IMS have shown that increasing the binding of an interdimer contact (IXI motif into pocket) destabilizes the AP interface within a dimer, arguing for a role for the C-terminal region in regulating quaternary dynamics and hence access to potential substrate binding sites (Hilton *et al.*, 2013).

In mammalian sHsps, oligomer assembly is sensitive to phosphorlyation, a modification that can be a marker of disease. Using an unbiased proteomics approach, comparing the range of phosphopeptides between Alzheimer's disease and normal brains, the pathology was associated with phosphorylated HspB1 Ser-82 and phosphorylated αB Ser-59, which were correlated with phosphopeptides of several kinases and phosphorylated other chaperones (Dammer *et al.*, 2015). In human Hsp27 (HspB1), biophysical techniques readily detect equilibrium between large oligomers and dimers, with phosphorylation of three serines in the N-terminal extension promoting dissociation (Mchaourab *et al.*, 2009). Hsp27 undergoes concentration-dependent dissociation in the μM range, and a combination of SEC and MS data shows that an aspartate phosphorlyation mimic, D3-Hsp27, predominantly exists as a disulfide-linked dimer under nonreducing conditions, which dissociates into monomers upon reduction (Jovcevski *et al.*, 2015). It now seems to be accepted that Hsp27 dimers have the greater affinity for denatured substrate (Hayes *et al.*, 2009, McDonald *et al.*, 2012; Jovcevski *et al.*, 2015). Spin labeling across the N-terminal extension of wild type human Hsp27 indicated its buried location, becoming exposed in D3-Hsp27 when the protein was largely dimeric (McDonald *et al.*, 2012). Representative spin-labeled D3-Hsp27 variants can bind destabilized T4/L mutants at room temperature which reassemble into complexes. Replacement of residues within the conserved motif 37-RLPEE-41 in the N-terminal domain (Fig. 4.7) reverses dissociation in D3-Hsp27, implying perhaps that these residues are there to destabilize the oligomer and/or present a conserved binding region for substrate in the disassembled form.

Subcellular Compartments

A common cellular response to stress is the sequestration of protein aggregates to subcellular compartments. The cytoplasmic site termed CytoQ has been characterized in yeast (Escusα-Toret *et al.*, 2013), although the correspondence of this deposition site to the mammalian aggresome, a perinuclear compartment with high autophagic activity, is unclear (Miller *et al.*, 2015). A specific role for sHsps in sequestration of misfolded proteins in yeast was indicated by the finding that *Saccharomyces cerevisiae* Hsp42 (Haslbeck *et al.*, 2004) but not Hsp26 played an essential role in the cellular machinery for formation of CytoQ, and that this activity largely resided in the N-terminal sequence (Specht *et al.*, 2011). The 240-residue N-terminal sequence of yeast Hsp42 shows little resemblance to the N-terminal extension of Hsp26, or the N-terminal extension of Hsp16.1 from *Schizosaccharomyces pombe;* nor is there an equivalent sequence to Hsp42 identified in the genome of *Schizosaccharomyces pombe.* However, Hsp42 does contain the motif 150-IGV-152 in the N-terminal extension.

Protein aggregates in deposition sites such as yeast CytoQ and mammalian aggresomes are marked for removal by UPS or autophagy, with Bag1 targeting proteins for degradation by the UPS, whereas Bag3 mediates degradation by macroautophagy (Kim *et al.*, 2013). Bag domains are nucleotide exchange factors in eukaryotic cells that make a direct interaction with the N-terminal domain of Hsp70. When Bag1 and Bag3 associate via their Bag domain with the nucleotide binding domain of Hsp70, they are connected to the Hsp70 machinery which triages misfolded proteins (Gamerdinger *et al.*, 2009; Kim *et al.*, 2013). A new cellular pathway that uses the specificity of Hsp70 to recognize misfolded proteins is regulated by Bag3, which also interacts with the microtubule motor dynein and selectively directs Hsp70 substrates to the motor and thereby to the aggresome (Gamerdinger *et al.*, 2011). The PxxP region, located upstream from the Bag domain, is a determinant of Bag3–dynein interaction. The ratio of Bag3/Bag1 increases with aging, especially in neuronal cells (Gamerdinger *et al.*, 2009). It is thought that this may be responsible for a switch in protein degradation toward autophagy, possibly in partnership with p62 (SQSTM1), as more highly modified proteins form bigger aggregates in aged cells. Bag3 is constitutively expressed in skeletal muscle cells and cardiomyocytes (Rosati *et al.*, 2011).

There is firm evidence for direct binding of Bag3 to vertebrate HspB8 (Carra *et al.*, 2008). Protein engineering supported linkage by IPV motifs in Bag3 full length protein into $\beta4/\beta8$ pockets of HspB8 (Fuchs *et al.*, 2010). Bag3 and HspB8 were shown to interact *in vitro* using SEC and AUC (Shemetov and Gusev, 2011). Bag3 can bind to HspB6, but with a lower affinity than HspB8 (Fuchs *et al.*, 2010). In principle, as all ACDs have $\beta4/\beta8$ pockets, potentially all could engage in this kind of interaction.

Bag3 and starvin (Q9vu81), the fly functional homologue of Bag3, both localize to the Z-disk (Arndt *et al.*, 2010). Both possess a Bag domain at their C-terminals, required for Hsc70 binding. In addition, Bag3 carries a WW and a proline-rich (PXXP) domain, while starvin has flanking low complexity and disordered regions. *Drosophila* HspB8 (CG14207) is specifically located in the Z-disks of insect flight muscle (Arndt *et al.*, 2010), where it is considered to reside in a complex with Bag3/Hspc7/CHIP, with a functional role of constantly dispatching exercise-damaged proteins toward destruction by autophagy. The sHsp sequence labeled HspB8 in *Drosophila* and its orthologue SiHsp21.4 in the rice pest moth have the set of conserved hydrophobics to form the $\beta4/\beta8$ pockets and both have standard IXI motifs in the C-terminal and N-terminal extensions, unlike vertebrate HspB8. Although the *Drosophila* Bag3 (starvin) sequence has not the number of IXI motifs in the sequence upstream of the Bag domain as in humans, there are two such motifs. Thus, insect Bag3 IPV motifs are available to insert into insect HspB8 pockets, and the insect IXI motifs in the HspB8 extensions are available to interact with their own or other sHsp $\beta4/\beta8$ pockets, whereas vertebrate HspB8 is less equipped to engage in exchange interactions with other sHsps.

In vertebrate muscle cells, sHsps are stored in the cytosol and then go to the Z-disk on mechanical stress, whereas Hsp90 and unc45 (myosin chaperones) are stored in the Z-disk and on stress go to the A-band (Smith *et al.*, 2014). During strenuous exercise HspB1 (Hsp27) and HspB5 (αB-crystallin) translocate to myofibrillar structures, where they are considered to carry out their protective role (Dubinskα-Magiera *et al.*, 2014). HspB5 and HspB1 bind to the titin immunoglobulin-like domain in the I-band region proximal to the Z-disk, where they may protect titin from aggregation (K€otter *et al.*, 2014). These authors review the findings that suggest a relationship among acidosis, sHSP induction and muscle elasticity. They show that HspB1 and HspB5 specifically protect the unfolded Ig regions of titin, but not the intrinsically disordered segments, from aggregating under acidic stress. HspB1 and HpsB5 were found to translocate preferentially to the titin springs in stretched sarcomeres as well as in diseased human skeletal and cardiac myocytes, with the mechanical effect of decreasing stiffness (K€otter *et al.*, 2014).

There are IXI/V motifs in the target region of human titin, such as in Ig-like domains 76, 77 and 79. Myofibrillar myopathies are associated with disintegration of the Z-disks, with most myopathy mutations mapping to Z-disk components such as desmin, αB-crystallin, myotilin, ZASP, filamin C and Bag3 (Selcen, 2011). A homozygous frameshift in exon1 of CRYAB or a homozygous frameshift in exon 3 within the ACD, in different human families, resulted in profound infantile muscular stiffness (Forrest *et al.*, 2011; Selcen, 2011). This suggests an essential role for αB in human skeletal muscle.

Conserved actinomyosin contractile machinery predates the evolution of muscle cells and metazoans. It is presumed to have performed cytoskeletal functions before gaining additional roles in muscle contraction in the evolution of animals and there is evidence that striated muscle evolved independently in cnidarians and bilaterians (Steinmetz *et al.*, 2012). The vertebrate Z-disk comprises 28 proteins, of which only 4 are shared with the insect Z-disk, among which is titin, which regulates sarcomeric length and integrity (Steinmetz *et al.*, 2012). Among animals with muscle cells (bilaterians, cnidarians, ctenophors), titin appears to be absent from the nonbilaterians, and is absent from placozoans, the simplest known animal (Steinmetz *et al.*, 2012). sHsp sequences (PF00011) are not detected in the genomes of ctenophors (comb jellies) and placozoans, but are found in cnidarian genomes (*Nematostella Vectensis*, Starlet sea anemone and *Hydra vulgaris*). The cnidarian AP interface is not properly formed (Fig. 4.2), so the shaping of the histidine-regulated AP interface may have been an adaptation to protect titin in bilaterian skeletal muscle.

Protein Partners: Folded Versus Unfolded Proteins

The most common folded partner of an sHsp is itself, with close family members able to coassemble to form hetero-oligomers, all using the IXI motif-into-pockets interaction. The interaction with Bag3 also uses this recognition device except now sHsp $\beta 4/\beta 8$ pockets are filled by an IXI motif embedded in a disordered region of a protein which has a separate domain with a well-defined binding partner. In this way, the sHsp is acting as a scaffold or component of complex machinery. It is less clear if sHsps engage in direct interactions with specific folded proteins, rather than acting as a holdase, when they would interact with unfolded or detabilized proteins. Sometimes clues to potential partners can be found from their sequence architectures. However, apart from the presence of an N-terminal ARID domain in sHsps from several plant species, sequences containing the α-crystallin domain are rarely accompanied by other domains. By contrast, the sequences of the ubiquitous CS domain have a range of architectures, spanning from complex multidomain proteins, often with an NUDC domain, or 2 CHORD domains, or a TPR repeat domain. The structure of human p23 (prostaglandin E synthase 3) (Q15185), a cochaperone which comprises solely of a single CS domain, has been resolved in isolation (pdb 1ejf; Weaver *et al.*, 2000), and the yeast homologue p23/Sba1 in complex with Hsp90 (pdb 2cg9; Ali *et al.*, 2006). It does not self-assemble. The third sHsp gene from the closest unicellular relative *Monosiga brevicollis*, MBrS8, is a 471-residue sequence with a single ACD near the N-terminal, with no IXI motif C-terminal to the ACD but rather a V-D-I sequence that precedes the domain. The sequence also has a pair of CBS domains near the C-terminal, known as a Bateman domain, which generally forms a nucleotide binding site (Bateman, 1997; Jin *et al.*, 2007), hinting perhaps that the ACD domain can work in partnership with other machines regulated by nucleotides.

Hsp27 is highly expressed in epithelial airways and there is evidence for Hsp27-mediated F508del CFTR degradation in mammalian cells via a pathway that involves substrate modification by SUMO followed by SUMO-dependent degradation by the ubiquitin–proteasome system (Ahner *et al.*, 2013). The authors proposed that Hsp27 made a physical interaction with Ubc9, though this was based on yeast two-hybrid studies and followup co-IP of Hsp27 and Ubc9 from coexpressing HEK cells, which does not prove a direct interaction. However, Ubc9 has an accessible 4-IAL-6 sequence at the N-terminal that could potentially promote binding when folded, and a 107-ITI-109 motif buried in the folded Ubc9 structure that would become accessible when unfolded.

There is some evidence that in mechanically stretched myocytes, the β4/β8 pockets of phopshorylated αB-crystallin can bind a folded α-helical protein, the FAK C-terminal domain (FAT domain) of focal adhesion kinase, based on pulldowns and immunochemistry (Periera *et al.*, 2014). Biophysical analysis of His-tagged αB and the FAK C-terminal domain indicated a dimer–dimer interaction of the complex deduced by a change in the elution profile of the two proteins when mixed. The structural basis of the direct interaction was tested using crosslinking in combination with mass spectrometry (CXMS) and computer modeling. The data supported a model where one of the helices of the four-helix bundle of the FAT domain docks into a β4/β8 pocket, based on crosslinked K-103 of αB. There was no biophysical or mutation data using the well-characterized αB ACD with FAK C-terminal to test this model. The authors propose that in the fully assembled state of αB the β4/β8 pockets are filled with C-terminal extensions, but on dissociation the dimer binds FAK FAT when they are empty. The general idea is that when FAK is bound to αB it is protected from calpain-mediated proteolysis, providing a nonchaperone role for αB in a cytoskeletal complex that enables cardiomyocytes to survive under stress conditions.

Conclusions

These small stress proteins, perhaps because they form few firm specific attachments, are free to duplicate and diversify, providing cytoprotection at modest genetic cost. In solution sHsps can populate a range of oligomeric distributions, and in cells they will be in dynamic equilibrium with close family members as well, where their surfaces and interfaces can explore the state of the proteome. Modern solution spectroscopy and spectrometry can monitor some of the conformational transitions sHsps perform as they dissociate and reassemble, under normal and stressful conditions. 3D imaging techniques have resolved a few static structures of symmetric sHsp oligomers that reveal some of the principles of assembly. However, the structural role of the N-terminal extension in animal sHsps is unknown, so the basic principle of assembly of, for example, α-crystallin is in doubt, even though it has been visible as spherical particles by EM for over 40 years. Progress in cryoEM is now rapid in terms of the microscope, the detectors and the *in silico* methods for processing data (Orlova and Saibil, 2011; Bai *et al.*, 2015). When the level of sample polydispersity is low and the difference in the species structure is pronounced, particle classification during processing and reconstruction can lead to several 3D structures resolved at high resolution from a single dataset. Just as crystallography of sHsps has been driven by opportunistic discovery of monodisperse forms that sometimes yield diffracting crystals, so for EM the goal must be to reduce sHsp polydispersity. This may allow some of the species to be resolved to higher resolution, perhaps in combination with single particle tomography to generate a correct but low resolution starting model. The hope is that sHsps will be resolved to near-atomic resolution by EM, as is achieved with other conformationally variable assemblies (Voorhees *et al.*, 2014).

Acknowledgments

CS is grateful for the financial support of the Medical Research Council, UK (grant G0801846), and the synchrotron beamline staff at the European Synchrotron Radiation Facility (Grenoble, France), the Diamond Light Source (Oxfordshire, UK) and Soleil, France (partially funded through BIOSTRUCT-X).

References

Ahner A, Gong XY, Schmidt BZ, *et al.* (2013). Small heat shock proteins target mutant cystic fibrosis transmembrane conductance regulator for degradation via a small ubiquitin-like modifier-dependent pathway. *Mol Biol Cell* **24:** 74–84.

Ali MMU, Roe SM, Vaughan CK, *et al.* (2006). Crystal structure of an Hsp90-nucleotide-p23/Sba1 closed chaperone complex. *Nature* **440:** 1013–1017.

Arndt V, Dick N, Tawo R, *et al.* (2010). Chaperone-assisted selective autophagy is essential for muscle maintenance. *Curr Biol*, **20:** 143–148.

Arrigo AP (2013). Human small heat shock proteins: protein interactomes of homo- and hetero-oligomeric complexes — an update. *Febs Lett*, **587:** 1959–1969.

Arrigo AP, Simon S, Gibert B, *et al.* (2007). Hsp27 (HspB1) and alpha B-crystallin (HspB5) as therapeutic targets. *Febs Lett*, **581:** 3665–3674.

Bagnéris C, Bateman OA, Naylor CE, *et al.* (2009). Crystal structures of alpha-crystallin domain dimers of alpha B-crystallin and Hsp20. *J Mol Biol* **392:** 1242–1252.

Bai XC, McMullan G, Scheres SHW (2015). How cryo-EM is revolutionizing structural biology. *Trends Biochem Sci*. **40:** 49–57.

Baldwin AJ, Hilton GR, Lioe H, *et al.* (2011a). Quaternary dynamics of alpha B-crystallin as a direct consequence of localised tertiary fluctuations in the C-terminus. *J Mol Biol* **413:** 310–320.

Baldwin AJ, Lioe H, Hilton GR, *et al.* (2011b). The polydispersity of alpha B-crystallin is rationalized by an interconverting polyhedral architecture. *Structure* **19:** 1855–1863.

Baldwin AJ, Walsh P, Hansen DF, *et al.* (2012). Probing dynamic conformations of the high-molecular-weight alpha B-crystallin heat shock protein ensemble by NMR spectroscopy. *J Am Chem Soc* **134:** 15343–15350.

Basha E, Jones C, Blackwell AE, *et al.* (2013). An unusual dimeric small heat shock protein provides insight into the mechanism of this class of chaperones. *J Mol Biol* **425:** 1683–1696.

Basha E, O'Neill H, Vierling E (2012). Small heat shock proteins and alpha-crystallins: dynamic proteins with flexible functions. *Trends Biochem Sci* **37:** 106–117.

Bateman A (1997). The structure of a domain common to archaebacteria and the homocystinuria disease protein. *Trends Biochem Sci* **22:** 12–13.

Beere HM (2005). Death versus survival: functional interaction between the apoptotic and stress-inducible heat shock protein pathways. *J Clin Investig* **115:** 2633–2639.

Bellaye PS, Burgy O, Colas J, *et al.* (2015). Antifibrotic role of αB-Crystallin inhibition in pleural and subpleural fibrosis. *Am J Resp Cell Mol Biol* **52:** 244–252.

Bepperling A, Alte F, Kriehuber T, *et al.* (2012). Alternative bacterial two-component small heat shock protein systems. *Proc Natl Acad Sci USA* **109:** 20407–20412.

Boncoraglio A, Minoia M, Carra S (2012). The family of mammalian small heat shock proteins (HSPBs): implications in protein deposit diseases and motor neuropathies. *Int J Biochem Cell Biol* **44:** 1657–1669.

Braun N, Zacharias M, Peschek J, *et al.* (2011). Multiple molecular architectures of the eye lens chaperone alpha B-crystallin elucidated by a triple hybrid approach. *Proc Natl Acad Sci USA* **108:** 20491–20496.

Carra S, Rusmini P, Crippa V, *et al.* (2013). Different anti-aggregation and pro-degradative functions of the members of the mammalian sHSP family in neurological disorders. *Philos Trans R Soc Lond B Biol Sci* **368:** article No. 20110409.

Carra S, Seguin SJ, Lambert H, Landry J (2008). HspB8 chaperone activity toward poly(Q)-containing proteins depends on its association with Bag3, a stimulator of macroautophagy. *J Biol Chem* **283:** 1437–1444.

Clark AR, Naylor CE, Bagnéris C, *et al.* (2011). Crystal structure of R120G disease mutant of human alphaB-crystallin domain dimer shows closure of a groove. *J Mol Biol* **408**:118–134.

Clark AR, Lubsen, NH, Slingsby, C (2012). sHSP in the eye lens: crystallin mutations, cataract and proteostasis. *Int J Biochem Cell Biol* **44:** 1687–1697.

Dammer EB, Lee AK, Duong DM, *et al.* (2015). Quantitative phosphoproteomics of Alzheimer's disease reveals cross-talk between kinases and small heat shock proteins. *Proteomics* **15:** 508–519.

Delbecq SP, Jehle S, Klevit R (2012). Binding determinants of the small heat shock protein, alpha B-crystallin: recognition of the "I × I" motif. *EMBO J* **31:** 4587–4594.

Delbecq SP, Klevit RE (2013). One size does not fit all: the oligomeric states of alpha B crystallin. *Febs Lett* **587:** 1073–1080.

De Jong WW, Caspers GJ, Leunissen JAM (1998). Genealogy of the alpha-crystallin–small heat-shock protein superfamily. *Int J Biol Macromol* **22:** 151–162.

de Thonel A, Le Mouel A, Mezger V (2012). Transcriptional regulation of small HSP-HSF1 and beyond. *Int J Biochem Cell Biol* **44:** 1593–1612.

Dubinska-Magiera M, Jablonska J, Saczko J, *et al.* (2014). Contribution of small heat shock proteins to muscle development and function. *Febs Lett* **588:** 517–530.

Esposito G, Garvey M, Alverdi V, *et al.* (2013). Monitoring the interaction between beta(2)-microglobulin and the molecular chaperone alpha B-crystallin by NMR and mass spectrometry: alpha B-crystallin dissociates Beta(2)-microglobulin oligomers. *J Biol Chem* **288:** 17844–17858.

Escusa-Toret S, Vonk WIM, Frydman J (2013). Spatial sequestration of misfolded proteins by a dynamic chaperone pathway enhances cellular fitness during stress. *Nat Cell Biol* **15:** 1231–1243.

Eyles SJ, Gierasch LM (2010). Nature's molecular sponges: small heat shock proteins grow into their chaperone roles. *Proc Natl Acad Sci USA* **107:** 2727–2728.

Finn RD, Bateman A, Clements J, *et al.* (2014). Pfam: the protein families database. *Nucl Acids Res* **42:** D222–D230.

Forrest KML, Al-Sarraj S, Sewry C, *et al.* (2011). Infantile onset myofibrillar myopathy due to recessive CRYAB mutations. *Neuromusc Dis* **21:** 37–40.

Franck E, Madsen O, van Rheede T, *et al.* (2004). Evolutionary diversity of vertebrate small heat shock proteins. *J Mol Evol* **59:** 792–805.

Fuchs M, Poirier DJ, Seguin SJ, *et al.* (2010). Identification of the key structural motifs involved in HspB8/HspB6–Bag3 interaction. *Biochem J* **425:** 245–255.

Gamerdinger M, Hajieva P, Kaya AM, *et al.* (2009). Protein quality control during aging involves recruitment of the macroautophagy pathway by BAG3. *EMBO J* **28:** 889–901.

Gamerdinger M, Kaya AM, Wolfrum U, *et al.* (2011). BAG3 mediates chaperone-based aggresome-targeting and selective autophagy of misfolded proteins. *EMBO Rep*, **12:** 149–156.

Gunst SJ, Zhang W.W (2008). Actin cytoskeletal dynamics in smooth muscle: a new paradigm for the regulation of smooth muscle contraction. *Am J Physiol Cell Physiol* **295:** C576–C587.

Haley DA, Bova MP, Huang QL, *et al.* (2000). Small heat-shock protein structures reveal a continuum from symmetric to variable assemblies. *J Mol Biol* **298:** 261–272.

Hanazono Y, Takeda K, Oka T, *et al.* (2013). Nonequivalence observed for the 16-meric structure of a small heat shock protein, SpHsp16.0, from *Schizosaccharomyces pombe*. *Structure* **21:** 220–228.

Haslbeck M, Braun N, Stromer T, *et al.* (2004). Hsp42 is the general small heat shock protein in the cytosol of *Saccharomyces cerevisiae*. *EMBO J* **23:** 638–649.

Haslbeck M, Franzmann T, Weinfurtner D, Buchner J (2005). Some like it hot: the structure and function of small heat-shock proteins. *Nat Struct Mol Biol* **12:** 842–846.

Haslbeck M, Vierling E (2015). A first line of stress defense: small heat shock proteins and their function in protein homeostasis. *J Mol Biol* **427:** 1537–1548.

Haslbeck M, Walke S, Stromer T, *et al.* (1999). Hsp26: a temperature-regulated chaperone. *EMBO J* **18:** 6744–6751.

Hayes D, Napoli V, Mazurkie A, *et al.* (2009). Phosphorylation dependence of Hsp27 multimeric size and molecular chaperone function. *J Biol Chem* **284:** 18801–18807.

Heirbaut M, Beelen S, Strelkov SV, Weeks SD (2014). Dissecting the functional role of the N-terminal domain of the human small heat shock protein HSPB6. *PLoS ONE* **9:** article No. e105892.

Hilario E, Martin FJM, Bertolini MC, Fan, L (2011). Crystal structures of *xanthomonas* small heat shock protein provide a structural basis for an active molecular chaperone oligomer. *J Mol Biol* **408:** 74–86.

Hilton GR, Hochberg GK, Laganowsky A, *et al.* (2013). C-terminal interactions mediate the quaternary dynamics of alphaB-crystallin. *Philos Trans R Soc Lond B Biol Sci* **368:** 20110405.

Hochberg GKA, Benesch JLP (2014). Dynamical structure of alpha B-crystallin. *Prog Biophys Mol Biol* **115:** 11–20.

Hochberg GKA, Ecroyd H, Liu C, *et al.* (2014). The structured core domain of alpha B-crystallin can prevent amyloid fibrillation and associated toxicity. *Proc Natl Acad Sci USA* **111:** E1562–E1570.

Horwitz J (1992). Alpha-crystallin can function as a molecular chaperone. *Proc Natl Acad Sci USA* **89:** 10449–10453.

Houlden H, Laura M, Wavrant-De Vrieze F, *et al.* (2008). Mutations in the HSP27 (HSPB1) gene cause dominant, recessive, and sporadic distal HMN/CMT type 2. *Neurology* **71:** 1660–1668.

Ingolia TD, Craig EA (1982). 4 Small *Drosophila* heat-shock proteins are related to each other and to mammalian alpha-crystallin. *Proc Natl Acad Sci USA* **79:** 2360–2364.

Jehle S, van Rossum B, Stout JR, *et al.* (2009). Alpha B-Crystallin: A hybrid solid-state/solution-state NMR investigation reveals structural aspects of the heterogeneous oligomer. *J Mol Biol* **385:** 1481–1497.

Jehle S, Rajagopal P, Bardiaux B, *et al.* (2010). Solid-state NMR and SAXS studies provide a structural basis for the activation of alpha B-crystallin oligomers. *Nat Struct Mol Biol* **17:** 1037–1042.

Jin XS, Townley R, Shapiro L (2007). Structural insight into AMPK regulation: ADP comes into play. *Structure* **15:** 1285–1295.

Jovcevski B, Kelly MA, Rote AP, *et al.* (2015). Phosphomimics destabilize hsp27 oligomeric assemblies and enhance chaperone activity. *Chem Biol* **22:** 186–195.

Kappé G, Franck E, Verschuure P, *et al.* (2003). The human genome encodes 10 alpha-crystallin-related small heat shock proteins: HspB1-10. *Cell Stress Chaperones* **8:** 53–61.

Kennaway CK, Benesch JLP, Gohlke U, *et al.* (2005). Dodecameric structure of the small heat shock protein Acr1 from *Mycobacterium* tuberculosis. *J Biol Chem* **280:** 33419–33425.

Kim YE, Hipp MS, Bracher A, *et al.* (2013). Molecular chaperone functions in protein folding and proteostasis. *Annu Rev Biochem* **82:** 323–355.

Kim KK, Kim R, Kim SH (1998). Crystal structure of a small heat-shock protein. *Nature* **394:** 595–599.

King N, Westbrook MJ, Young SL, *et al.* (2008). The genome of the choanoflagellate *Monosiga brevicollis* and the origin of metazoans. *Nature*, **451:** 783–788.

Koetter S, Unger A, Hamdani N, *et al.* (2014). Human myocytes are protected from titin aggregation-induced stiffening by small heat shock proteins. *J Cell Biol* **204:** 187–202.

Kourtis N, Nikoletopoulou V, Nikoletopoulou V, Tavernarakis N (2012). Small heat-shock proteins protect from heat-stroke-associated neurodegeneration. *Nature* **490:** 213–218.

Kriehuber T, Rattei T, Weinmaier T, *et al.* (2010). Independent evolution of the core domain and its flanking sequences in small heat shock proteins. *FASEB J* **24:** 3633–3642.

Laganowsky A, Eisenberg D (2010). Non-3D domain swapped crystal structure of truncated zebrafish alphaA crystallin. *Prot Sci* **19:** 1978–1984.

Laganowsky A, Benesch JLP, Landau M, *et al.* (2010). Crystal structures of truncated alphaA and alphaB crystallins reveal structural mechanisms of polydispersity important for eye lens function. *Prot Sci* **19:** 1031–1043.

Liu L, Chen JY, Yang B, Wang YH (2015). Crystal structure and function of an unusual dimeric Hsp20.1 provide insight into the thermal protection mechanism of small heat shock proteins. *Biochem Biophys Res Comm* **458:** 429–434.

Lutomiah J, Musila L, Makio A, *et al.* (2014). Ticks and tick-borne viruses from livestock hosts in arid and semiarid regions of the eastern and northeastern parts of Kenya. *J Med Entomol* **51:** 269–277.

Mainz A, Peschek J, Stavropoulou M *et al.* (2015). The chaperone αB-crystallin uses different interfaces to capture an amorphous and an amyloid client. *Nature Struct Mol Biol* **22:** 898–905.

Makley LN, McMenimen KA, DeVree BT *et al.* (2015). Pharmacological chaperone for α-crystallin partially restores transparency in cataract models. *Science* **350:** 674–677.

Mathieson W, Wilson R (2010). A comparative proteomic study of the undeveloped and developed *Schistosoma mansoni* egg and its contents: the miracidium, hatch fluid and secretions. *Int J Parasitol* **40:** 617–628.

McDonald ET, Bortolus M, Koteiche HA, Mchaourab HS (2012). Sequence, structure, and dynamic determinants of Hsp27 (HspB1) equilibrium dissociation are encoded by the N-terminal domain. *Biochemistry* **51:** 1257–1268.

Mchaourab HS, Godar JA, Stewart PL (2009). Structure and mechanism of protein stability sensors: chaperone activity of small heat shock proteins. *Biochemistry* **48:** 3828–3837.

Mchaourab HS, Lin YL, Spiller BW (2012). Crystal structure of an activated variant of small heat shock protein Hsp16.5. *Biochemistry* **51:** 5105–5112.

Mickelbart MV, Hasegawa PM, Bailey-Serres J, (2015). Genetic mechanisms of abiotic stress tolerance that translate to crop yield stability. *Nat Rev Genet* **16:** 237–251.

Miller SBM, Mogk A, Bukau B (2015). Spatially organized aggregation of misfolded proteins as cellular stress defense strategy. *J Mol Biol* **427:** 1564–1574.

Moyano JV, Evans JR, Chen F, *et al.* (2006). Alpha B-crystallin is a novel oncoprotein that predicts poor clinical outcome in breast cancer. *J Clinic Investig* **116:** 261–270.

Mymrikov EV, Seit-Nebi AS, Gusev NB (2011). Large potentials of small heat shock proteins. *Physiol Rev* **91:** 1123–1159.

Mymrikov EV, Seit-Nebi AS, Gusev NB (2012). Heterooligomeric complexes of human small heat shock proteins. *Cell Stress Chaperones* **17:** 157–169.

Orlova EV, Saibil HR (2011). Structural analysis of macromolecular assemblies by electron microscopy. *Chem Rev* **111:** 7710–7748.

Pereira MBM, Santos AM, Goncalves DC, (2014). αB-crystallin interacts with and prevents stress-activated proteolysis of focal adhesion kinase by calpain in cardiomyocytes. *Nat Commun* **5:** article No. 5159.

Poulain P, Gelly JC, Flatters D (2010). Detection and architecture of small heat shock protein monomers. *PLoS One* **5:** article No. e9990.

Quinlan RA, Zhang Y, Lansbury A, *et al.* (2013). Changes in the quaternary structure and function of MjHSP16.5 attributable to deletion of the IXI motif and introduction of the substitution, R107G, in the α-crystallin domain. *Philos Trans R Soc Lond B Biol Sci* **368(1617):** 20120327.

Rajagopal P, Tse E, Borst AJ, *et al.* (2015). A conserved histidine modulates HSPB5 structure to trigger chaperone activity in response to stress-related acidosis. *ELIFE* **4:** article No. UNSP e07304.

Reddy VS, Reddy GB (2015). Emerging role for alpha B-crystallin as a therapeutic agent: pros and cons. *Curr Mol Med* **15:** 47–61.

Richter K, Haslbeck H, Buchner J (2010). The heat shock response: life on the verge of death. *Mol Cell* **40:** 253–266.

Rosati A, Graziano V, De Laurenzi V, *et al.* (2011). Bag3: a multifaceted protein that regulates major cell pathways. *Cell Death Dis* **2:** e141.

Selcen D (2011). Myofibrillar myopathies. *Neuromusc Dis* **21:** 161–171.

Shemetov AA, Gusev NB (2011). Biochemical characterization of small heat shock protein HspB8 (Hsp22)–Bag3 interaction. *Arch Biochem Biophys* **513:** 1–9.

Shi J, Koteiche HA, McDonald ET, *et al.* (2013). Cryoelectron microscopy analysis of small heat shock protein 16.5 (Hsp16.5) complexes with T4 lysozyme reveals the structural basis of multimode binding. *J. Biol. Chem.* **288:** 4819–4830.

Slingsby C, Clark AR (2013). Flexible nanoassembly for sequestering non-native proteins. *Structure* **21:** 193–194.

Slingsby C, Wistow GJ (2014). Functions of crystallins in and out of lens: roles in elongated and post-mitotic cells. *Prog Biophys Mol Biol* **115:** 52–67.

Slingsby C, Wistow GJ, Clark AR (2013). Evolution of crystallins for a role in the vertebrate eye lens. *Prot Sci* **22:** 367–380.

Smith DA. Carland CR, Guo YM, Bernstein SI (2014). Getting folded: chaperone proteins in muscle development, maintenance and disease. *Anat Rec: Advances in Integrative Anatomy and Evolutionary Biology* **297:** 1637–1649.

Specht S, Miller SBM, Mogk A, Bukau B (2011). Hsp42 is required for sequestration of protein aggregates into deposition sites in *Saccharomyces cerevisiae*. *J Cell Biol* **195:** 617–629.

Stamler R, Kappe G, Boelens W, Slingsby C (2005). Wrapping the alpha-crystallin domain fold in a chaperone assembly. *J Mol Biol* **353:** 68–79.

Steinmetz PRH, Kraus JEM, Larroux C, *et al*. (2012). Independent evolution of striated muscles in cnidarians and bilaterians. *Nature*, **487:** 231–234.

Stengel F, Baldwin AJ, Painter AJ, *et al*. (2010). Quaternary dynamics and plasticity underlie small heat shock protein chaperone function. *Proc Natl Acad Sci USA* **107:** 2007–2012.

Stengel F, Baldwin AJ, Bush MF, *et al*. (2012). Dissecting heterogeneous molecular chaperone complexes using a mass spectrum deconvolution approach. *Chem Biol* **19:** 599–607.

Sun M, Lu MX, Tang XT, Du YZ (2014). Characterization and expression of genes encoding three small heat shock proteins in *Sesamia inferens* (*Lepidoptera*: *Noctuidae*). *Int J Mol Sci* **15:** 23196–23211.

van Montfort RLM, Basha E, Friedrich KL, *et al*. (2001). Crystal structure and assembly of a eukaryotic small heat shock protein. *Nat Struc Biol* **8:** 1025–1030.

van Noort JM, Bsibsi M, Gerritsen WH, *et al*. (2010). Alpha B-crystallin is a target for adaptive immune responses and a trigger of innate responses in preactive multiple sclerosis lesions. *J Neuropathol Exp Neurol* **69:** 694–703.

Voorhees RM, Fernández IS, Scheres SH, Hegde RS, (2014). Structure of the mammalian ribosome–Sec61 complex to 3.4 Å resolution. *Cell* **157:** 1632–1643.

Walther DM, Kasturi P, Zheng M, *et al.* (2015). Widespread proteome remodeling and aggregation in aging *C. elegans*. *Cell* **161:** 919–932.

Waters ER (2013). The evolution, function, structure, and expression of the plant sHSPs. *J Exp Botany* **64:** 391–403.

Weaver AJ, Sullivan WP, Felts SJ, *et al.* (2000). Crystal structure and activity of human p23, a heat shock protein 90 co-chaperone. *J Biol Chem* **275:** 23045–23052.

Weeks SD, Baranova EV, Heirbaut M, *et al.* (2014). Molecular structure and dynamics of the dimeric human small heat shock protein HSPB6. *J Struct Biol* **185:** 342–354.

White HE, Orlova EV, Chen SX, *et al.* (2006). Multiple distinct assemblies reveal conformational flexibility in the small heat shock protein Hsp26. *Structure* **14:** 1197–1204.

Wojtowicz I, Jablonska J, Zmojdzian M, *et al.* (2015). *Drosophila* small heat shock protein CryAB ensures structural integrity of developing muscles, and proper muscle and heart performance. *Development* **142:** 994–1005.

Zhang KM, Ezemaduka AN, Wang Z, *et al.* (2015). A novel mechanism for small heat shock proteins to function as molecular chaperones. *Sci Rep* **5:** article No. 8811.

Chapter 5

Structure and Mechanism of AAA+ Chaperones

Shannon M. Doyle, Andrea N. Kravats and Sue Wickner

Introduction

Clp/Hsp100 molecular chaperones belong to the AAA+ superfamily of ATPases (ATPases Associated with various cellular Activities) and are found in the cytoplasm of bacteria and fungi and in the organelles of higher eukaryotes. They are energy-dependent protein remodeling machines that participate in maintaining the integrity of the cellular proteome by facilitating protein remodeling, disaggregation, reactivation or degradation of misfolded and inactive proteins. Clp/Hsp100 chaperones function in protein remodeling and degradation in a similar manner; they recognize, bind and unfold proteins (Fig. 5.1A). In this chapter we will cover the recent developments in the field that advance our understanding of the structures and mechanism of Clp/Hsp100 chaperones, focusing on the best-studied members of the family. We will discuss how these chaperones function as protein disaggregating machines, in particular the activities of ClpB of bacteria and Hsp104 of yeast. We will also cover the role of Clp/Hsp100 proteins, including ClpA, ClpX and HslU, in protein degradation in collaboration with proteolytic components.

Laboratory of Molecular Biology, National Cancer Institute, National Institutes of Health, Bethesda 20892, USA

Correspondence should be addressed to: Sue Wickner, 37 Convent Drive, Room 5144, NIH, Bethesda, MD 20892, USA; e-mail: wickners@mail.nih.gov or Shannon M. Doyle, 37 Convent Drive, Room 5144, NIH, Bethesda, MD; 20892, USA; e-mail: doyles@mail.nih.gov.

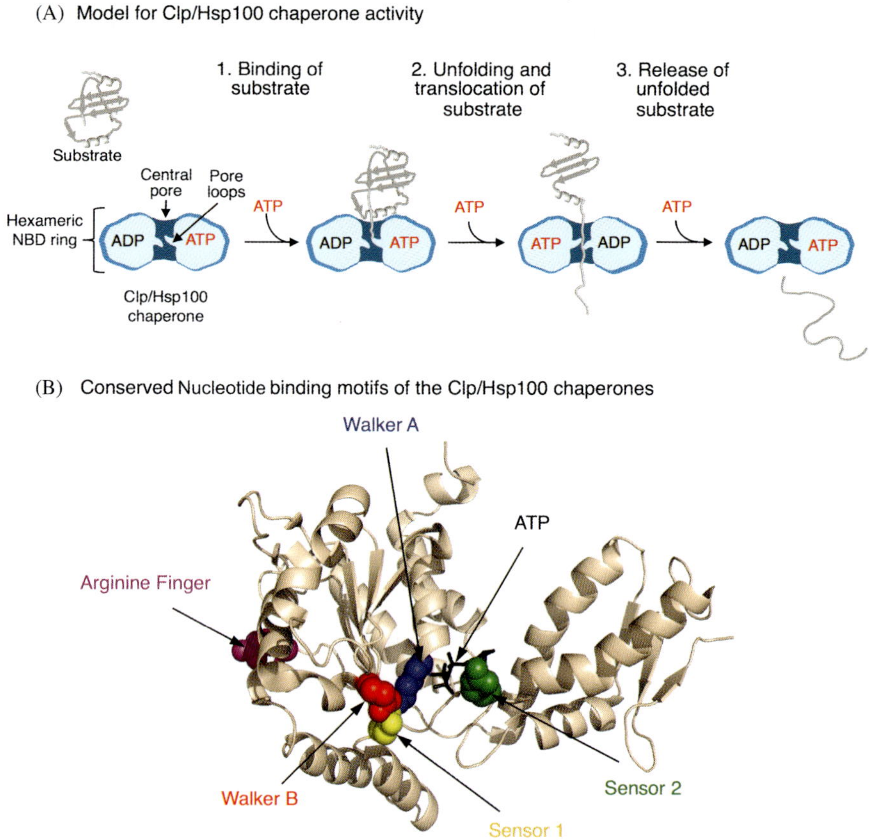

Fig. 5.1 Protein remodeling by Clp/Hsp100 molecular chaperones. (A) Model for protein remodeling by Clp/Hsp100 molecular chaperones. In the absence of cochaperones or proteases, substrate proteins with exposed unstructured regions interact with the chaperone via the pore loop tyrosines and in some cases the N-terminal domain (not shown) (step 1). ATP stabilizes the Clp/Hsp100 hexamer as well as its interactions with substrates. Once engaged, substrates are unfolded by forcible translocation through the central channel using the energy of ATP hydrolysis (step 2). The unfolded polypeptides are released (step 3) — a step induced by ATP hydrolysis — and refold to the native state, misfold or, when a protease component is present, are degraded. **(B)** Structure of NBD-2 from *T. thermophilus* ClpB (PDB 1qvr – chain C) (Lee *et al.*, 2003), showing amino acid residues from conserved motifs involved in nucleotide binding. The Walker A (blue), Walker B (red), sensor 1 (yellow), sensor 2 (green) and arginine finger (purple) are shown as CPK models; AMP-PNP is shown as a stick model and is in black. The image in **B** was prepared using PYMOL (www.pymol.org).

Clp/Hsp100 proteins have diverse functions *in vivo*. For example, yeast Hsp104 and ClpB from bacteria are homologous Clp/Hsp100 chaperones that are essential for growth following extreme stress conditions, such as heat (Doyle *et al.*, 2013; Hodson *et al.*, 2012; Mogk *et al.*, 2015; Mogk *et al.*, 1999; Parsell *et al.*, 1994; Sanchez and Lindquist, 1990; Sanchez *et al.*, 1992; Squires *et al.*, 1991; Winkler *et al.*, 2012a). They promote survival by participating in the disaggregation and reactivation of large insoluble protein aggregates that form during exposure to cellular stresses. However, for growth under nonstress conditions ClpB and Hsp104 are not essential. Hsp104 plays an additional role; it is required for the transmission of specific amyloid protein aggregates, namely prions, in yeast (Chernoff, 2007; Romanova and Chernoff, 2009; Tuite *et al.*, 2011; Tuite and Serio, 2010; Wickner *et al.*, 2011; Winkler *et al.*, 2012a). When highly expressed, Hsp104 also promotes curing (elimination) of the [PSI^+] prion (Hung and Masison, 2006; Park *et al.*, 2014). Several other Clp/Hsp100 proteins that are the chaperone components of proteases, such as ClpX of *Caulobacter* and *Bacillus subtilis*, are essential in these prokaryotes where they regulate replication and cell division by modulation of essential proteins through direct degradation or via the degradation of key transcription factors (Jenal and Fuchs, 1998; Vass and Chien, 2013; Weart *et al.*, 2005). However, in *E. coli*, protease-associated Clp/Hsp100 chaperones, ClpA, ClpX and HslU/ClpY, are not essential.

Mutation of a number of human Clp/Hsp100 chaperones has been linked with diseases. Very recently, multiple studies have shown that mutation or truncation of human mitochondrial ClpB leads to the development of 3-methylglutaconic aciduria, cataracts, neutropenia and other disorders (Capo-Chichi *et al.*, 2015; Kanabus *et al.*, 2015; Saunders *et al.*, 2015; Wortmann *et al.*, 2015). Mutation of another Clp/Hsp100 protein, Torsin A, causes early onset torsin dystonia, a disease associated with dystonic movements (Granata and Warner, 2010; Liang *et al.*, 2014). Single amino acid substitutions in the human p97/VCP chaperone, a eukaryotic Clp/Hsp100 homolog involved in many cellular processes (Meyer *et al.*, 2012; Meyer and Weihl, 2014), appear to cause defects in multiple protein degradation pathways (Tang and Xia, 2012; Zhang and Ye, 2014), which lead to inclusion body myopathy associated with Paget's disease of the bone and frontotemporal dementia (IBMPFD), and ultimately death (Meyer and Weihl, 2014; Nalbandian *et al.*, 2011; Tang and Xia, 2012; Zhang and Ye, 2014). Thus, understanding the structure and mechanisms of action of these molecular chaperones and how they contribute to regulating protein quality control is of significant therapeutic interest.

Architecture of Clp/Hsp100 Chaperones

Clp/Hsp100 chaperones belong to the AAA+ protein superfamily (Neuwald *et al.*, 1999). The signature feature of Clp/Hsp100 proteins is the ATP-binding module of ~230 amino acids that is common to all AAA+ family members. This module contains conserved Walker A and Walker B motifs, sensor 1 and sensor 2 motifs and an arginine finger motif (Fig. 1B). The AAA+ module has two subdomains, with residues from both subdomains contributing to the nucleotide-binding site (Erzberger and Berger, 2006; Hanson and Whiteheart, 2005; Ogura *et al.*, 2004; Wendler *et al.*, 2012). Some Clp/Hsp100 chaperones, such as HslU and ClpX, have a single AAA+ nucleotide-binding domain (NBD) per monomer, while others, including the ClpB and Hsp104 disaggregases and the ClpA component of the ClpAP protease, have two AAA+ modules, NBD-1 and NBD-2. ATP binding and hydrolysis by these domains provide the energy essential for protein remodeling (Fig. 5.2A).

In addition to differing in the number of AAA+ modules present, Clp/Hsp100 proteins differ in the number of extra domains they possess (Figs. 2B–2F). Many Clp/Hsp100 chaperones contain a poorly conserved N-terminal domain (N-domain) that is involved in interaction with specific substrates, but is not essential under many conditions (Chow *et al.*, 2005; Chowdhury *et al.*, 2010; Gur *et al.*, 2013; Hinnerwisch

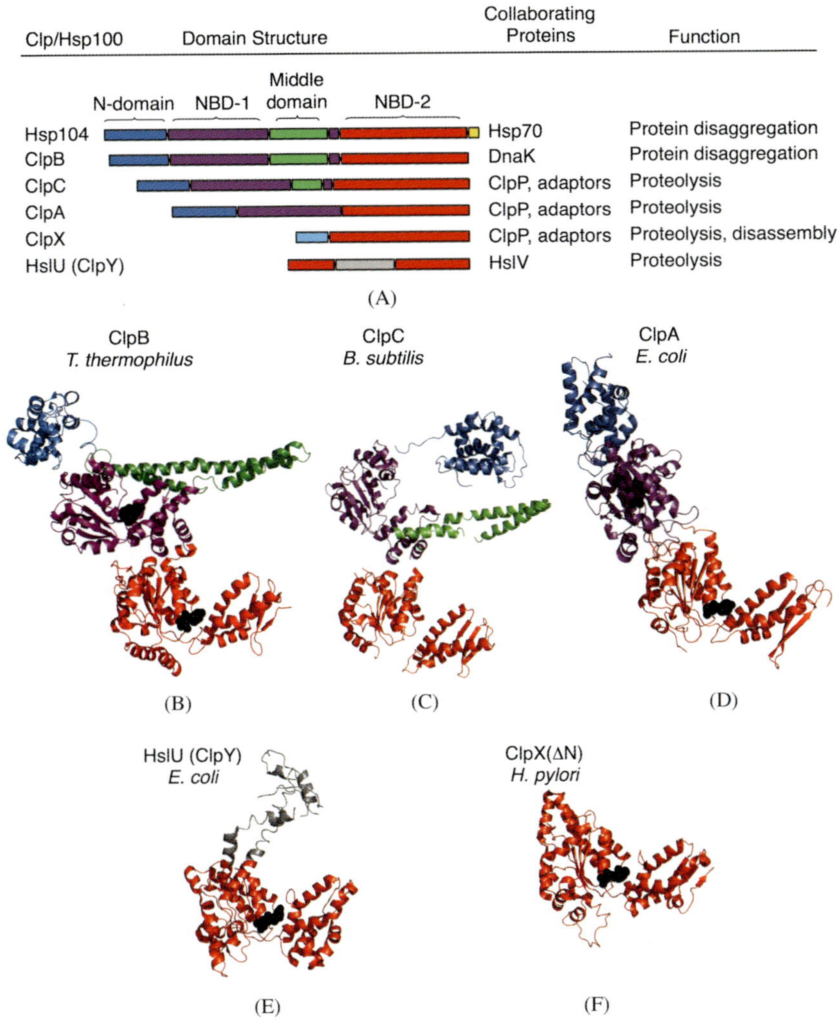

Fig. 5.2 Domain organization of Clp/Hsp100 protomers. (A) Domain arrangement of members of the Clp/Hsp100 family. The Clp/Hsp100 family has homologs in fungi, bacteria, plants and eukaryotes. Some members have two nucleotide-binding domains (NBDs), whereas others have a single NBD. NBDs are highly conserved and are shown in purple (NBD-1) and red (NBD-2), while the N-terminal domain [N-domain (blue)] is poorly conserved. Some Clp/Hsp100 proteins have a coiled-coil middle domain [M-domain (green)] or an I-domain (gray) inserted within an NBD. Hsp104 has a unique C-terminal domain (yellow). Hsp104 and ClpB function in protein disaggregation and collaborate with the Hsp70/DnaK chaperone system. ClpC, ClpA, ClpX and HslU interact with the compartmentalized proteases, ClpP or HslV, and form ClpCP, ClpAP,

Fig. 5.2 (*Continued*) ClpXP and HslUV protease complexes. **(B)** Protomer structure of the AMP-PNP bound *T. thermophilus* ClpB (PDB 1qvr – chain C) (Lee *et al.*, 2003). **(C)** Protomer structure of *B. subtilis* ClpC (Glu280Ala/Glu618Ala, with loop deletions 247–251, 281–292, 587–596 and 665–685) in complex with an adaptor protein, MecA (not shown) (PDB 3pxi – chain A) (Wang *et al.*, 2011). **(D)** Protomer structure of *E. coli* ClpA with ADP bound (PDB 1r6b) (Xia *et al.*, 2004). **(E)** Protomer structure of *E. coli* HslU with AMP-PNP bound (PDB 1e94 – chain F) (Song *et al.*, 2000). **(F)** Protomer structure of *H. pylori* ClpX with an N-terminal deletion (ΔN) with ADP bound (PDB 1um8) (Kim and Kim, 2003). In **B** and **D–F**, nucleotide is shown in black as a CPK model. The images in **B–F** were prepared using PYMOL (www.pymol.org).

et al., 2005b; Nagy *et al.*, 2010) (Figs. 5.2B–5.2D). In yeast the Hsp104 N-domain is not essential for thermotolerance or prion propagation, but is required for prion curing (Hung and Masison, 2006; Sweeny *et al.*, 2015). In bacteria the ClpB N-domain is not essential for thermotolerance (Chow *et al.*, 2005; Mogk *et al.*, 2003). Some Clp/Hsp100 chaperones, including Hsp104 and ClpB, also possess a coiled-coil middle domain (M-domain) composed of four α-helices (Lee *et al.*, 2007; Lee *et al.*, 2003; Wang *et al.*, 2011). The M-domain is inserted within NBD-1 and is required for protein disaggregation (Lee *et al.*, 2007; Lee *et al.*, 2010; Lee *et al.*, 2003; Wang *et al.*, 2011) (Figs. 5.2B and 5.2C). The M-domains are of variable size among Clp/Hsp100 family members and poorly conserved (Lee *et al.*, 2010; Lee *et al.*, 2003; Wang *et al.*, 2011). HslU contains an intermediate domain (I-domain) embedded in its single NBD that is involved in substrate binding (Sousa *et al.*, 2000; Sundar *et al.*, 2012) (Fig. 5.2E). In addition, a few Clp/Hsp100 proteins contain a short N-terminal domain. In Hsp104, the C-domain is involved in hexamerization and is important in thermotolerance (Mackay *et al.*, 2008).

Asymmetric Quaternary Structure of Clp/Hsp100 Chaperones

The biologically active structural unit of Clp/Hsp100 molecular chaperones is an oligomeric ring comprising six protomers (Doyle *et al.*, 2013; Erzberger and Berger, 2006; Mogk *et al.*, 2015) (Figs. 5.3A–5.3C). Several Clp/Hsp100 proteins have been crystallized as symmetric hexamers,

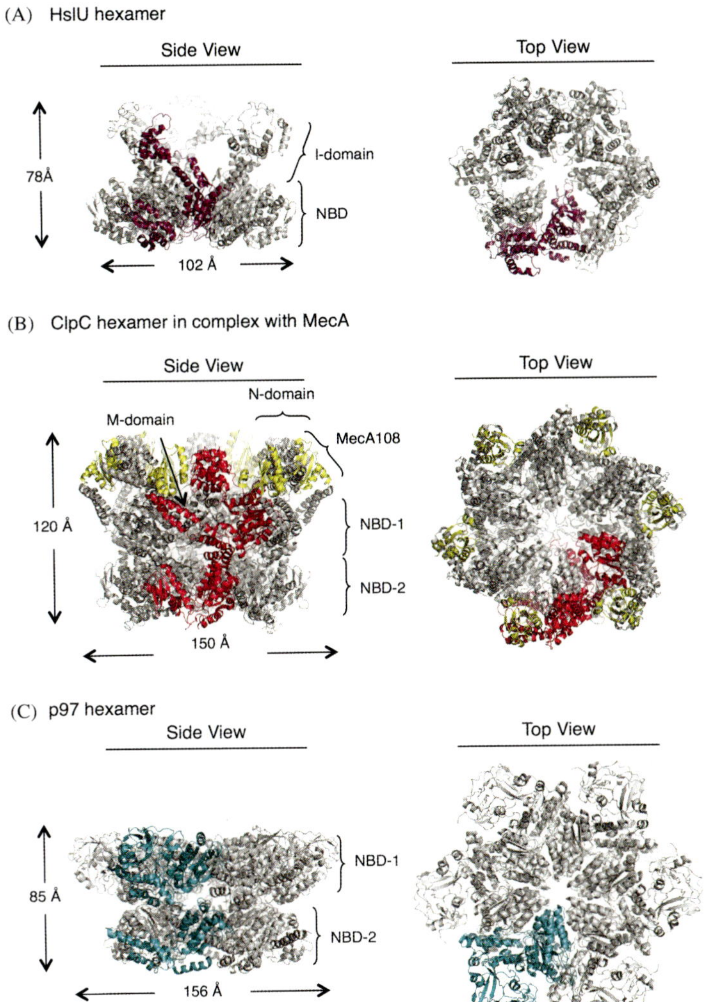

Fig. 5.3 Hexamer structures of Clp/Hsp100 proteins. **(A)** The HslU hexamer structure from *E. coli* (ATP) (PDB 1do2) (Bochtler *et al.*, 2000). Each HslU protomer comprises a nucleotide-binding domain (NBD) and an I-domain with one protomer shown in purple. Side (left) and top (right) views are shown. **(B)** The *B. subtilis* ClpC hexamer structure (Glu280Ala/Glu618Ala) in complex with MecA108 (PDB 3j3u) (Liu *et al.*, 2013). Each ClpC protomer comprises an N-domain, two nucleotide-binding domains (NBD-1 and NBD-2) and an M-domain with a single protomer in pink and the six MecA molecules in yellow. Side (left) and top (right) views are shown. **(C)** Hexameric structure of *M. musculus* p97 (ADP-AlF$_x$) (PDB 3cf1) (Davies *et al.*, 2008). Each p97 protomer comprises two nucleotide-binding domains (NBD-1 and NBD-2) with a single protomer shown in blue. Side (left) and top (right) views are shown. The images in **A–C** were prepared using PYMOL (www.pymol.org).

including HslU, ClpC and p97 (DeLaBarre and Brunger, 2003; Sousa *et al.*, 2000; Wang *et al.*, 2011); however, recent results strongly suggest that active hexamers form asymmetric rings (Baker and Sauer, 2012; Carroni *et al.*, 2014; DeSantis *et al.*, 2012; Doyle *et al.*, 2007b; Glynn *et al.*, 2009; Glynn *et al.*, 2012; Hoskins *et al.*, 2009). While one crystal structure of HslU showed a symmetric hexamer, additional HslU crystal structures indicated that only three or four nucleotide-binding sites were occupied with nucleotide in the hexamer, indicating a potential functional asymmetry (Wang *et al.*, 2001b; Yakamavich *et al.*, 2008). The crystal structures of other Clp/Hsp100 chaperones directly show asymmetry (Glynn *et al.*, 2009; Guo *et al.*, 2002b; Lee *et al.*, 2003). For example, *E. coli* ClpA and *T. thermophilus* ClpB ($ClpB_{T. th}$) formed spirals with six and three subunits, respectively (Guo *et al.*, 2002b; Lee *et al.*, 2003). For ClpB, each of the three subunits in the spiral was shown to have a distinct conformation (Lee *et al.*, 2003). Additionally, using covalently linked *E. coli* ClpX trimers, Sauer, Baker and colleagues showed that ClpX crystallized as an asymmetric hexamer (Glynn *et al.*, 2009; Glynn *et al.*, 2012). These observations, as well as results from the many biochemical experiments mentioned below, suggest that structural asymmetry might be relevant and important (Baker and Sauer, 2012; Carroni *et al.*, 2014; DeSantis *et al.*, 2012; Doyle *et al.*, 2007b; Hoskins *et al.*, 2009). Nevertheless, symmetric hexameric models have been generated for ClpA, ClpB and Hsp104 using the symmetric crystal structures of HslU and p97 as templates and have proven useful in visualizing the assembly that closely resembles the biologically active hexamer (DeLaBarre and Brunger, 2003; Diemand and Lupas, 2006; Sousa *et al.*, 2000). More recently asymmetric hexamer models of ClpB and Hsp104 have been built to more closely reflect the functional structures of the molecules (Carroni *et al.*, 2014; Doyle *et al.*, 2015).

The crystal structures of the Clp/Hsp100 chaperones showed that the nucleotide-binding sites are located at the interfaces of two neighboring protomers, with the Walker A and B motifs and sensor 1 and 2 motifs supplied by one protomer and the arginine finger motif supplied by the adjacent protomer (DeLaBarre and Brunger, 2003; Kim and Kim, 2003; Lee *et al.*, 2003; Ogura *et al.*, 2004; Sousa *et al.*, 2000; Wendler *et al.*, 2012) (Fig. 5.4A). This arrangement allows protomers within the hexamer to communicate with each other throughout the reaction cycle.

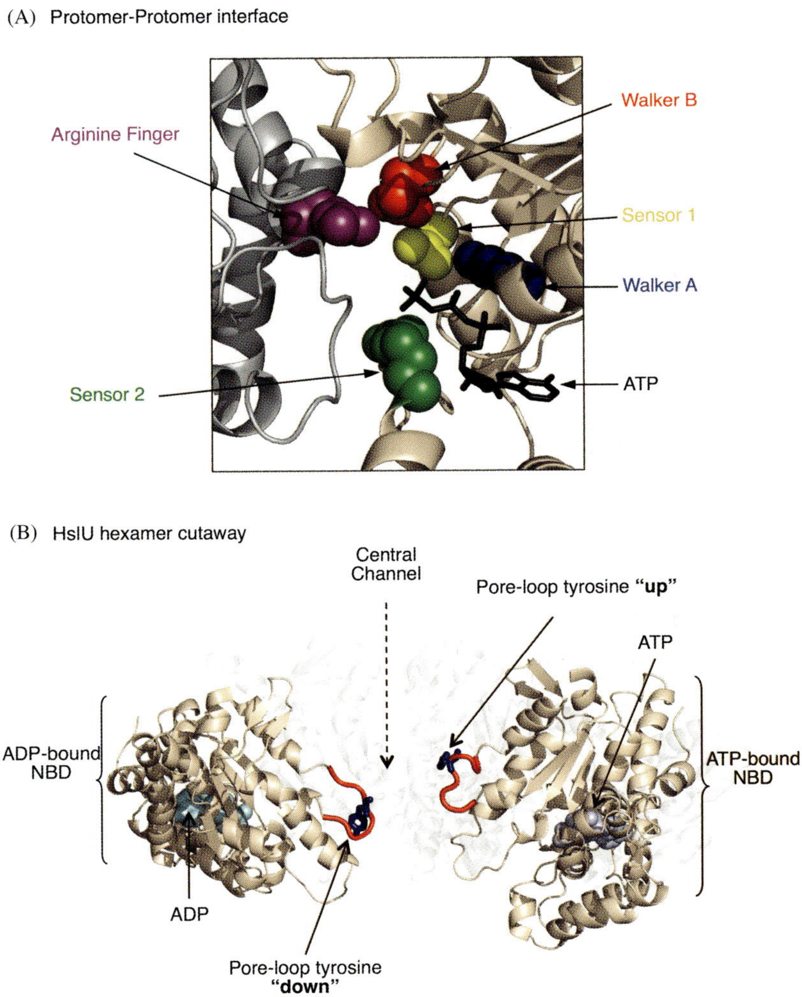

Fig. 5.4 Conserved motifs of Clp/Hsp100 nucleotide-binding domains. (A) ClpB nucleotide-binding domain 2 (NBD-2) protomer–protomer interface. A close-up view of the subunit interface of NBD-2 from a *T. thermophilus* ClpB hexamer model (Diemand and Lupas, 2006), showing a conserved arginine finger residue (purple) from one

Fig. 5.4 (*Continued*) protomer (gray) participating in ATP (black sticks) binding in the neighboring protomer (wheat). Additional conserved residues in NBD-2, Walker A (blue), Walker B (red), sensor 1 (yellow) and sensor 2 (green) are shown as CPK models. **(B)** Model of the Clp/Hsp100 central channel. A model of four HslU subunits without I-domains (residues 108–243 were excluded) is shown. The HslU structures (PDB 1do2 and PDB 1do0) (Bochtler *et al.*, 2000) were oriented to minimize the RMSD of the NBD of HslU protomers in the ATP-bound (open) and the ADP bound (closed) state, respectively. The left subunit has ADP bound (PDB 1do0) and the right subunit has ATP bound (PDB 1do2). To illustrate the central pore and ring structure, two subunits in the back are shown in light gray while the front two subunits have been removed for clarity. Pore loops in the "down" (ADP-bound) and "up" (ATP-bound) positions are shown in red and the pore loop tyrosines are shown in blue as stick models. In **A** and **B**, the images were prepared using PYMOL (www.pymol.org).

Nucleotide binding and hydrolysis cause local conformational changes in individual subunits. These changes are transduced to adjoining domains within the protomer and to adjacent protomers in the hexamer, leading to global changes in the hexameric ring that are required for protein unfolding and translocation (Biter *et al.*, 2012a; Biter *et al.*, 2012b; Franzmann *et al.*, 2011; Glynn *et al.*, 2009; Glynn *et al.*, 2012).

The structures of hexameric assemblies of Clp/Hsp100 proteins, including ClpB and Hsp104, have also been explored using cryoelectron microscopy (Figs. 5.5A and 5.5B). These images have indicated the positions of protomers within the hexamers and in some cases the positions of individual domains within protomers (Carroni *et al.*, 2014; Lee *et al.*, 2007; Lee *et al.*, 2010; Lee *et al.*, 2003). The $ClpB_{T.\ th}$ hexamer model was reconstructed in work by Tsai and colleagues by fitting X-ray crystallographic structures into the cryo-EM map and imposing sixfold symmetry (Lee *et al.*, 2003). The resulting model showed two stacked rings, composed of six NBD-1 domains and six NBD-2 domains, with a channel at the center of the hexamer (Lee *et al.*, 2007; Lee *et al.*, 2003). A ring composed of N-domains was not visible in the EM images of ClpB, likely because the N-domains are connected to NBD-1 by flexible linkers (Lee *et al.*, 2007; Lee *et al.*, 2003). The M-domains of ClpB were located on the hexamer surface, projecting outward from the ring

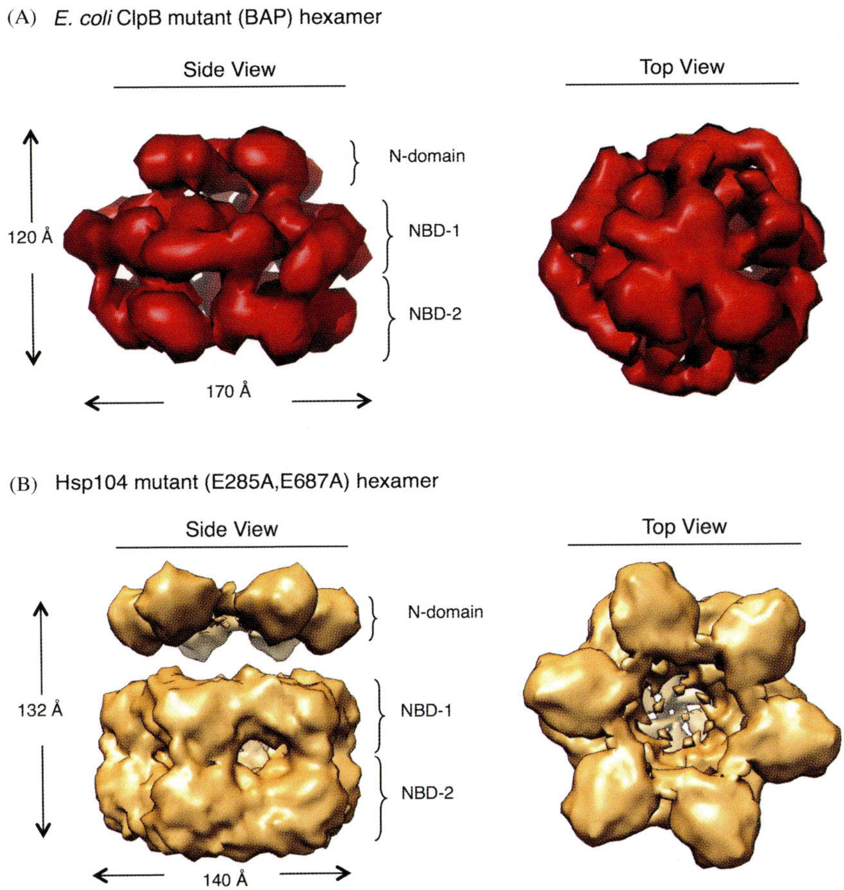

Fig. 5.5 Oligomeric structure and models of ClpB and Hsp104. (A) Three-dimensional asymmetric cryo-EM reconstruction (21 Å) of an *E. coli* ClpB mutant, BAP [ClpB with the ClpP interaction motif of ClpA transplanted to the homologous region of ClpB [(Weibezahn *et al.*, 2004)] in complex with ClpP (but not included in the figure for clarity) and ATPγS (EMD 2558) (Carroni *et al.*, 2014). Each protomer in the hexamer comprises an N-domain, two nucleotide-binding domains (NBD-1 and NBD-2) and an M-domain. The density of only five N-domains is visible due to the inherent flexibility of the domain and the asymmetry of the model. Side (left) and top (right) views are shown and were prepared using Chimera (www.cgl.ucsf.edu/chimera). **(B)** Three-dimensional symmetric cryo-EM reconstruction (11 Å) of *S. cerevisiae* Hsp104 (Glu285Ala, Glu687Ala) with ATP bound (EMD 1631) (Lee *et al.*, 2010). Each protomer in the hexamer comprises an N-domain, two nucleotide-binding domains (NBD-1 and NBD-2) and an M-domain although the M-domains are not visible in this model due to the inherent flexibility of the domain. Side (left) and top (right) views are shown and were prepared using Chimera (www.cgl.ucsf.edu/chimera).

of NBD-1 domains (Lee *et al.*, 2007; Lee *et al.*, 2003). Only portions of each M-domain were visible in the reconstructions, due to the intrinsic flexibility of the domain. Conformational differences were observed under various nucleotide conditions and were reflected in the diameter of the central pore and position adopted by the M-domains based on the nucleotide bound (Carroni *et al.*, 2014; Lee *et al.*, 2007; Lee *et al.*, 2010; Lee *et al.*, 2003). Recent asymmetric reconstructions of ClpB from Saibil and colleagues show that the M-domains adopt different positions in a single ClpB hexamer and may more closely represent the true structure of these complex machines (Carroni *et al.*, 2014) (Fig. 5.5A). These new findings suggest that individual subunit conformations may be important for chaperone activity.

In contrast to the results of the cryo-EM work on $ClpB_{T. th}$, the initial Hsp104 cryo-EM studies showed a much larger hexamer, indicating an unusual arrangement of AAA+ modules (Wendler *et al.*, 2007; Wendler *et al.*, 2009). The prominent coiled-coil domain was embedded between the AAA+ domains, and the central channel was a very large cavity (Wendler *et al.*, 2007; Wendler *et al.*, 2009). More recently, the discrepancies in the results have been resolved: it is now clear that the overall structure of Hsp104 is similar to ClpB in size and it is agreed upon that the M-domains are located on the exterior of the hexameric molecule (Carroni *et al.*, 2014; Lee *et al.*, 2010) (Fig. 5.5B).

Arrangement of Clp/Hsp100 Chaperones in Proteolytic Complexes

Clp/Hsp100 proteins associated with proteases are arranged with the Clp/Hsp100 chaperone ring located at one or both ends of a proteolytic core (Baker and Sauer, 2012; Gur *et al.*, 2013; Kress *et al.*, 2009; Sauer and Baker, 2011; Striebel *et al.*, 2009). The proteolytic core is made up of two stacked rings comprising six or seven subunits, which together form a chamber with the active sites of the protease on the interior surface of the cavity (Alexopoulos *et al.*, 2012; Liu *et al.*, 2014) (Figs. 5.6B and 5.6C).

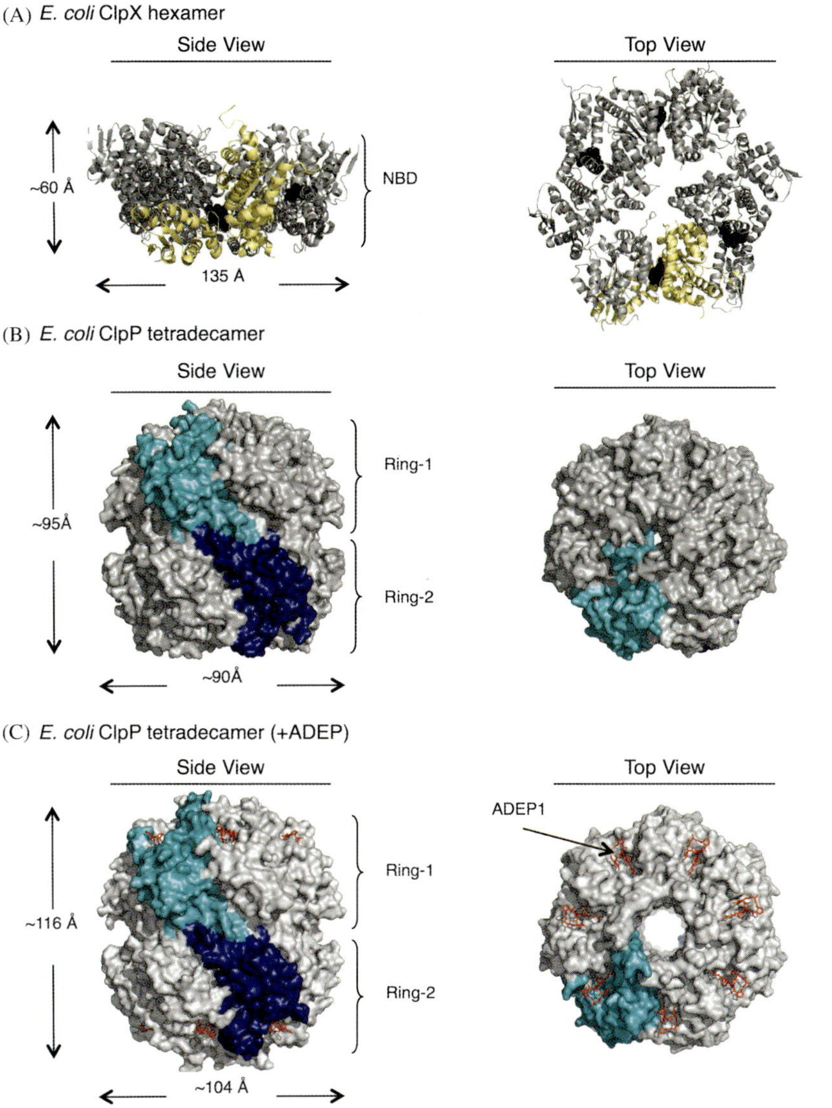

Fig. 5.6 The ClpX–ClpP degradation machine. (A) Atomic model of the asymmetric crystal structure of the *E. coli* ClpX (ΔN) hexamer comprising a dimer of covalently linked ClpX trimers (PDB 3hws) (Glynn *et al.*, 2009). Each protomer in the hexamer is a ClpX nucleotide-binding domain (NBD). One protomer is shown in yellow. Nucleotide is bound by four of the six protomers and is shown in black as CPK models. Side (left) and top (right) views are shown and were prepared using PYMOL (www.pymol.org). **(B)** Surface rendering of the crystal structure of the *E. coli* ClpP tetradecamer comprising two stacked

Fig. 5.6 (*Continued*) heptameric rings (Ring-1 and Ring-2) (PDB 1yg6) (Bewley *et al.*, 2006). One protomer in Ring-1 (cyan) and Ring-2 (blue) is in color. Side (left) and top (right) views are shown and were prepared using PYMOL (www.pymol.org). **(C)** Surface rendering of the crystal structure of the *E. coli* ClpP tetradecamer comprising two stacked, heptameric rings (Ring-1 and Ring-2) with ADEP1, an acyldepsipeptide antibiotic, bound to the N-terminal loops of each ClpP protomer (PDB 3mt6) (Li *et al.*, 2010). One protomer in Ring-1 (cyan) and Ring-2 (blue) is in color. The 14 ADEP1 molecules are shown in red as stick models. Side (left) and top (right) views are shown and were prepared using PYMOL (www.pymol.org).

Access to the proteolytic chamber is through narrow pores at the ends of the core's stacked rings. These pores are not large enough to allow the entry of folded proteins. As a result, the core component by itself is unable to degrade proteins, although in the case of ClpP it can degrade peptides (Alexopoulos *et al.*, 2012; Liu *et al.*, 2014; Nagpal *et al.*, 2013). The interaction between the Clp/Hsp100 chaperone and the proteolytic core causes structural rearrangements that open the narrow pores and allow polypeptides that have been unfolded by the chaperone component to enter the protease cavity and be degraded (Alexopoulos *et al.*, 2013; Alexopoulos *et al.*, 2012; Liu *et al.*, 2014). The rearrangements caused by the chaperone component can be mimicked by an antibiotic, acyldepsipeptide (ADEP), which binds to ClpP and promotes degradation by ClpP independent of ClpA or ClpX (Alexopoulos *et al.*, 2013; Alexopoulos *et al.*, 2012; Brotz-Oesterhelt *et al.*, 2005; Gersch *et al.*, 2015; Kirstein *et al.*, 2009a; Li *et al.*, 2010). Crystallographic, electron-microscopic and small angle X-ray scattering studies show that ADEP binds to the apical surface of ClpP and causes enlargement of the apical pore (Alexopoulos *et al.*, 2013; Alexopoulos *et al.*, 2012; Gersch *et al.*, 2015; Li *et al.*, 2010) (Fig. 5.6C).

While the Clp/Hsp100 chaperone component of the protease is hexameric, the proteolytic core component comprises either two hexameric (HslV) or two heptameric (ClpP) rings of protomers (Alexopoulos *et al.*, 2012; Baker and Sauer, 2012; Gur *et al.*, 2013; Kress *et al.*, 2009; Liu *et al.*, 2014; Sauer and Baker, 2011; Striebel *et al.*, 2009). The HslUV protease is a symmetric complex composed of a hexameric ring of HslU protomers and an HslV dodecamer. The crystal structure reveals that two HslU hexamers interact with one HslV particle; one bound to each proximal face of the stacked hexameric HslV rings via conserved loops in the C-domain of HslU (Kress *et al.*, 2009; Sousa *et al.*, 2000; Striebel *et al.*, 2009; Wang *et al.*, 2001a). Cryoelectron-microscopic studies have shown that ClpAP and ClpXP are arranged similarly, although there is a symmetry mismatch between the hexameric ClpA or ClpX and the ClpP tetradecamer. ClpX and ClpA contact ClpP through conserved tripeptide motifs, referred to as IGF loops, which are displayed on the distal surface of the Clp/Hsp100 chaperone (Alexopoulos *et al.*, 2013; Alexopoulos *et al.*, 2012; Baker and Sauer, 2012; Sauer and Baker, 2011; Striebel *et al.*, 2009). These loops interact with a hydrophobic pocket on the apical surface of ClpP (Alexopoulos *et al.*, 2013; Alexopoulos *et al.*, 2012; Baker and Sauer, 2012; Kress *et al.*, 2009; Sauer and Baker, 2011; Striebel *et al.*, 2009). In addition, there is a weaker and more dynamic interaction between pore loops on the distal surface of the Clp/Hsp100 protein and loops in the N-terminal region of the ClpP protomers (Martin *et al.*, 2007). The consequences of the asymmetry between the rings of the Clp/Hsp100 component and the proteolytic component are not yet understood.

Mechanism of Clp/Hsp100 Chaperone Action

Regardless of whether the Clp/Hsp100 chaperone is disaggregating an aggregate, reactivating an inactive protein or delivering a substrate to a protease for degradation, one common theme among these chaperones is that the substrate must be recognized by the chaperone, unfolded and then threaded through the central pore in an energy-dependent manner (Fig. 5.1A). The mechanism of several Clp/Hsp100 proteins has been explored in great detail *in vitro* under conditions where the chaperones are able to act alone, without cochaperones or protease components (Baker and Sauer, 2012; Doyle *et al.*, 2013; Doyle *et al.*, 2007b; Mogk *et al.*, 2015; Tessarz *et al.*, 2008; Weber-Ban *et al.*, 1999; Weibezahn *et al.*, 2004). While the specifics of the mechanisms vary among members of the Clp/Hsp100 family, the mechanisms are remarkably similar.

Substrate Binding

Substrate specificity varies for Clp/Hsp100 chaperones and substrates can be in the form of aggregated, unfolded, misfolded, abnormal or native proteins. In most cases the region recognized by the chaperone is unstructured, with exposed hydrophobic residues. Unfolded and aggregated proteins are recognized nonspecifically through unstructured regions that become accessible following heat or other stress (Chowdhury *et al.*, 2010; Hoskins *et al.*, 2002; Li *et al.*, 2015; Li and Lucius, 2013; Mogk *et al.*, 1999; Neher *et al.*, 2006; Schlieker *et al.*, 2004; Sundar *et al.*, 2010). Abnormal polypeptides resulting from ribosome stalling become substrates for degradation by ClpAP and ClpXP through ClpX and ClpA recognition of an 11-amino-acid tag, coded by *ssrA*. The tag is cotranslationally added to the C-terminal of the polypeptide (Himeno *et al.*, 2014). Other Clp/Hsp100 substrates can be native proteins, whose remodeling or degradation is regulated by Clp/Hsp100 chaperones and proteases (Baker and Sauer, 2012; Doyle *et al.*, 2013; Kress *et al.*, 2009; Mogk *et al.*, 2015; Sauer and Baker, 2011; Striebel *et al.*, 2009). Most native protein substrates are recognized through short peptide tags at either the N- or C-terminal of the protein. Proteomic studies have identified more than 50 natural substrates of *E. coli* ClpXP and approximately 100 potential substrates of both *Caulobacter* ClpXP and *S. aureus* ClpC (Chien *et al.*, 2007; Flynn *et al.*, 2003; Graham *et al.*, 2013; Williams *et al.*, 2014). In addition, several plasmid rep proteins, phage Mu transposase and phage lambda Oh protein are substrates of ClpX and ClpXP (Burton and Baker, 2005; Hoskins *et al.*, 2002; Zabrocka *et al.*, 2014). For some native substrates, internal sequence tags are recognized and act as secondary binding sites for Clp/Hsp100. Utilization of duel recognition sites allows for increased stability of the substrate–Clp/Hsp100 complex and presentation of the N- or C-terminal of the substrate to the chaperone (Burton *et al.*, 2001; Haslberger *et al.*, 2008; Hoskins *et al.*, 2002; Lee *et al.*, 2002; Ling *et al.*, 2015).

Substrate polypeptides interact with Clp/Hsp100 chaperones via highly conserved tyrosine residues located in a motif most often consisting of GYVG (Hinnerwisch *et al.*, 2005a; Martin *et al.*, 2008b; Schlieker *et al.*, 2004; Siddiqui *et al.*, 2004; Weibezahn *et al.*, 2004) (Fig. 4B). In the Clp/Hsp100 hexamer, the critical tyrosines, one per NBD, reside on short unstructured loops that protrude into the central channel (Lee *et al.*, 2007; Wang *et al.*, 2001a). For the substrate to stably interact with the pore loops, ATP binding is required (Glynn *et al.*, 2009; Hinnerwisch *et al.*, 2005a; Lee *et al.*, 2007; Martin *et al.*, 2008b; Wang *et al.*, 2001b; Weibezahn *et al.*, 2004). Moreover, only a single substrate protein is bound per Clp/Hsp100 hexamer (Martin *et al.*, 2008b; Pak and Wickner, 1997; Piszczek *et al.*, 2005). Not only do the pore loops participate in the initial engagement of the substrate (Martin *et al.*, 2008a,b,c; Park *et al.*, 2005; Weibezahn *et al.*, 2004), but they are involved in translocation.

Some but not all substrates require additional interactions with the N-domain or I-domain of the Clp/Hsp100 protein. For example, substrate binding by ClpB is enhanced and stabilized by the N-domain (Doyle *et al.*, 2012; Nagy *et al.*, 2010; Zhang *et al.*, 2012). The N-terminal domain of Hsp104 is not required for stable substrate binding or the majority of disaggregation activities *in vitro* and prion propagation *in vivo* (Hung and Masison, 2006; Lum *et al.*, 2008; Tipton *et al.*, 2008), but it is involved in prion curing (Hung and Masison, 2006; Sweeny *et al.*, 2015). Deletion of the N-domain of ClpX decreases recognition and degradation of specific substrates (Baker and Sauer, 2012; Dougan *et al.*, 2003; Gur *et al.*, 2013; Kirstein *et al.*, 2009b; Kress *et al.*, 2009; Wojtyra *et al.*, 2003). Similarly, deletion of the N-terminal domain of ClpA results in a decrease in the rate of degradation of some substrates; however, the deletion also increases the rate of degradation of others (Cranz-Mileva *et al.*, 2008; Guo *et al.*, 2002a; Gur *et al.*, 2013; Kirstein *et al.*, 2009b), and deletion of the I-domain of HslU alters the degradation of some HslUV substrates (Kravats *et al.*, 2011; Sundar *et al.*, 2012).

Other substrates require adaptor proteins: proteins that facilitate recognition of specific substrates by Clp/Hsp100 chaperones (Figs. 5.7A–5.7E) (Baker and Sauer, 2012; Battesti and Gottesman, 2013;

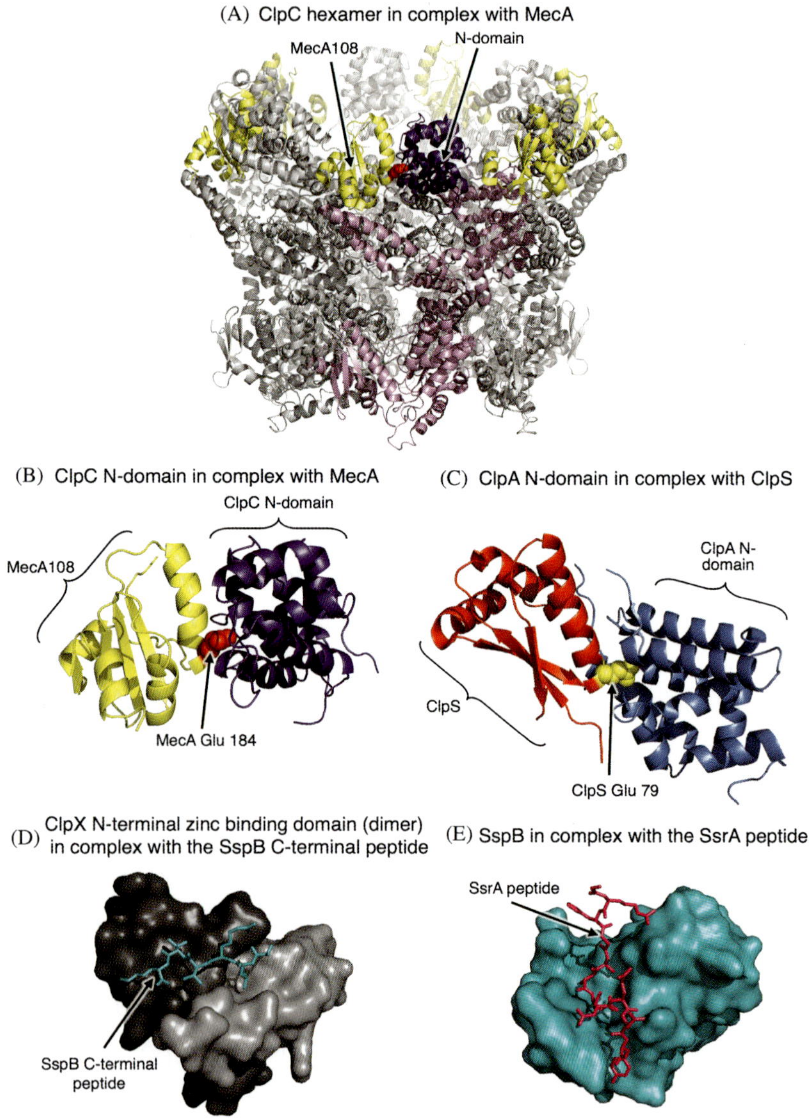

Fig. 5.7 Adaptor proteins directly interact with the Clp/Hsp100 chaperone. (A) The *B. subtilis* ClpC hexamer structure (Glu280Ala/Glu618Ala) in complex with MecA108 (PDB 3j3u) (Liu *et al.*, 2013). A single protomer is shown in lavender, with the N-terminal

Fig. 5.7 (*Continued*) domain in purple and the six MecA molecules in yellow. Side (left) and top (right) views are shown. **(B)** Structure of a single *B. subtilis* ClpC N-domain in complex with MecA108 (PDB 3j3u) (Liu *et al*., 2013) in the same orientation as in the hexamer in **A**. The critical Glu 184 residue is shown in red as a CPK model. **(C)** Structure of a single *E. coli* ClpA N-domain in complex with ClpS (H66A) (PDB 1mg9) (Zeth *et al*., 2002) in an orientation similar to the ClpC–MecA structure shown in **B**. The ClpA N-domain is in blue and the ClpS monomer is in red with the critical Glu 79 residue shown in yellow as a CPK model. **(D)** Structure of the *E. coli* N-terminal zinc-binding domain of ClpX (residues 9–51) as a homodimer (surface representation) in complex with an SspB C-terminal peptide (residues 160–165) represented with sticks (PDB 2ds8) (Park *et al*., 2007). Each monomer of ClpX is shown in light or dark gray and the SspB peptide is cyan. **(E)** Structure of a monomer of the *H. influenzae* SspB adaptor protein (residues 5–111) in association with the SsrA peptide (PDB 1ou8) (Levchenko *et al*., 2003). The SspB monomer is shown as a surface representation in cyan and the 11-amino-acid SsrA peptide is in magenta and shown as sticks. In **A–E**, the images were prepared using PYMOL (www.pymol.org).

Dougan *et al*., 2003; Gur *et al*., 2013; Kirstein *et al*., 2009b; Kress *et al*., 2009; Nagpal *et al*., 2013; Wojtyra *et al*., 2003). Adaptor proteins directly interact with the substrate and simultaneously interact with the N-domain of the Clp/Hsp100 chaperone. In this way, low affinity substrates are specifically targeted for degradation (Battesti and Gottesman, 2013; Gur *et al*., 2013; Kirstein *et al*., 2009b; Nagpal *et al*., 2013). For example, by binding to the SsrA tag and the N-domain of ClpX, SspB increases the effective local concentration of SsrA-tagged substrate proteins, which in turn stimulates the rate of degradation (Figs. 5.7D and 5.7E). It is thought that a single SspB dimer interacts with a ClpX hexamer during substrate degradation despite potential binding sites for three dimers (Levchenko, 2002). Another ClpXP adaptor protein, RssB (also known as SprE or MviA), specifically targets σ^S, the stationary phase sigma factor of *E. coli*, for degradation during exponential growth (Battesti and Gottesman, 2013; Gur *et al*., 2013; Kirstein *et al*., 2009b). RssB interacts with the N-domain of ClpX and simultaneously with σ^S to facilitate σ^S degradation. While SspB increases the rate of substrate degradation, RssB is essential for the recognition of σ^S by ClpX (Zhou *et al*., 2001). ClpA has a specific adaptor, ClpS, that binds substrates containing Tyr, Phe, Trp or

Leu at the N-terminal (referred to as N-end rule substrates) and simultaneously tethers them to ClpA, thereby delivering the substrates to ClpA and enhancing substrate degradation (Battesti and Gottesman, 2013; Gur *et al.*, 2013; Kirstein *et al.*, 2009b; Roman-Hernandez *et al.*, 2011). An α-helix in the C-terminal domain of ClpS contains a Glu residue that binds in a conserved pocket of the ClpA N-domain; other residues in the C-terminal region of ClpS interact with substrate (Dougan *et al.*, 2002; Guo *et al.*, 2002a; Zeth *et al.*, 2002) (Fig. 7C). Similarly, the C-terminal region of MecA from *B. subtilis* interacts with the N-terminal domain of ClpC via a Glu residue that resides in the interacting α-helix of MecA (Fig. 7B) (Liu *et al.*, 2013). In contrast to other adaptors, which only require the N-terminal domain of the Clp protein for binding, MecA also requires the M-domain of ClpC for docking and substrate delivery (Fig. 7A) (Kirstein *et al.*, 2006; Liu *et al.*, 2013; Wang *et al.*, 2011).

Protein Unfolding and Threading

In classic experiments by Horwich and colleagues, it was shown that Clp chaperones, possess the ability to unfold proteins by forcibly translocating the polypeptides through the central channel of the chaperone (Weber-Ban *et al.*, 1999). These observations explained the basic mechanism of action of Clp/Hsp100 chaperones (Hoskins *et al.*, 2000; Kim *et al.*, 2000; Lee *et al.*, 2001; Ortega *et al.*, 2000; Singh *et al.*, 2000; Weber-Ban *et al.*, 1999). ClpB and Hsp104 also possess the innate ability to unfold and translocate substrates despite the fact that cochaperones are required *in vivo* and in many *in vitro* disaggregation assays (Doyle *et al.*, 2007b; Tessarz *et al.*, 2008; Weibezahn *et al.*, 2004). This activity for ClpB and Hsp104 was first observed in experiments perfomed by Bukau and colleagues where the conserved ClpP protease interaction motif of ClpA, a helix–loop–helix, was transplanted to the homologous region of ClpB or Hsp104 (Tessarz *et al.*, 2008; Weibezahn *et al.*, 2004). These modified ClpB and Hsp104 proteins were able to interact with the ClpP protease and degrade proteins (Tessarz *et al.*,

2008; Weibezahn *et al.*, 2004). By the use of this tool, it was demonstrated that ClpB and Hsp104 bind and translocate unfolded substrates through the central pore into the ClpP chamber for degradation (Tessarz *et al.*, 2008; Weibezahn *et al.*, 2004). Additionally, in the absence of cochaperones and ClpP, ClpB and Hsp104 were shown to unfold and remodel proteins using conditions that restrict ATP hydrolysis (Doyle *et al.*, 2007b; Hoskins *et al.*, 2009). Together, these studies demonstrated that ClpB and Hsp104 act via a mechanism similar to that of other Clp/Hsp100 proteins.

Stable substrate interactions with Clp/Hsp100 central pore loops are required for efficient substrate unfolding and translocation (Hinnerwisch *et al.*, 2005a; Martin *et al.*, 2008b; Weibezahn *et al.*, 2004). Pore loop mutations in ClpX, HslU, ClpA, ClpB and the ATPase domain of FtsH lower the affinity for substrate and severely impair the ability of the Clp/Hsp100 to process substrate proteins (Doyle *et al.*, 2012; Farbman *et al.*, 2008; Hinnerwisch *et al.*, 2005a; Lum *et al.*, 2008; Martin *et al.*, 2008b; Park *et al.*, 2005; Weibezahn *et al.*, 2004). Work by Martin, Baker and Sauer showed that mutation of specific pore loop tyrosines in single chain ClpX hexamers reduces substrate unfolding and translocation activity (Martin *et al.*, 2008b). For example, a ClpX hexamer with two mutated pore tyrosines used approximately threefold more ATP than wild-type ClpX during substrate unfolding, indicating that the tyrosine is important for efficient substrate binding and translocation (Martin *et al.*, 2008b). Additionally, a recent study showed that the ClpX pore loops function synergistically, and the number of wild-type pore loops required for efficient substrate degradation depends on the thermodynamic stability of the protein substrate (Cordova *et al.*, 2014; Iosefson *et al.*, 2015; Kenniston *et al.*, 2003; Lee *et al.*, 2001; Matouschek and Bustamante, 2003). In ClpB, substitutions of pore loop tyrosines cause defects in substrate binding, unfolding and processing (Doyle *et al.*, 2012; Schlieker *et al.*, 2004; Weibezahn *et al.*, 2004). Moreover, subunit mixing experiments showed that, as seen with ClpX, the number of wild-type ClpB pore loops required for substrate remodeling was substrate-dependent (DeSantis *et al.*, 2012; Doyle *et al.*, 2012).

Following substrate engagement by the pore loop tyrosines, ATP hydrolysis induces conformational changes that move the pore loops down into the axial channel, generating mechanical force via a power-stroke mechanism (Hinnerwisch *et al.*, 2005a; Iosefson *et al.*, 2015; Koga *et al.*, 2009; Lee *et al.*, 2007; Lum *et al.*, 2008; Martin *et al.*, 2008b; Schlieker *et al.*, 2004; Sen *et al.*, 2013; Weibezahn *et al.*, 2004) (Fig. 4B). Successive paddling motions of the pore loops promote unfolding and translocation of the substrate (Aubin-Tam *et al.*, 2011; Hinnerwisch *et al.*, 2005a; Iosefson *et al.*, 2015; Lum *et al.*, 2008; Maillard *et al.*, 2011; Martin *et al.*, 2008b,c; Sen *et al.*, 2013). Moreover, the successive motions of the pore loops facilitate substrate handover from subunit to subunit as well as prevent substrate backsliding or premature release (Glynn *et al.*, 2009; Lyubimov *et al.*, 2011; Nyquist and Martin, 2014).

ATP binding and hydrolysis is a universal requirement for protein unfolding and translocation by Clp/Hsp100 chaperones. However, demonstrating how ATP hydrolysis is coordinated and ordered within the hexameric ring has been challenging. To elucidate the mechanism of ATP utilization by ClpX, Baker, Sauer and colleagues constructed and studied a single chain variant of ClpX (Glynn *et al.*, 2009; Martin *et al.*, 2005, 2008a–c). Using this variant, amino acid substitutions were introduced into specific ATP–binding sites within the hexamer and the mutated hexamers were tested for chaperone activity. For example, when a ClpX hexamer had only a single protomer capable of ATP hydrolysis, substrate unfolding and translocation was supported, although the activity was significantly reduced (Martin *et al.*, 2005). These results suggest a random or probabilistic sequence of ATP hydrolysis and rule out strictly concerted (where all six active sites of a ring simultaneously bind ATP, hydrolyze ATP and release ADP) or sequential (where activity depends upon an endless cycle of ATP utilization around the ring) mechanisms of ATP hydrolysis (Cordova *et al.*, 2014; Martin *et al.*, 2005). However, some inter-subunit coordination was indicated; the arrangement of active and inactive subunits in the ClpX hexamer showed differences in overall activity, suggesting that a strictly probabilistic mechanism is not likely (Aubin-Tam *et al.*, 2011; Cordova *et al.*, 2014; Maillard *et al.*, 2011; Martin *et al.*, 2005; Sen *et al.*, 2013).

Understanding how the ATP hydrolytic cycle of Hsp104 and ClpB is coordinated has been more difficult as each protomer has two nucleotide-binding sites able to bind and hydrolyze ATP. As suggested by the asymmetry observed for the ClpB hexamer, nucleotide binding does not occur at all 12 NBDs at the same time; only 8 of 12 nucleotide-binding sites bind nucleotide concurrently (Carroni *et al.*, 2014; Fernandez-Higuero *et al.*, 2011). Moreover, studies have indicated that four ClpB subunits make up a cooperative unit (del Castillo *et al.*, 2011; Fernandez-Higuero *et al.*, 2011), and cooperativity within the ClpB ring is maintained by highly conserved arginine residues, which interact with ATP bound by the neighboring protomer (Biter *et al.*, 2012a; Mogk *et al.*, 2003; Yamasaki *et al.*, 2011; Zeymer *et al.*, 2014). Interestingly, the coordination of ATP hydrolysis for ClpB and Hsp104 has also been seen to vary based on the substrate (DeSantis *et al.*, 2012; Doyle *et al.*, 2012; Hoskins *et al.*, 2009). For example, with small soluble aggregates the disaggregase utilizes a random mechanism of ATP hydrolysis (DeSantis *et al.*, 2012; Doyle *et al.*, 2012; Hoskins *et al.*, 2009). However, for large insoluble aggregates or prions, disaggregation requires more cooperation between subunits: a sequential or semisequential mechanism of ATP hydrolysis is employed (DeSantis *et al.*, 2012; Hoskins *et al.*, 2009; Werbeck *et al.*, 2008). The ability of Hsp104 and ClpB to utilize substrate-adaptive ATP hydrolysis mechanisms likely explains how Hsp104 and ClpB can act on such diverse substrates.

In recent work, substrate unfolding and translocation by ClpX and ClpA have been monitored using single molecule studies (Aubin-Tam *et al.*, 2011; Cordova *et al.*, 2014; Maillard *et al.*, 2011). In studies using ClpXP, the experiments suggested stochastic unfolding and translocation by ClpX (Cordova *et al.*, 2014). ClpX was seen to translocate polypeptides in physical steps of ~1–4 nm in length and the rates of substrate translocation were dependent on the sequence of the unfolded substrate. Similar single molecule experiments using ClpAP showed that

ClpA, which has two nucleotide-binding rings, unfolds most protein substrates substantially faster than ClpX, which has a single ring (Olivares *et al.*, 2014). It was further shown that ClpA translocates polypeptides slower than ClpX by taking significantly smaller and more regular steps of ~1–2 nm.

Collaboration between Clp/Hsp100 and Hsp70 Chaperones in Protein Disaggregation

The Clp/Hsp100 disaggregases, ClpB and Hsp104, collaborate with the Hsp70 chaperone system in the disaggregation of large insoluble protein aggregates both *in vivo* and *in vitro* (Glover and Lindquist, 1998; Goloubinoff *et al.*, 1999; Motohashi *et al.*, 1999; Zolkiewski, 1999). Hsp70 in eukaryotes and the prokaryotic homolog, DnaK, are highly conserved ATP-dependent molecular chaperones (Hartl *et al.*, 2011; Mayer, 2010; Mayer and Bukau, 2005; Meimaridou *et al.*, 2009). They function in the folding of nascent protein and remodeling of proteins under nonstress conditions as well as the refolding and reactivation of damaged proteins under stress conditions. Hsp70 functions with two cochaperones: a J-domain protein, such as Hsp40 in eukaryotes or DnaJ in prokaryotes, and a nucleotide exchange factor (NEF, GrpE) for protein remodeling activity (Acebron *et al.*, 2008; Genevaux *et al.*, 2007; Harrison, 2003; Kampinga and Craig, 2010). Hsp70 comprises an N-terminal nucleotide-binding domain (NBD) connected by a flexible linker to a C-terminal substrate-binding domain (CTD) (Bertelsen *et al.*, 2009; Kityk *et al.*, 2012). Nucleotide binding in the NBD allosterically regulates substrate binding in the CTD (Qi *et al.*, 2013; Schlecht *et al.*, 2011; Schweizer *et al.*, 2011; Zhuravleva *et al.*, 2012; Zhuravleva and Gierasch, 2011; Zuiderweg *et al.*, 2013). ADP binding promotes high affinity substrate binding at the CTD, while ATP binding triggers substrate release. Structural studies have elucidated the ADP as well as the ATP conformation of Hsp70 (Bertelsen *et al.*, 2009; Kityk *et al.*, 2012). The ATP hydrolytic cycle of Hsp70 is regulated by the cochaperones, Hsp40 and NEF. Hsp40 stimulates ATP hydrolysis by Hsp70, promoting stable substrate binding by the ADP form of Hsp70 (Genevaux *et al.*, 2007; Kampinga and Craig, 2010). It is also involved in presenting some substrates to Hsp70 (Acebron *et al.*, 2008; Genevaux *et al.*, 2007; Kampinga and Craig, 2010). NEF stimulates the exchange of ADP with ATP by Hsp70, which triggers substrate release from Hsp70 (Genevaux *et al.*, 2007; Harrison, 2003). The Hsp70 system alone has limited ability to act on aggregates, and ClpB and Hsp104 are unable to disaggregate aggregates; together they form a powerful disaggregase.

Physical Interaction between Hsp104 (ClpB) and Hsp70 (DnaK)

The observation by Glover and Lindquist that Hsp104 acts in a species-specific manner with Hsp70 suggested early on that the functional interaction involved a physical interaction between the two proteins (Glover and Lindquist, 1998). The work of many groups has demonstrated the species-specific collaboration between Hsp104 (ClpB) and Hsp70 (DnaK), and defined the protein–protein interaction sites between the chaperones (Doyle *et al.*, 2015; Glover and Lindquist, 1998; Krzewska *et al.*, 2001; Lee *et al.*, 2013; Miot *et al.*, 2011; Reidy *et al.*, 2012; Rosenzweig *et al.*, 2013; Seyffer *et al.*, 2012; Sielaff and Tsai, 2010; Tipton *et al.*, 2008).

The regions of Hsp104 and ClpB that contact Hsp70 and DnaK, respectively, are located on helices 2 and 3 of the M-domain (Miot *et al.*, 2011; Rosenzweig *et al.*, 2013; Seyffer *et al.*, 2012; Sielaff and Tsai, 2010; Tipton *et al.*, 2008). Biochemical and structural studies show that ClpB binds to the DnaK NBD utilizing a region overlapping the GrpE binding site (Doyle *et al.*, 2015; Harrison, 2003; Rosenzweig *et al.*, 2013) (Figs. 5.8A and 5.8B). The region of the ClpB M-domain that interacts with DnaK also forms intraprotomer interactions with NBD-1 of ClpB (Carroni *et al.*, 2014; Haslberger *et al.*, 2007; Lee *et al.*, 2003; Oguchi

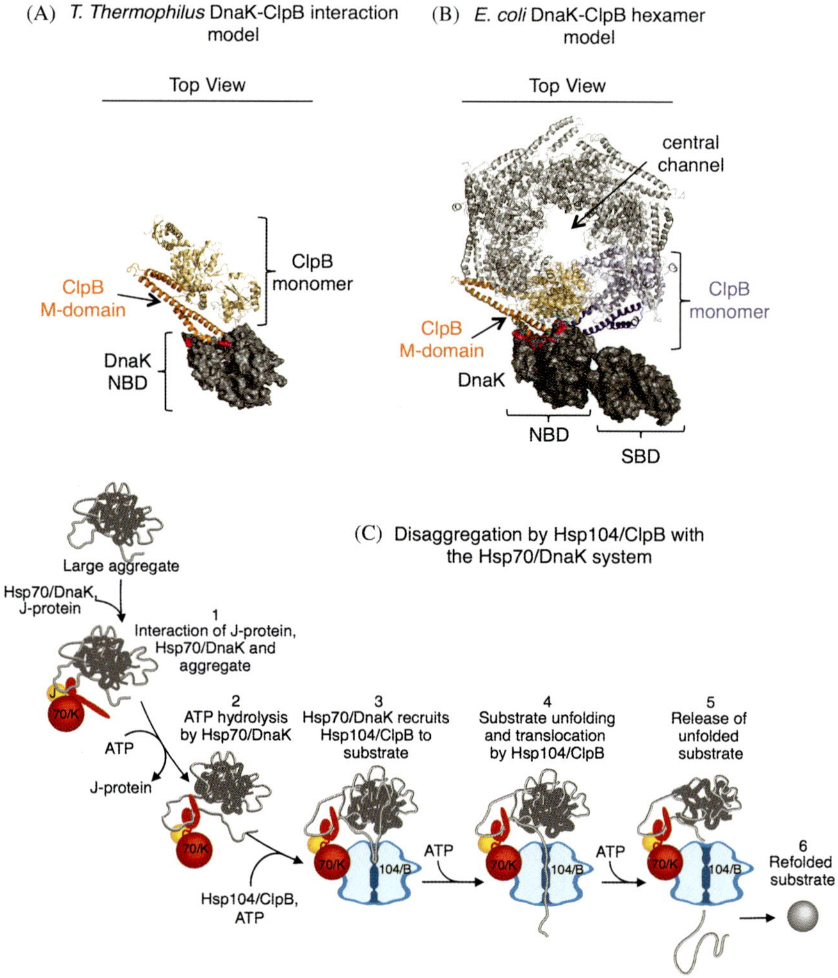

Fig. 5.8 Interaction and collaboration between ClpB/Hsp104 and DnaK/Hsp70 in protein disaggregation. (A) Docking model of the *T. thermophilus* ClpB–DnaK interaction (Rosenzweig *et al.*, 2013) based on NMR experiments of a complex formed by the NBD-1 and M-domain of a ClpB monomer and the NBD of DnaK. A surface rendering of the DnaK NBD is shown in gray, with residues identified as interacting shown in pink (Rosenzweig *et al.*, 2013); ClpB (ΔN) is shown as an atomic model based on the *T. thermophilus* crystal structure (PDB 1qvr) with NBD-1 and NBD-2 in wheat, the M-domain in orange and residues identified as interacting with DnaK in cyan (Lee *et al.*, 2003). The model was prepared using PYMOL (www.pymol.org). **(B)** Computational model of the interaction between *E. coli* DnaK and a ClpB hexamer (Doyle *et al.*, 2015). Model of a full-length ClpB hexamer shown as a top view bound to full-length DnaK shown as a surface-rendered model in gray. ClpB protomers that interact with DnaK are in blue

Fig. 5.8 (*Continued*) and orange, and specific residues identified as interacting with DnaK in cyan. Residues in the DnaK NBD specifically identified as interacting with ClpB are shown in pink. The NBD and substrate-binding domain (SBD) of DnaK are shown, as are the central channel and one M-domain of the ClpB hexamer. The orientation of the DnaK NBD and the ClpB M-domain is as in **A** and the model was prepared using PYMOL (www.pymol.org). **(C)** Model for the mechanism of disaggregation by ClpB/Hsp104 and the DnaK/Hsp70 chaperone system. Initially, the J-protein (yellow) interacts with aggregated substrate and recruits DnaK/Hsp70 (red) to the aggregate, where initial loosening of the aggregate can occur. The DnaK/Hsp70 NBD directly engages the M-domain of ClpB/Hsp104 (blue) and likely presents substrate to ClpB/Hsp104. Once substrate is engaged by the pore loops and in some cases the N-domains of ClpB/Hsp104, DnaK/Hsp70 coordinates ATP hydrolysis by ClpB/Hsp104, which leads to activation of the ClpB/Hsp104 disaggregase activity. ClpB/Hsp104 extracts polypeptides from the aggregate by forcibly unfolding and translocating polypeptides through the central channel using the power of ATP hydrolysis. The disaggregated polypeptides are released in an unfolded conformation, providing them with the opportunity to refold spontaneously or with the assistance of other chaperones, including the DnaK/Hsp70 system.

et al., 2012; Watanabe *et al.*, 2009). Mutations that stabilize the interaction between the M-domain with NBD-1 decrease ClpB unfolding activity (Oguchi *et al.*, 2012) and those that destabilize the interaction between the M-domain and NBD-1 stimulate the ATPase and unfolding activities of ClpB (Oguchi *et al.*, 2012). The stimulation in ClpB activity observed upon destabilization of the M-domain interaction with NBD-1 mimics the effects of DnaK binding to ClpB (Carroni *et al.*, 2014; Doyle *et al.*, 2007a; Doyle *et al.*, 2015; Oguchi *et al.*, 2012). One possibility is that DnaK binding to the M-domain disrupts the inhibitory interactions between the M-domain and NBD-1, leading to stimulation of the ATPase and disaggregation activity of ClpB in the presence of DnaK (Doyle *et al.*, 2007a; Doyle *et al.*, 2015; Rosenzweig *et al.*, 2013; Seyffer *et al.*, 2012). In parallel studies, the Hsp70 NBD was shown to interact with the Hsp104 M-domain (Lee *et al.*, 2013). Further, modeling studies of the *E. coli* ClpB and DnaK interaction correlated well with experimental results and indicate that many of the residues of the GrpE binding region of DnaK interact with residues in the M-domain of ClpB

(Carroni *et al*., 2014; Doyle *et al*., 2015). The modeling study also suggested that DnaK interacts with NBD-1 residues of ClpB (Doyle *et al*., 2015).

Current Model for Disaggregation by ClpB or Hsp104 and the Hsp70 System

Many groups have contributed to our current understanding of protein disaggregation and the mechanism of collaboration between DnaK and ClpB (Barends *et al*., 2010; Doyle and Wickner, 2009; Mogk *et al*., 2015; Winkler *et al*., 2012a; Zolkiewski *et al*., 2012). Altogether the results support a mechanism in which the Hsp70 system acts first on the aggregated substrate with the J-domain protein making the initial interaction and then recruiting Hsp70 to the aggregate (Acebron *et al*., 2008; Winkler *et al*., 2012b; Zietkiewicz *et al*., 2004; Zietkiewicz *et al*., 2006) (Fig. 8C). Protein remodeling is then initiated by the Hsp70 system and involves the disruption of interactions between unstructured hydrophobic regions on the aggregate surface.

The M-domain of Hsp104 (ClpB) directly interacts with the NBD of Hsp70 (DnaK), bringing Hsp104 (ClpB) in close proximity to the aggregate (Acebron *et al*., 2009; Doyle *et al*., 2015; Miot *et al*., 2011; Rosenzweig *et al*., 2013; Seyffer *et al*., 2012; Sielaff and Tsai, 2010; Tipton *et al*., 2008; Winkler *et al*., 2012b). This leads to the interaction between an unstructured region from the aggregate with Hsp104 (ClpB) central pore loop tyrosines and the N-domain (when necessary) (Doyle *et al*., 2012; Nagy *et al*., 2010; Sweeny *et al*., 2015; Tessarz *et al*., 2008; Weibezahn *et al*., 2004). Moreover, Hsp70 (DnaK) likely stabilizes the substrate–Hsp104 (ClpB) interaction by maintaining a ternary complex with substrate (Doyle *et al*., 2012).

Following substrate engagement by the Hsp104 (ClpB) pore loops, polypeptides are extracted from the aggregate and unfolded via ATP-dependent translocation through the central pore (Lum *et al*., 2008; Schlieker *et al*., 2004; Tessarz *et al*., 2008; Weibezahn *et al*., 2004). By directly binding to a region of the Hsp104 (ClpB) M-domain located near NBD-1, Hsp70 (DnaK) functions at this step to coordinate ATP hydrolysis by Hsp104 (ClpB) and drive conformational changes in the disaggregase that are required for polypeptide unfolding and translocation (Doyle *et al.*, 2007a; Doyle *et al.*, 2015; Hoskins *et al.*, 2009; Lee *et al.*, 2013; Miot *et al.*, 2011; Rosenzweig *et al.*, 2013; Seyffer *et al.*, 2012; Sielaff and Tsai, 2010). The unfolded polypeptides are released and either refold spontaneously or are refolded by other molecular chaperones (Doyle *et al.*, 2013; Mogk *et al.*, 2015; Zolkiewski *et al.*, 2012).

Ordered Aggregate Disassembly and Prion Propagation

In yeast, Hsp104 is essential for the propagation of prions. Its role is to sever prions, which are large amyloid fibers, into smaller fibers (Chernoff, 2007; Romanova and Chernoff, 2009; Tuite *et al.*, 2011; Tuite and Serio, 2010; Winkler *et al.*, 2012a). Only small fibers are able to partition between mother and daughter cells upon cell division, and act as templates for further prion growth (Chernoff *et al.*, 1999; Helsen and Glover, 2012a,b; Liebman and Chernoff, 2012; Winkler *et al.*, 2012a). Inheritance of some prions also requires the participation of yeast Hsp70, Hsp40 and NEF (Chernoff, 2007; Helsen and Glover, 2012b; Liebman and Chernoff, 2012; Masison *et al.*, 2009; Moran *et al.*, 2013; Sharma *et al.*, 2009; Winkler *et al.*, 2012b). For the [*PSI*$^+$] prion, high levels of Hsp104 expression can cure the cell of the prion (Helsen and Glover, 2012b; Hung and Masison, 2006). Because of the unique ability of Hsp104 to disassemble prion amyloids, it has been proposed that Hsp104 is a more powerful disaggregase than ClpB (DeSantis *et al.*, 2012). However, it is now clear that Hsp104 and ClpB possess very similar capabilities. For example, ClpB, in collaboration with the DnaK system, is able to promote prion propagation in yeast lacking Hsp104 (Reidy *et al.*, 2012). Moreover, when yeast prions are introduced into *E. coli*, the prions are propagated in a ClpB–dependent fashion, indicating that ClpB can act in prion propagation similarly to Hsp104 (Garrity *et al.*, 2010; Yuan *et al.*, 2014).

In vitro studies have suggested that prion amyloids are disassembled by Hsp104 alone in an ATP-dependent manner (DeSantis *et al.*, 2012; DeSantis and Shorter, 2012; Shorter and Lindquist, 2004, 2006), although aspects of these studies have not been reproduced (Glover and Lum, 2009; Inoue *et al.*, 2004; Krzewska and Melki, 2006; Savistchenko *et al.*, 2008). Other *in vitro* studies, performed under different conditions, indicated that prion disassembly required the combination of Hsp104 and the yeast Hsp70 system (DeSantis *et al.*, 2012; Shorter and Lindquist, 2004, 2006, 2008). Further studies are required to provide a clearer understanding of the structures of the many prion amyloids and the mechanism of prion disassembly.

Summary and Perspectives

Many recent advances have provided insight into the structures and mechanisms of action of Clp/Hsp100 chaperones. Yet, many questions remain due to the complexity of the Clp/Hsp100 chaperones, their dependence on the Hsp70 chaperone system, and the involvement of adaptor proteins. The prevalence of neurodegenerative disorders, including Alzheimer's, Huntington's and Parkinson's disease, which involve aggregated proteins that share a common amyloid fold, implores us to expand our understanding of how molecular chaperones attenuate and potentially reverse these disease states. Therefore, a stronger understanding of the structure and function of these Clp/Hsp100 molecular chaperones may ultimately lead to therapies that act on the myriad of protein misfolding and aggregation diseases.

Acknowledgments

We thank Joel Hoskins for critical reading of this work and the many helpful discussions. This research was supported by the Intramural Research Program of the NIH, NCI, Center for Cancer Research.

References

Acebron SP, Fernandez-Saiz V, Taneva SG, *et al.* (2008). DnaJ recruits DnaK to protein aggregates. *J Biol Chem* **283:** 1381–1390.

Acebron SP, Martin I, del Castillo U, *et al.* (2009). DnaK-mediated association of ClpB to protein aggregates: A bichaperone network at the aggregate surface. *FEBS Lett* **583:** 2991–2996.

Alexopoulos J, Ahsan B, Homchaudhuri L, *et al.* (2013). Structural determinants stabilizing the axial channel of ClpP for substrate translocation. *Mol Microbiol* **90:** 167–180.

Alexopoulos JA, Guarne A, Ortega J (2012). ClpP: a structurally dynamic protease regulated by AAA+ proteins. *J Struct Biol* **179:** 202–210.

Aubin-Tam ME, Olivares AO, Sauer RT, *et al.* (2011). Single-molecule protein unfolding and translocation by an ATP-fueled proteolytic machine. *Cell* **145:** 257–267.

Baker TA, Sauer RT (2012). ClpXP, an ATP-powered unfolding and protein-degradation machine. *Biochim Biophys Acta* **1823:** 15–28.

Barends TR, Werbeck ND, Reinstein J (2010). Disaggregases in 4 dimensions. *Curr Opin Struct Biol* **20:** 46–53.

Battesti A, Gottesman S (2013). Roles of adaptor proteins in regulation of bacterial proteolysis. *Curr Opin Microbiol* **16:** 140–147.

Bertelsen EB, Chang L, Gestwicki JE, Zuiderweg ER (2009). Solution conformation of wild-type *E. coli* Hsp70 (DnaK) chaperone complexed with ADP and substrate. *Proc Nat Acad Sci USA* **106:** 8471–8476.

Bewley MC, Graziano V, Griffin K, Flanagan JM (2006). The asymmetry in the mature amino-terminal of ClpP facilitates a local symmetry match in ClpAP and ClpXP complexes. *J Struct Biol* **153:** 113–128.

Biter AB, Lee J, Sung N, *et al.* (2012a). Functional analysis of conserved cis- and trans-elements in the Hsp104 protein disaggregating machine. *J Struct Biol* **179:** 172–180.

Biter AB, Lee S, Sung N, Tsai FT (2012b). Structural basis for intersubunit signaling in a protein disaggregating machine. *Proc Nat Acad Sci USA* **109:** 12515–12520.

Bochtler M, Hartmann C, Song HK, *et al.* (2000). The structures of HsIU and the ATP-dependent protease HsIU–HsIV. *Nature* **403:** 800–805.

Brotz-Oesterhelt H, Beyer D, Kroll HP, *et al.* (2005). Dysregulation of bacterial proteolytic machinery by a new class of antibiotics. *Nat Medi* **11:** 1082–1087.

Burton BM, Baker TA, *et al.* (2005). Remodeling protein complexes: insights from the AAA+ unfoldase ClpX and Mu transposase. Protein Science: *A Publication of the Protein Society* **14:** 1945–1954.

Burton RE, Siddiqui SM, Kim YI, *et al.* (2001). Effects of protein stability and structure on substrate processing by the ClpXP unfolding and degradation machine. *EMBO J* **20:** 3092–3100.

Capo-Chichi JM, Boissel S, Brustein E, *et al.* (2015). Disruption of CLPB is associated with congenital microcephaly, severe encephalopathy and 3-methylglutaconic aciduria. *J Med Genet* **52:** 303–311.

Carroni M, Kummer E, Oguchi Y, *et al.* (2014). Head-to-tail interactions of the coiled-coil domains regulate ClpB activity and cooperation with Hsp70 in protein disaggregation. *eLife* **3:** e02481.

Chernoff YO (2007). Stress and prions: lessons from the yeast model. *FEBS Lett* **581:** 3695–3701.

Chernoff YO, Newnam GP, Kumar J, *et al.* (1999). Evidence for a protein mutator in yeast: role of the Hsp70-related chaperone ssb in formation, stability, and toxicity of the [PSI] prion. *Mol Cell Biol* **19:** 8103–8112.

Chien P, Perchuk BS, Laub MT, *et al.* (2007). Direct and adaptor-mediated substrate recognition by an essential AAA+ protease. *Proc Nat Acad Sci USA* **104:** 6590–6595.

Chow IT, Barnett ME, Zolkiewski M, Baneyx F (2005). The N-terminal domain of *Escherichia coli* ClpB enhances chaperone function. *FEBS Lett* **579:** 4242–4248.

Chowdhury T, Chien P, Ebrahim S, *et al.* (2010). Versatile modes of peptide recognition by the ClpX N domain mediate alternative adaptor-binding specificities in different bacterial species. *Protein Science: A Publication of the Protein Society* **19:** 242–254.

Cordova JC, Olivares AO, Shin Y, *et al.* (2014). Stochastic but highly coordinated protein unfolding and translocation by the ClpXP proteolytic machine. *Cell* **158:** 647–658.

Cranz-Mileva S, Imkamp F, Kolygo K, *et al.* (2008). The flexible attachment of the N-domains to the ClpA ring body allows their use on demand. *J Mol Biol* **378:** 412–424.

Davies JM, Brunger AT, Weis WI (2008). Improved structures of full-length p97, an AAA ATPase: implications for mechanisms of nucleotide-dependent conformational change. *Structure* **16:** 715–726.

del Castillo U, Alfonso C, Acebron SP, *et al.* (2011). A quantitative analysis of the effect of nucleotides and the M domain on the association equilibrium of ClpB. *Biochemistry* **50:** 1991–2003.

DeLaBarre B, Brunger AT (2003). Complete structure of p97/valosin-containing protein reveals communication between nucleotide domains. *Nat Struct Biol* **10:** 856–863.

DeSantis ME, Leung EH, Sweeny EA, *et al.* (2012). Operational plasticity enables hsp104 to disaggregate diverse amyloid and nonamyloid clients. *Cell* **151:** 778–793.

DeSantis ME, Shorter J (2012). Hsp104 drives "protein-only" positive selection of Sup35 prion strains encoding strong [PSI(+)]. *Chem & Biol* **19:** 1400–1410.

Diemand AV, Lupas AN (2006). Modeling AAA+ ring complexes from monomeric structures. *J Struct Biol* **156:** 230–243.

Dougan DA, Reid BG, Horwich AL, Bukau B (2002). ClpS, a substrate modulator of the ClpAP machine. *Mol Cell* **9:** 673–683.

Dougan DA, Weber-Ban E, Bukau B (2003). Targeted delivery of an ssrA-tagged substrate by the adaptor protein SspB to its cognate AAA+ protein ClpX. *Mol Cell* **12:** 373–380.

Doyle SM, Genest O, Wickner S (2013). Protein rescue from aggregates by powerful molecular chaperone machines. *Nat Rev Mol Cell Biol* **14:** 617–629.

Doyle SM, Hoskins JR, Wickner S (2007a). Collaboration between the ClpB AAA+ remodeling protein and the DnaK chaperone system. *Proc Nat Acad Sci USA* **104:** 11138–11144.

Doyle SM, Hoskins JR, Wickner S (2012). DnaK chaperone-dependent disaggregation by caseinolytic peptidase B (ClpB) mutants reveals functional overlap in the N-terminal domain and nucleotide-binding domain-1 pore tyrosine. *J Biol Chem* **287:** 28470–28479.

Doyle SM, Shastry S, Kravats AN, *et al.* (2015). Interplay between *E. coli* DnaK, ClpB and GrpE during protein disaggregation. *J Mol Biol* **427:** 312–327.

Doyle SM, Shorter J, Zolkiewski M, *et al.* (2007b). Asymmetric deceleration of ClpB or Hsp104 ATPase activity unleashes protein-remodeling activity. *Nat Struct Mol Biol* **14:** 114–122.

Doyle SM, Wickner S (2009). Hsp104 and ClpB: protein disaggregating machines. *Trends Biochem Sci* **34:** 40–48.

Erzberger JP, Berger JM (2006). Evolutionary relationships and structural mechanisms of AAA+ proteins. *Annu Rev Biophys Biomol Struct* **35:** 93–114.

Farbman ME, Gershenson A, Licht S (2008). Role of a conserved pore residue in the formation of a prehydrolytic high substrate affinity state in the AAA+ chaperone ClpA. *Biochemistry* **47:** 13497–13505.

Fernandez-Higuero JA, Acebron SP, Taneva SG, *et al.* (2011). Allosteric communication between the nucleotide binding domains of caseinolytic peptidase B. *J Biol Chem* **286:** 25547–25555.

Flynn JM, Neher SB, Kim YI, *et al.* (2003). Proteomic discovery of cellular substrates of the ClpXP protease reveals five classes of ClpX-recognition signals. *Mol Cell* **11:** 671–683.

Franzmann TM, Czekalla A, Walter SG (2011). Regulatory circuits of the AAA+ disaggregase Hsp104. *J Biol Chem* **286:** 17992–18001.

Garrity SJ, Sivanathan V, Dong J, *et al.* (2010). Conversion of a yeast prion protein to an infectious form in bacteria. *Proc Nat Acad Sci USA* **107:** 10596–10601.

Genevaux P, Georgopoulos C, Kelley WL (2007). The Hsp70 chaperone machines of *Escherichia coli*: A paradigm for the repartition of chaperone functions. *Mol Microbiol* **66:** 840–857.

Gersch M, Famulla K, Dahmen M, *et al.* (2015). AAA+ chaperones and acyldepsipeptides activate the ClpP protease via conformational control. *Nat Commun* **6:** 6320.

Glover JR, Lindquist S (1998). Hsp104, Hsp70, and Hsp40: a novel chaperone system that rescues previously aggregated proteins. *Cell* **94:** 73–82.

Glover JR, Lum R. (2009). Remodeling of protein aggregates by Hsp104. *Protein Pept Lett* **16:** 587–597.

Glynn SE, Martin A, Nager AR, *et al.* (2009). Structures of asymmetric ClpX hexamers reveal nucleotide-dependent motions in a AAA+ protein-unfolding machine. *Cell* **139:** 744–756.

Glynn SE, Nager AR, Baker TA, Sauer RT (2012). Dynamic and static components power unfolding in topologically closed rings of a AAA+ proteolytic machine. *Nat Struct Mol Biol* **19:** 616–622.

Goloubinoff P, Mogk A, Zvi AP, *et al.* (1999). Sequential mechanism of solubilization and refolding of stable protein aggregates by a bichaperone network. *Proc Nat Acad Sci USA* **96:** 13732–13737.

Graham JW, Lei MG, Lee CY (2013). Trapping and identification of cellular substrates of the *Staphylococcus aureus* ClpC chaperone. *J Bacteriol* **195:** 4506–4516.

Granata A, Warner TT (2010). The role of torsin A in dystonia. *European Journal of Neurology: The official journal of the European Federation of Neurological Societies* **17**(Suppl 1): 81–87.

Guo F, Esser L, Singh SK, *et al.* (2002a). Crystal structure of the heterodimeric complex of the adaptor, ClpS, with the N-domain of the AAA+ chaperone, ClpA. *J Biol Chem* **277:** 46753–46762.

Guo F, Maurizi MR, Esser L, Xia D (2002b). Crystal structure of ClpA, an Hsp100 chaperone and regulator of ClpAP protease. *J Biol Chem* **277:** 46743–46752.

Gur E, Ottofueling R, Dougan DA (2013). Machines of destruction — AAA+ proteases and the adaptors that control them. *Subcell Biochem* **66:** 3–33.

Hanson PI, Whiteheart SW (2005). AAA+ proteins: have engine, will work. *Nat Rev Mol Cell Biology* **6:** 519–529.

Harrison C (2003). GrpE, a nucleotide exchange factor for DnaK. *Cell Stress & Chaperones* **8:** 218–224.

Hartl FU, Bracher A, Hayer-Hartl M (2011). Molecular chaperones in protein folding and proteostasis. *Nature* **475:** 324–332.

Haslberger T, Weibezahn J, Zahn R, *et al.* (2007). M domains couple the ClpB threading motor with the DnaK chaperone activity. *Mol Cell* **25:** 247–260.

Haslberger T, Zdanowicz A, Brand I, *et al.* (2008). Protein disaggregation by the AAA+ chaperone ClpB involves partial threading of looped polypeptide segments. *Nat Struct Mol Biol* **15:** 641–650.

Helsen CW, Glover JR (2012a). Insight into molecular basis of curing of [PSI+] prion by overexpression of 104-kDa heat shock protein (Hsp104). *J Biol Chem* **287:** 542–556.

Helsen CW, Glover JR (2012b). A new perspective on Hsp104-mediated propagation and curing of the yeast prion [PSI (+)]. *Prion* **6:** 234–239.

Himeno H, Kurita D, Muto A (2014). tmRNA-mediated trans-translation as the major ribosome rescue system in a bacterial cell. *Front Genet* **5:** 66.

Hinnerwisch J, Fenton WA, Furtak KJ *et al.* (2005a). Loops in the central channel of ClpA chaperone mediate protein binding, unfolding, and translocation. *Cell* **121:** 1029–1041.

Hinnerwisch J, Reid BG, Fenton, WA, Horwich AL (2005b). Roles of the N-domains of the ClpA unfoldase in binding substrate proteins and in stable complex formation with the ClpP protease. *J Biol Chem* **280:** 40838–40844.

Hodson S, Marshall JJ, Burston SG (2012). Mapping the road to recovery: the ClpB/Hsp104 molecular chaperone. *J Struct Biol* **179:** 161–171.

Hoskins JR, Doyle SM, Wickner S. (2009). Coupling ATP utilization to protein remodeling by ClpB, a hexameric AAA+ protein. *Proc Nat Acad Sci USA* **106:** 22233–22238.

Hoskins JR, Singh SK, Maurizi MR, Wickner S (2000). Protein binding and unfolding by the chaperone ClpA and degradation by the protease ClpAP. *Proc Nat Acad Sci USA* **97:** 8892–8897.

Hoskins JR, Yanagihara K, Mizuuchi K, Wickner S (2002). ClpAP and ClpXP degrade proteins with tags located in the interior of the primary sequence. *Proc Nat Acad Sci USA* **99:** 11037–11042.

Hung GC, Masison DC (2006). N-terminal domain of yeast Hsp104 chaperone is dispensable for thermotolerance and prion propagation but necessary for curing prions by Hsp104 overexpression. *Genetics* **173:** 611–620.

Inoue Y, Taguchi H, Kishimoto A, Yoshida M (2004). Hsp104 binds to yeast Sup35 prion fiber but needs other factor(s) to sever it. *J Biol Chem* **279:** 52319–52323.

Iosefson O, Nager AR, Baker TA, Sauer RT (2015). Coordinated gripping of substrate by subunits of a AAA+ proteolytic machine. *Nat Chem Biol* **11:** 201–206.

Jenal U, Fuchs T (1998). An essential protease involved in bacterial cell-cycle control. *EMBO J* **17:** 5658–5669.

Kampinga HH, Craig EA (2010). The HSP70 chaperone machinery: J proteins as drivers of functional specificity. *Nat Rev Mol Cell Biol* **11:** 579–592.

Kanabus M, Shahni R, Saldanha JW, *et al.* (2015). Bi-allelic CLPB mutations cause cataract, renal cysts, nephrocalcinosis and 3-methylglutaconic aciduria, a novel disorder of mitochondrial protein disaggregation. *J Inherit Metab Dis* **38:** 211–219.

Kenniston JA, Baker TA, Fernandez JM, Sauer RT (2003). Linkage between ATP consumption and mechanical unfolding during the protein processing reactions of an AAA+ degradation machine. *Cell* **114:** 511–520.

Kim DY, Kim KK (2003). Crystal structure of ClpX molecular chaperone from *Helicobacter pylori*. *J Biol Chem* **278:** 50664–50670.

Kim YI, Burton RE, Burton BM, *et al.* (2000). Dynamics of substrate denaturation and translocation by the ClpXP degradation machine. *Mol Cell* **5:** 639–648.

Kirstein J, Hoffmann A, Lilie H, Schmidt R, *et al.* (2009a). The antibiotic ADEP reprogrammes ClpP, switching it from a regulated to an uncontrolled protease. *EMBO Mol Medi* **1:** 37–49.

Kirstein J, Moliere N, Dougan DA, Turgay K (2009b). Adapting the machine: adaptor proteins for Hsp100/Clp and AAA+ proteases. *Nat Rev Microbiol* **7:** 589–599.

Kirstein J, Schlothauer T, Dougan DA, *et al.* (2006). Adaptor protein controlled oligomerization activates the AAA+ protein ClpC. *EMBO J* **25:** 1481–1491.

Kityk R, Kopp J, Sinning I, Mayer MP (2012). Structure and dynamics of the ATP-bound open conformation of Hsp70 chaperones. *Mol Cell* **48,** 863–874.

Koga N, Kameda T, Okazaki K, Takada S (2009). Paddling mechanism for the substrate translocation by AAA+ motor revealed by multiscale molecular simulations. *Proc Nat Acad Sci USA* **106:** 18237–18242.

Kravats A, Jayasinghe M, Stan G (2011). Unfolding and translocation pathway of substrate protein controlled by structure in repetitive allosteric cycles of the ClpY ATPase. *Proc Nat Acad Sci USA* **108:** 2234–2239.

Kress W, Maglica Z, Weber-Ban E (2009). Clp chaperone-proteases: structure and function. *Res Microbiol* **160:** 618–628.

Krzewska J, Langer T, Liberek K (2001). Mitochondrial Hsp78, a member of the Clp/Hsp100 family in *Saccharomyces cerevisiae*, cooperates with Hsp70 in protein refolding. *FEBS Lett* **489:** 92–96.

Krzewska J, Melki R (2006). Molecular chaperones and the assembly of the prion Sup35p, an *in vitro* study. *EMBO J* **25:** 822–833.

Lee C, Prakash S, Matouschek A (2002). Concurrent translocation of multiple polypeptide chains through the proteasomal degradation channel. *J Biol Chem* **277:** 34760–34765.

Lee C, Schwartz MP, Prakash S, *et al.* (2001). ATP-dependent proteases degrade their substrates by processively unraveling them from the degradation signal. *Mol Cell* **7:** 627–637.

Lee J, Kim JH, Biter AB, *et al.* (2013). Heat shock protein (Hsp) 70 is an activator of the Hsp104 motor. *Proc Nat Acad Sci USA* **110:** 8513–8518.

Lee S, Choi JM, Tsai FT (2007). Visualizing the ATPase cycle in a protein disaggregating machine: structural basis for substrate binding by ClpB. *Mol Cell* **25:** 261–271.

Lee S, Sielaff B, Lee J, Tsai FT (2010). CryoEM structure of Hsp104 and its mechanistic implication for protein disaggregation. *Proc Nat Acad Sci USA* **107:** 8135–8140.

Lee S, Sowa ME, Watanabe YH, *et al.* (2003). The structure of ClpB: a molecular chaperone that rescues proteins from an aggregated state. *Cell* **115:** 229–240.

Levchenko I, Grant RA, Wah DA, *et al.* (2003). Structure of a delivery protein for an AAA+ protease in complex with a peptide degradation tag. *Mol Cell* **12:** 365–372.

Li DH, Chung YS, Gloyd M, *et al.* (2010). Acyldepsipeptide antibiotics induce the formation of a structured axial channel in ClpP: A model for the ClpX/ClpA-bound state of ClpP. *Chem & Biol* **17:** 959–969.

Li T, Lin J, Lucius AL (2015). Examination of polypeptide substrate specificity for *Escherichia coli* ClpB. *Proteins* **83:** 117–134.

Li T, Lucius AL (2013). Examination of the polypeptide substrate specificity for *Escherichia coli* ClpA. *Biochemistry* **52:** 4941–4954.

Liang CC, Tanabe LM, Jou S, *et al.* (2014). TorsinA hypofunction causes abnormal twisting movements and sensorimotor circuit neurodegeneration. *J Clin Invest* **124:** 3080–3092.

Liebman SW, Chernoff YO (2012). Prions in yeast. *Genetics* **191:** 1041–1072.

Ling L, Montano SP, Sauer RT *et al.* (2015). Deciphering the roles of multicomponent recognition signals by the AAA+ unfoldase ClpX. *J Mol Biol* **427:** 2966–2982.

Liu J, Mei Z, Li N, *et al.* (2013). Structural dynamics of the MecA–ClpC complex: a type II AAA+ protein unfolding machine. *J Biol Chem* **288:** 17597–17608.

Liu K, Ologbenla A, Houry WA (2014). Dynamics of the ClpP serine protease: a model for self-compartmentalized proteases. *Crit Rev Biochem Mol Biol* **49:** 400–412.

Lum R, Niggemann M, Glover JR (2008). Peptide and protein binding in the axial channel of Hsp104: insights into the mechanism of protein unfolding. *J Biol Chem* **283:** 30139–30150.

Lyubimov AY, Strycharska M, Berger JM (2011). The nuts and bolts of ring-translocase structure and mechanism. *Curr Opin Struct Biol* **21:** 240–248.

Mackay RG, Helsen CW, Tkach JM, Glover JR (2008). The C-terminal extension of *Saccharomyces cerevisiae* Hsp104 plays a role in oligomer assembly. *Biochemistry* **47:** 1918–1927.

Maillard RA, Chistol G, Sen M, *et al.* (2011). ClpX(P) generates mechanical force to unfold and translocate its protein substrates. *Cell* **145:** 459–469.

Martin A, Baker TA, Sauer RT (2005). Rebuilt AAA+ motors reveal operating principles for ATP-fuelled machines. *Nature* **437:** 1115–1120.

Martin A, Baker TA, Sauer RT (2007). Distinct static and dynamic interactions control ATPase-peptidase communication in a AAA+ protease. *Mol Cell* **27:** 41–52.

Martin A, Baker TA, Sauer RT (2008a). Diverse pore loops of the AAA+ ClpX machine mediate unassisted and adaptor-dependent recognition of ssrA-tagged substrates. *Mol Cell* **29:** 441–450.

Martin A, Baker TA, Sauer RT (2008b). Pore loops of the AAA+ ClpX machine grip substrates to drive translocation and unfolding. *Nat Struct Mol Biol* **15:** 1147–1151.

Martin A, Baker T.A, Sauer RT (2008c). Protein unfolding by a AAA+ protease is dependent on ATP-hydrolysis rates and substrate energy landscapes. *Nat Struct Mol Biol* **15:** 139–145.

Masison DC, Kirkland PA, Sharma D (2009). Influence of Hsp70s and their regulators on yeast prion propagation. *Prion* **3:** 65–73.

Matouschek A, Bustamante C (2003). Finding a protein's Achilles heel. *Nat Struct Biol* **10:** 674–676.

Mayer MP (2010). Gymnastics of molecular chaperones. *Mol Cell* **39:** 321–331.

Mayer MP, Bukau B (2005). Hsp70 chaperones: cellular functions and molecular mechanism. *Cell Mol Life Sci* **62:** 670–684.

Meimaridou E, Gooljar SB, Chapple JP (2009). From hatching to dispatching: the multiple cellular roles of the Hsp70 molecular chaperone machinery. *J Mol Endocrinol* **42:** 1–9.

Meyer H, Bug M, Bremer S (2012). Emerging functions of the VCP/p97 AAA-ATPase in the ubiquitin system. *Nat Cell Biol* **14:** 117–123.

Meyer H, Weihl CC (2014). The VCP/p97 system at a glance: connecting cellular function to disease pathogenesis. *J Cell Sci* **127:** 3877–3883.

Miot M, Reidy M, Doyle SM, *et al.* (2011). Species-specific collaboration of heat shock proteins (Hsp) 70 and 100 in thermotolerance and protein disaggregation. *Proc Nat Acad USA* **108:** 6915–6920.

Mogk A, Kummer E, Bukau B (2015). Cooperation of Hsp70 and Hsp100 chaperone machines in protein disaggregation. *Front Mol Biosci* **2:** 22.

Mogk A, Schlieker C, Strub C, *et al.* (2003). Roles of individual domains and conserved motifs of the AAA+ chaperone ClpB in oligomerization, ATP hydrolysis, and chaperone activity. *J Biol Chem* **278:** 17615–17624.

Mogk A, Tomoyasu T, Goloubinoff P, *et al.* (1999). Identification of thermolabile *Escherichia coli* proteins: prevention and reversion of aggregation by DnaK and ClpB. *EMBO J* **18:** 6934–6949.

Moran C, Kinsella GK, Zhang ZR, *et al.* (2013). Mutational analysis of Sse1 (Hsp110) suggests an integral role for this chaperone in yeast prion propagation *in vivo*. *G3* **:** 1409–1418.

Motohashi K, Watanabe Y, Yohda M, Yoshida M (1999). Heat-inactivated proteins are rescued by the DnaK.J-GrpE set and ClpB chaperones. *Proc Nat Acad Sci USA* **96:** 7184–7189.

Nagpal J, Tan JL, Truscott KN, *et al.* (2013). Control of protein function through regulated protein degradation: biotechnological and biomedical applications. *J Mol Microbiol Biotechnol* **23:** 335–344.

Nagy M, Guenther I, Akoyev V, *et al.* (2010). Synergistic cooperation between two ClpB isoforms in aggregate reactivation. *J Mol Biol* **396:** 697–707.

Nalbandian A, Donkervoort S, Dec E, *et al.* (2011). The multiple faces of valosin-containing protein-associated diseases: inclusion body myopathy with Paget's disease of bone, frontotemporal dementia, and amyotrophic lateral sclerosis. *J Mol Neurosci: MN* **45:** 522–531.

Neher SB, Villen J, Oakes EC, *et al.* (2006). Proteomic profiling of ClpXP substrates after DNA damage reveals extensive instability within SOS regulon. *Mol Cell* **22:** 193–204.

Neuwald AF, Aravind L, Spouge JL, Koonin EV (1999). AAA+: a class of chaperone-like ATPases associated with the assembly, operation, and disassembly of protein complexes. *Genome Res* **9:** 27–43.

Nyquist K, Martin A (2014). Marching to the beat of the ring: polypeptide translocation by AAA+ proteases. *Trends Biochem Sci* **39:** 53–60.

Oguchi Y, Kummer E, Seyffer F, *et al.* (2012). A tightly regulated molecular toggle controls AAA+ disaggregase. *Nat Struct Mol Biol* **19:** 1338–1346.

Ogura T, Whiteheart SW, Wilkinson AJ (2004). Conserved arginine residues implicated in ATP hydrolysis, nucleotide-sensing, and inter-subunit interactions in AAA and AAA+ ATPases. *J Struct Biol* **146:** 106–112.

Olivares AO, Nager AR, Iosefson O, *et al.* (2014). Mechanochemical basis of protein degradation by a double-ring AAA+ machine. *Nat Struct Mol Biol* **21:** 871–875.

Ortega J, Singh SK, Ishikawa T, *et al.* (2000). Visualization of substrate binding and translocation by the ATP-dependent protease, ClpXP. *Mol Cell* **6:** 1515–1521.

Pak M, Wickner S (1997). Mechanism of protein remodeling by ClpA chaperone. *Proc Nat Acad Sci USA* **94:** 4901–4906.

Park E, Rho YM, Koh OJ, *et al.* (2005). Role of the GYVG pore motif of HslU ATPase in protein unfolding and translocation for degradation by HslV peptidase. *J Biol Chem* **280:** 22892–22898.

Park SH, Bolender N, Eisele F, *et al.* (2007). The cytoplasmic Hsp70 chaperone machinery subjects misfolded and endoplasmic reticulum import-incompetent proteins to degradation via the ubiquitin-proteasome system. *Mol Biol Cell* **18:** 153–165.

Park YN, Zhao X, Yim YI, *et al.* (2014). Hsp104 overexpression cures *Saccharomyces cerevisiae* [PSI+] by causing dissolution of the prion seeds. *Eukaryotic Cell* **13:** 635–647.

Parsell DA, Kowal AS, Singer MA, Lindquist S (1994). Protein disaggregation mediated by heat-shock protein Hsp104. *Nature* **372:** 475–478.

Piszczek G, Rozycki J, Singh SK, *et al.* (2005). The molecular chaperone, ClpA, has a single high affinity peptide binding site per hexamer. *J Biol Chem* **280:** 12221–12230.

Qi R, Sarbeng EB, Liu Q, *et al.* (2013). Allosteric opening of the polypeptide-binding site when an Hsp70 binds ATP. *Nat Struct Mol Biol* **20:** 900–907.

Reidy M, Miot M, Masison DC (2012). Prokaryotic chaperones support yeast prions and thermotolerance and define disaggregation machinery interactions. *Genetics* **192:** 185–193.

Roman-Hernandez G, Hou JY, Grant RA, *et al.* (2011). The ClpS adaptor mediates staged delivery of N-end rule substrates to the AAA+ ClpAP protease. *Mol Cell* **43:** 217–228.

Romanova NV, Chernoff YO (2009). Hsp104 and prion propagation. *Protein Pept Lett* **16:** 598–605.

Rosenzweig R, Moradi S, Zarrine-Afsar A, *et al.* (2013). Unraveling the mechanism of protein disaggregation through a ClpB–DnaK interaction. *Science* **339:** 1080–1083.

Sanchez Y, Lindquist SL (1990). HSP104 required for induced thermotolerance. *Science* **248:** 1112–1115.

Sanchez Y, Taulien J, Borkovich KA, Lindquist S (1992). Hsp104 is required for tolerance to many forms of stress. *EMBO J* **11:** 2357–2364.

Sauer RT, Baker TA (2011). AAA+ proteases: ATP-fueled machines of protein destruction. *Annu Rev Biochem* **80:** 587–612.

Saunders C, Smith L, Wibrand F, *et al.* (2015). CLPB variants associated with autosomal-recessive mitochondrial disorder with cataract, neutropenia, epilepsy, and methylglutaconic aciduria. *Am J Hum Genet* **96:** 258–265.

Savistchenko J, Krzewska J, Fay N, Melki R (2008). Molecular chaperones and the assembly of the prion Ure2p *in vitro*. *J Biol Chem* **283:** 15732–15739.

Schlecht R, Erbse AH, Bukau B, Mayer MP (2011). Mechanics of Hsp70 chaperones enables differential interaction with client proteins. *Nat Struct Mol Biol* **18:** 345–351.

Schlieker C, Weibezahn J, Patzelt H, *et al.* (2004). Substrate recognition by the AAA+ chaperone ClpB. *Nat Struct Mol Biol* **11:** 607–615.

Schweizer RS, Aponte RA, Zimmermann S, *et al.* (2011). Fine tuning of a biological machine: DnaK gains improved chaperone activity by altered allosteric communication and substrate binding. *Chembiochem* **12:** 1559–1573.

Sen M, Maillard RA, Nyquist K, *et al.* (2013). The ClpXP protease unfolds substrates using a constant rate of pulling but different gears. *Cell* **155:** 636–646.

Seyffer F, Kummer E, Oguchi Y, *et al.* (2012). Hsp70 proteins bind Hsp100 regulatory M domains to activate AAA+ disaggregase at aggregate surfaces. *Nat Struct Mol Biol* **19:** 1347–1355.

Sharma D, Stanley RF, Masison DC (2009). Curing of yeast [URE3] prion by the Hsp40 cochaperone Ydj1p is mediated by Hsp70. *Genetics* **181:** 129–137.

Shorter J, Lindquist S (2004). Hsp104 catalyzes formation and elimination of self-replicating Sup35 prion conformers. *Science* **304:** 1793–1797.

Shorter J, Lindquist S (2006). Destruction or potentiation of different prions catalyzed by similar Hsp104 remodeling activities. *Mol Cell* **23:** 425–438.

Shorter J, Lindquist S (2008). Hsp104, Hsp70 and Hsp40 interplay regulates formation, growth and elimination of Sup35 prions. *EMBO J* **27:** 2712–2724.

Siddiqui SM, Sauer RT, Baker TA (2004). Role of the processing pore of the ClpX AAA+ ATPase in the recognition and engagement of specific protein substrates. *Genes Devt* **18:** 369–374.

Sielaff B, Tsai FT (2010). The M-domain controls Hsp104 protein remodeling activity in an Hsp70/Hsp40-dependent manner. *J Mol Biol* **402:** 30–37.

Singh SK, Grimaud R, Hoskins JR, *et al.* (2000). Unfolding and internalization of proteins by the ATP-dependent proteases ClpXP and ClpAP. *Proc Nat Acad Sci USA* **97:** 8898–8903.

Song HK, Hartmann C, Ramachandran R, *et al.* (2000). Mutational studies on HslU and its docking mode with HslV. *Proc Nat Acad Sci USA* **97:** 14103–14108.

Sousa MC, Trame CB, Tsuruta H, *et al.* (2000). Crystal and solution structures of an HslUV protease–chaperone complex. *Cell* **103:** 633–643.

Squires CL, Pedersen S, Ross BM, Squires C (1991). ClpB is the *Escherichia coli* heat shock protein F84.1. *J Bacteriol* **173:** 4254–4262.

Striebel F, Kress W, Weber-Ban E (2009). Controlled destruction: AAA+ ATPases in protein degradation from bacteria to eukaryotes. *Curr Opin Struct Biol* **19:** 209–217.

Sundar S, Baker TA, Sauer RT (2012). The I domain of the AAA+ HslUV protease coordinates substrate binding, ATP hydrolysis, and protein degradation. *Protein Science: A Publication of the Protein Society* **21:** 188–198.

Sundar S, McGinness KE, Baker TA, Sauer RT (2010). Multiple sequence signals direct recognition and degradation of protein substrates by the AAA+ protease HslUV. *J Mol Biol* **403:** 420–429.

Sweeny EA, Jackrel ME, Go MS, *et al.* (2015). The hsp104 N-terminal domain enables disaggregase plasticity and potentiation. *Mol Cell* **57:** 836–849.

Tang WK, Xia D (2012). Structural and functional deviations in disease-associated p97 mutants. *J Struct Biol* **179:** 83–92.

Tessarz P, Mogk A, Bukau B (2008). Substrate threading through the central pore of the Hsp104 chaperone as a common mechanism for protein disaggregation and prion propagation. *Mol Microbiol* **68:** 87–97.

Tipton KA, Verges KJ, Weissman JS. (2008). *In vivo* monitoring of the prion replication cycle reveals a critical role for Sis1 in delivering substrates to Hsp104. *Mol Cell* **32:** 584–591.

Tuite MF, Marchante R, Kushnirov V (2011). Fungal prions: structure, function and propagation. *Top Curr Chem* **305:** 257–298.

Tuite MF, Serio TR (2010). The prion hypothesis: from biological anomaly to basic regulatory mechanism. *Nat Rev Mol Cell Biol* **11:** 823–833.

Vass RH, Chien P (2013). Critical clamp loader processing by an essential AAA+ protease in Caulobacter crescentus. *Proc Nat Acad Sci USA* **110:** 18138–18143.

Wang F, Mei Z, Qi Y, *et al.* (2011). Structure and mechanism of the hexameric MecA–ClpC molecular machine. *Nature* **471:** 331–335.

Wang J, Song JJ, Franklin MC, *et al.* (2001a). Crystal structures of the HslVU peptidase–ATPase complex reveal an ATP-dependent proteolysis mechanism. *Structure* **9:** 177–184.

Wang J, Song JJ, Seong IS, *et al.* (2001b). Nucleotide-dependent conformational changes in a protease-associated ATPase HsIU. *Structure* **9:** 1107–1116.

Watanabe YH, Nakazaki Y, Suno R, Yoshida M (2009). Stability of the two wings of the coiled-coil domain of ClpB chaperone is critical for its disaggregation activity. *Biochem J* **421:** 71–77.

Weart RB, Nakano S, Lane BE, *et al.* (2005). The ClpX chaperone modulates assembly of the tubulin-like protein FtsZ. *Mol Microbiol* **57:** 238–249.

Weber-Ban EU, Reid BG, Miranker AD, Horwich AL (1999). Global unfolding of a substrate protein by the Hsp100 chaperone ClpA. *Nature* **401:** 90–93.

Weibezahn J, Tessarz P, Schlieker C, *et al.* (2004). Thermotolerance requires refolding of aggregated proteins by substrate translocation through the central pore of ClpB. *Cell* **119:** 653–665.

Wendler P, Ciniawsky S, Kock M, Kube S (2012). Structure and function of the AAA+ nucleotide binding pocket. *Biochim Biophys Acta* **1823:** 2–14.

Wendler P, Shorter J, Plisson C, *et al.* (2007). A typical AAA+ subunit packing creates an expanded cavity for disaggregation by the protein-remodeling factor Hsp104. *Cell* **131:** 1366–1377.

Wendler P, Shorter J, Snead D, *et al.* (2009). Motor mechanism for protein threading through Hsp104. *Mol Cell* **34:** 81–92.

Werbeck ND, Schlee S, Reinstein J (2008). Coupling and dynamics of subunits in the hexameric AAA+ chaperone ClpB. *J Mol Biol* **378:** 178–190.

Wickner RB, Edskes HK, Bateman D, *et al.* (2011). The yeast prions [PSI+] and [URE3] are molecular degenerative diseases. *Prion* **5:** 258–262.

Williams B, Bhat N, Chien P, Shapiro L (2014). ClpXP and ClpAP proteolytic activity on divisome substrates is differentially regulated following the *Caulobacter* asymmetric cell division. *Mol Microbiol* **93:** 853–866.

Winkler J, Tyedmers J, Bukau B, Mogk A (2012a). Chaperone networks in protein disaggregation and prion propagation. *J Struct Biol* **179:** 152–160.

Winkler J, Tyedmers J, Bukau B, Mogk A (2012b). Hsp70 targets Hsp100 chaperones to substrates for protein disaggregation and prion fragmentation. *J Cell Biol* **198:** 387–404.

Wojtyra UA, Thibault G, Tuite A, Houry WA (2003). The N-terminal zinc binding domain of ClpX is a dimerization domain that modulates the chaperone function. *J Biol Chem* **278:** 48981–48990.

Wortmann SB, Zietkiewicz S, Kousi M, *et al.* (2015). CLPB mutations cause 3-methylglutaconic aciduria, progressive brain atrophy, intellectual disability, congenital neutropenia, cataracts, movement disorder. *Am J Hum Genet* **96:** 245–257.

Xia D, Esser L, Singh SK, *et al.* (2004). Crystallographic investigation of peptide binding sites in the N-domain of the ClpA chaperone. *J Struct Biol* **146:** 166–179.

Yakamavich JA, Baker TA, Sauer RT (2008). Asymmetric nucleotide transactions of the HslUV protease. *J Mol Biol* **380:** 946–957.

Yamasaki T, Nakazaki Y, Yoshida M, Watanabe YH (2011). Roles of conserved arginines in ATP-binding domains of AAA+ chaperone ClpB from *Thermus thermophilus*. *FEBS J* **278:** 2395–2403.

Yuan AH, Garrity SJ, Nako E, Hochschild A (2014). Prion propagation can occur in a prokaryote and requires the ClpB chaperone. *eLife* **3:** e02949.

Zabrocka E, Wegrzyn K, Konieczny I (2014). Two replication initiators — one mechanism for replication origin opening? *Plasmid* **76C:** 72–78.

Zeth K, Ravelli RB, Paal K, *et al.* (2002). Structural analysis of the adaptor protein ClpS in complex with the N-terminal domain of ClpA. *Nat Struct Biol* **9:** 906–911.

Zeymer C, Fischer S, Reinstein J (2014). Trans-Acting arginine residues in the AAA+ chaperone ClpB allosterically regulate the activity through inter- and intradomain communication. *J Biol Chem* **289:** 32965–32976.

Zhang T, Ploetz EA, Nagy M, *et al.* (2012). Flexible connection of the N-terminal domain in ClpB modulates substrate binding and the aggregate reactivation efficiency. *Proteins* **80:** 2758–2768.

Zhang T, Ye Y (2014). The final moments of misfolded proteins *en route* to the proteasome. *DNA Cell Biol* **33:** 477–483.

Zhou Y, Gottesman S, Hoskins JR, *et al.* (2001). The RssB response regulator directly targets sigma(S) for degradation by ClpXP. *Genes Devt* **15:** 627–637.

Zhuravleva A, Clerico EM, Gierasch LM (2012). An interdomain energetic tug-of-war creates the allosterically active state in Hsp70 molecular chaperones. *Cell* **151:** 1296–1307.

Zhuravleva A, Gierasch LM (2011). Allosteric signal transmission in the nucleotide-binding domain of 70-kDa heat shock protein (Hsp70) molecular chaperones. *Proc Nat Acad Sci USA* **108:** 6987–6992.

Zietkiewicz S, Krzewska J, Liberek K (2004). Successive and synergistic action of the Hsp70 and Hsp100 chaperones in protein disaggregation. *J Biol Chem* **279:** 44376–44383.

Zietkiewicz S, Lewandowska A, Stocki P, Liberek K (2006). Hsp70 chaperone machine remodels protein aggregates at the initial step of Hsp70–Hsp100-dependent disaggregation. *J Biol Chem* **281:** 7022–7029.

Zolkiewski M (1999). ClpB cooperates with DnaK, DnaJ, and GrpE in suppressing protein aggregation: A novel multi-chaperone system from *Escherichia coli*. *J Biol Chem* **274:** 28083–28086.

Zolkiewski M, Zhang T, Nagy M (2012). Aggregate reactivation mediated by the Hsp100 chaperones. *Arch Biochem Biophys* **520:** 1–6.

Zuiderweg ER, Bertelsen EB, Rousaki A, *et al.* (2013). Allostery in the Hsp70 chaperone proteins. *Top Curr Chem* **328:** 99–153.

Chapter 6

Structural Insights into Hsp90 Function

*Elaine Kirschke, Daniel Elnatan, Kliment Verba, and David A. Agard**

Abstract

Hsp90 is a ubiquitous ATP-dependent molecular chaperone that accounts for 1–2% of the cytosolic proteins and is essential for viability in eukaryotes. It interacts with as much as 10% of proteome and facilitates the late stages of protein folding. It is unique among the chaperones, in that it is generally not required for *de novo* folding of most proteins, but facilitates the folding and activation of a specific subset of "client" proteins. A subset of these clients continue to be regulated by Hsp90 in their mature form, with Hsp90 being required for stability and the continual maintenance of their active state. ATP hydrolysis by Hsp90 is essential in eukaryotes, and is required for client activation. This review focuses on Hsp90 structure, its dramatic conformational rearrangements and the mechanistic details of ATP hydrolysis, and how all these potentially relate to client binding and activation.

Hsp90 is a highly dynamic homodimer, with large structural changes resulting from rotations of the mostly rigid three domains within each protomer. Many distinct conformations of Hsp90 make up its conformational ensemble, with the equilibrium between these states uniquely tuned to the individual homologues, and are only biased by nucleotide. In the eukaryotic cytosol, Hsp90 is highly

*Howard Hughes Medical Institute and Department of Biochemistry and Biophysics, University of California, San Francisco, CA 94158, USA

regulated by a set of cochaperones that stabilize and promote specific conformations during the chaperone cycle, as well as acting as an adaptor for particular classes of clients. How these conformational states and the energy available from ATP hydrolysis are utilized for client activation is still not clear, but recent insights provided by the crystal structure for an asymmetric ATP bound state of the mitochondrion-specific Hsp90 (TRAP1) provides new models for how ATP hydrolysis could be utilized to directly promote client remodeling.

Only recently have the principles governing client interactions begun to emerge. Hsp90 preferentially interacts with proteins in a nonnative partially folded state, with the degree to which a protein accesses these partially folded states as the major determinant for the strength of the interaction. The identification of the client binding surfaces on Hsp90 reveals a substrate interaction mode distinct from other chaperones, in which Hsp90 utilizes large exposed surfaces areas that are modulated by changes in domain orientation, with interaction strengths only weakly coupled to the nucleotide states. While this suggests that ATP hydrolysis is not likely utilized to govern client binding and release, it is clearly utilized for collaboration with the Hsp70 system, in which Hsp70 and Hsp90 act synergistically to refold and regulate client function and activity in an entirely ATP-hydrolysis-dependent process. The high degree of biochemical and structural coordination between the Hsp90 and Hsp70 cycles provides new insights into how the two chaperones collaborate in protein quality control.

Introduction

Hsp90 is one of the most highly conserved heat shock proteins, present from bacteria to mammals (but absent from archaea). Under normal conditions, it is highly expressed and is generally one of the most abundant proteins in eukaryotic cytosols (~1–2%). Under stress conditions, Hsp90 levels are upregulated, approximately two fold in eukaryotes, with Hsp90 playing an essential role in cell survival under heat shock (Borkovich *et al.*, 1989). However, Hsp90's function extends well beyond stress tolerance. Unlike the Hsp70 and Hsp60 chaperones, Hsp90 is generally not required for *de novo* folding of most proteins, but instead is required by a specific subset of proteins referred to as "clients," for which Hsp90 facilitates the late stages of folding (Nathan *et al.*, 1997) and activation to interact with ligands or other proteins (Pratt *et al.*, 2008; Richter and Buchner, 2001; Young, 2001).

In bacteria, Hsp90 is nonessential, but growing evidence indicates that it plays a role in a variety of host processes, including ribosomal biogenesis, DNA replication, and CRISPR function in *E. coli* and phycobilisome assembly in cyanobacteria (Grudniak *et al.*, 2015; Motojima-Miyazaki *et al.*, 2010; Sato *et al.*, 2010; Yosef *et al.*, 2011). By contrast, Hsp90 is essential in eukaryotes and, with the assistance of numerous cofactors often referred to as cochaperones, appears necessary for the activation of perhaps 10% of the proteome (Zhao *et al.*, 2005). While Hsp90's clients are numerous and diverse, they are disproportionately involved in signal transduction, including many kinases, steroid hormone receptors and E3 ligases. Additionally, its clients are involved in other critical cellular processes, such as cell cycle progression, telomere maintenance, apoptosis, protein trafficking and innate immunity, placing Hsp90 at a hub of cellular homeostasis (Taipale *et al.*, 2010). At least one way Hsp90 facilitates the late stages of folding is by receiving partially folded intermediates from Hsp70. In coordinating with the Hsp70 system, Hsp90 also facilitates protein quality control by assisting in the targeting of proteins for degradation by the proteasome (Edkins, 2015; Theodoraki and Caplan, 2012). Furthermore, at an organismal level Hsp90 plays a unique role as a "capacitor" for evolution. By stabilizing quasistable near native states, it can buffer the effects of structurally destabilizing mutations, therefore allowing the acquisition of new traits (Queitsch *et al.*, 2002; Rutherford and Lindquist, 1998).

Like Hsp60/GroEL and Hsp70 chaperones, Hsp90 binds and hydrolyzes ATP, with nucleotides promoting dramatic conformational rearrangements in Hsp90 (Plate 6.1B). In eukaryotes, ATP hydrolysis is

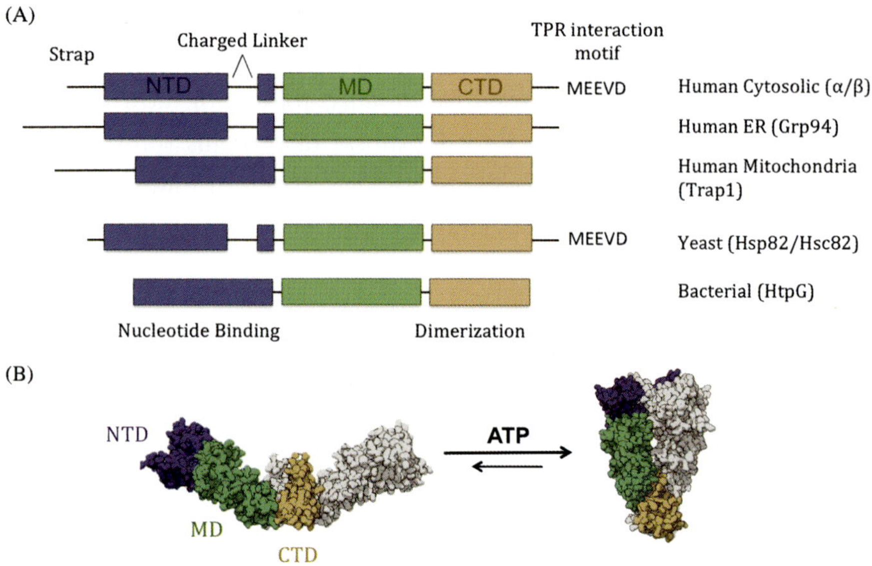

Plate 6.1. ***Architecture of Hsp90*****.** **A**. Domain architecture of different Hsp90 homologues. The NTD is in blue, the MD in green, and the CTD in brown. **B**. Hsp90 forms high affinity homodimers through CTD dimerization. ATP binding to the NTDs stabilizes a secondary dimerization interface between the NTDs.

essential for client activation *in vivo* (Obermann *et al.*, 1998; Panaretou *et al.*, 1998), with Hsp90's ATP cycle heavily regulated by cochaperones (Li *et al.*, 2012). Given that many Hsp90 clients are oncoproteins, in particular p53, Cdk4, c-Src/v-Src and HER-2/HER-3, Hsp90's nucleotide binding pocket has been intensely targeted by small molecules as potential cancer therapeutics (Trepel *et al.*, 2010). Hsp90's implication in other diseases, such as Alzheimer's, makes it an even more attractive drug target; however, full utility of Hsp90 inhibitors has been limited by our poor understanding of how Hsp90 interacts with clients, hampering the development of more client-specific inhibitors.

In vitro, Hsp90 moderately suppresses thermally induced aggregation, supporting Hsp90's role in heat shock. However, it does so by binding partially folded intermediates in an ATP-independent manner (Jakob *et al.*, 1995; Wiech *et al.*, 1992). The lack of ATP dependence suggests that this aggregation-suppressing function does not fully emulate the essential ATP-dependent client activation functionality required *in vivo*. It is generally expected that energy from ATP hydrolysis is somehow focused on driving client conformational rearrangements required for activation. However, this has yet to be explicitly demonstrated, due to the inherent instability of Hsp90 clients and the extreme challenges associated with such *in vitro* investigations. As discussed in more detail below, recent studies unexpectedly show that hydrolysis-dependent reactivation of a steroid hormone receptor results from Hsp90 reversing the inhibitory effect of Hsp70 (Kirschke *et al*., 2014). This reveals ATP hydrolysis as being integral to regulating client transfer from Hsp70, which in some cases is a critical aspect of client activation.

Cellular Localization

Hsp90 is found in almost every compartment of eukaryotic cells. Most eukaryotic genomes encode for multiple organelle-specific Hsp90 proteins. Humans possess four Hsp90 isoforms: two cytosolic, (Hsp90 α/β), a mitochondrion-specific Hsp90 (TRAP1) and an ER-specific Hsp90 (Grp94). Cytosolic Hsp90 is considered the canonical isoform, and although predominantly found in the cytoplasm, it is transported to other compartments of the cell, including the extracellular matrix where it facilitates tumor cell invasiveness (Eustace *et al*., 2004). In addition, while Hsp90 does not possess a nuclear localization sequence, about 2–3% of cellular Hsp90 is found in the nucleus, where it is thought to be cotransported with clients. Hsp90 is now emerging as a central player in many nuclear functions, such as telomere maintenance, pol II pausing and regulation of transcriptional complexes (DeZwaan and Freeman, 2010; Freeman and Yamamoto, 2002; Sawarkar and Paro, 2013).

Architecture of Hsp90

Hsp90 exists as a constitutive homodimer, with dimerization promoting ATP hydrolysis, and is essential for proper *in vivo* function (Wayne and Bolon, 2007; Wegele *et al*., 2003). Each monomer of ~90kDa consists of three major conserved domains: an N-terminal nucleotide binding domain (NTD), a middle domain (MD) and a C-terminal dimerization domain (CTD) (Plate 6.1A). The high affinity of the CTD dimerization interface ensures that Hsp90 exists as a dimer under physiological conditions with Hsp90 monomers coming together to form a V-shape (Harris *et al.*, 2004; Shiau *et al.*, 2006). ATP binding stabilizes a "closed" state in which the NTDs form a secondary dimerization interface (Plate 6.1B) (Ali *et al.*, 2006). Besides the highly conserved core domains, different Hsp90 species and organellar specific homologues have evolved additional features that confer additional functionality and distinct regulation. Such homologue-specific features include the length of the NTD extension referred to as the "strap" which allows for tuning of Hsp90's hydrolysis cycle, and a C-terminal extension that contains an MEEVD motif utilized in the eukaryotic cytosolic Hsp90s to interact with numerous TPR-domain-containing cochaperones (Plates 6.1A and 6.2B). These features will be discussed later in more detail. Additionally, most homologues contain a charged flexible loop inserted near the end of the NTD domain, referred to as the charged linker. The length and sequence of the charged linker is highly variable, ranging from 10 residues in *E. coli* to 56 in yeast, 63 in human and 95 in *Plasmodium falciparum*, with most of its length being nonessential in yeast (Hainzl *et al.*, 2009). The function of the charged linker is not well understood, although altering it does perturb chaperone activity (Jahn *et al.*, 2014; Tsutsumi *et al.*, 2012) and in yeast it imparts a solubility-enhancing function for particularly aggregation-prone clients during ATP-dependent Hsp90 maturation (Pursell *et al.*, 2012).

Structurally, Hsp90 belongs to a class of functionally divergent but structurally related split ATPases referred to as the GHKL (gyrase, Hsp90, histindine kinase, MutL) superfamily, which contain a Bergerat ATP-binding fold (Dutta and Inouye, 2000). A distinct feature of this protein class is the relatively long and flexible lid that covers the nucleotide binding pocket (Plates 6.2A and 6.2B). For Hsp90, the conserved lid is open in apo crystal structures and occludes NTD dimerization

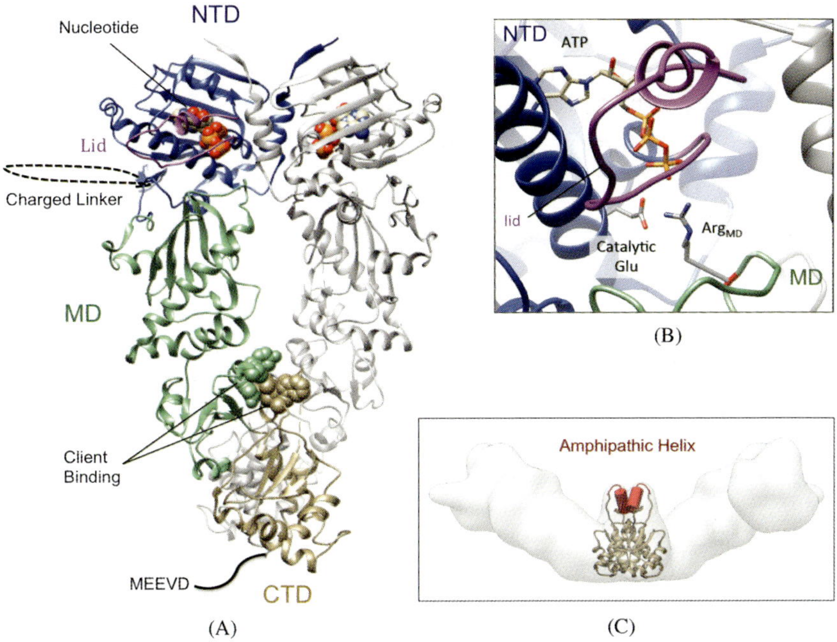

Plate 6.2. ***Structure of Hsp90*****. A**. Crystal structure of the closed ATP-bound state of the Hsp90 dimer, with one monomer in gray and the opposing monomer colored by domain as in Plate 1 (Ali *et al.*, 2006). The NTD nucleotide lid is highlighted in purple. Nucleotide and client binding residues are rendered as spheres. The approximate location of the charged linker, which was excluded from the crystallographic construct, is shown by the black dotted line. Also indicated is the approximate location of the flexible MEEVD C-terminal tail motif not resolved in the crystal structure. **B**. A zoom of the nucleotide binding pocket bound with the nonhydrolysable ATP analogue (AMPPNP), coordinated by the NTD catalytic glutamate along with Arg_{MD} (yeast-R380), which directly contacts the γ-phosphate. **C**. Highlighted is the CTD amphipathic helix (red) protruding into the cleft of Hsp90 for the extended state of Hsp90.

(Richter *et al.*, 2006) but, as shown in the closed Hsp90 crystal structure, lid closure permits NTD dimerization, which in turn stabilizes the lid in a closed conformation such that the lid wraps around the β- and γ-phosphates and makes direct contact with the ribose sugar of the bound ATP (Plates 6.2B and 6.3). Another signature feature of the GHKL ATPases is the conserved positively charged residue in the adjacent domain, referred to as the MD arginine (Arg_{MD}) for Hsp90, which extends up

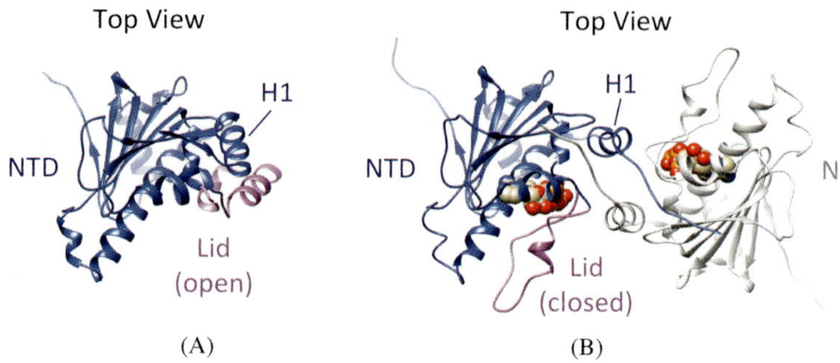

Plate 6.3. ***NTD dimerization stabilizes lid closure***. **A**. The crystal structure for the NTD of apo Hsp90 with the nucleotide lid in the open conformation inhibits dimerization (Prodromou *et al*., 1997). **B**. The NTD–dimer formed in the closed AMP-PNP-bound Hsp90 structure (Ali *et al*., 2006), with the opposing monomer in gray. Lid closure permits dimerization about helix1 (H1) and the NTD dimerization contacts stabilize the lid in the closed conformation.

and projects into the nucleotide binding pocket to make critical contact with the ATP γ-phosphate (Plate 6.2B).

Despite extensive structural information, the mechanism for client recognition by Hsp90 remains mostly enigmatic, with surfaces utilized for client binding only recently beginning to emerge (Karagöz and Rüdiger, 2015; Street *et al*., 2011b, 2014). Unlike other ATP-dependent chaperones, Hsp90's structure does not reveal an obvious location for client binding. Hsp90's elongated surface lacks any substantial pocket to enclose a substrate protein, and hydrophobic regions are spread relatively uniformly across the surface. The most suspicious feature suggesting substrate interactions is a short amphipathic helix that protrudes into the central cleft (Plate 6.2C) (Harris *et al*., 2004). As later discussed, this CTD amphipathic helix, and a distinct region in the MD, have since been confirmed as key determinants in client binding (Plate 6.2A) (Genest *et al*., 2013; Street *et al*., 2011b). In addition, clients have been documented to make extensive contacts across Hsp90, with client contacts reported to extend along the MD and NTD (Karagöz *et al*., 2014).

Conformational Dynamics of Hsp90

While the first crystal structure of Hsp90's NTD was reported in 1997 (Prodromou *et al*., 1997; Stebbins *et al*., 1997), obtaining full-length structures frustrated structural biologists for nearly a decade before they began to emerge. Intriguingly, the structures obtained from diverse homologues with various nucleotides revealed dramatically different conformations (Ali *et al*., 2006; Dollins *et al*., 2007; Shiau *et al*., 2006). The difficulty in obtaining full-length structures is further highlighted by it being nearly another decade before the next structure was revealed (Lavery *et al*., 2014). It is now clear that despite the differences in homologues, the various structures provide snapshots of distinct, universally conserved states Hsp90 assumes during its nucleotide-driven chaperone cycle, and that Hsp90's remarkable structural flexibility contributes to the considerable difficulty in attaining full-length crystal structures (Southworth and Agard, 2008). The *tour de force* crystallography from multiple labs combined with structural information obtained from small angle X-ray scattering (SAXS) and electron microscopy (EM) has yielded a more complete picture for the trajectory of conformational changes that occur during Hsp90 chaperone cycles (Plate 6.4). These large structural rearrangements mostly result from rigid body rotations of the individual domains about the NTD–MD or MD–CTD interface, indicating relatively weak coupling at the primary domain boundaries. More subtle changes can also be seen within the MD, suggesting that it is not a monolithic structure but that it should be divided into upper and lower subdomains (Lavery *et al*., 2014). This is mirrored by the strongest conservation within the GHKL superfamily covering the NTD and the upper part of the MD (Meyer *et al*., 2003).

For the apo state of Hsp90, solution studies by SAXS and negative stain EM revealed that Hsp90 favors a more extended state than observed in the apo state crystal structure, and that in the absence of NTD dimerization Hsp90 exhibits significant flexibility around the MD–CTD interface (Krukenberg *et al*., 2008; Shiau *et al*., 2006). In its most

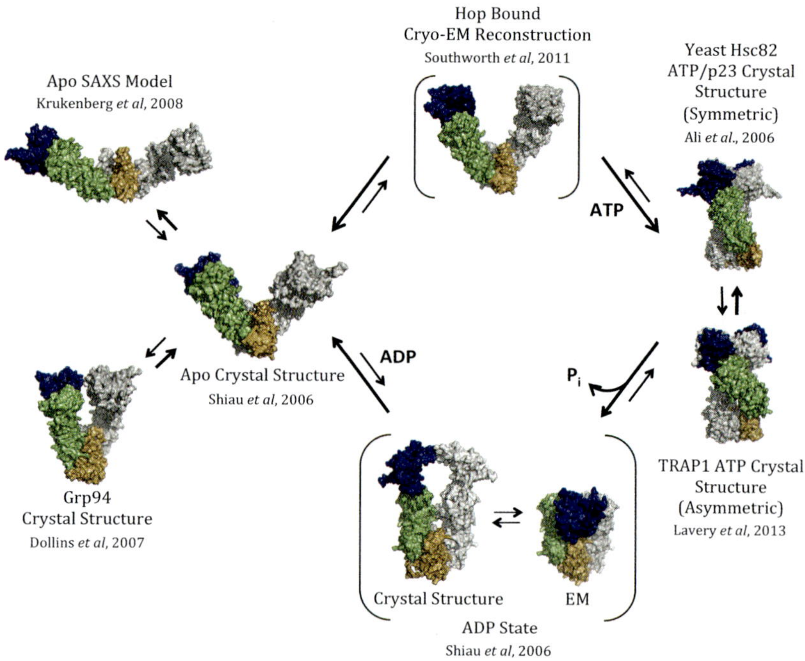

Plate 6.4. ***Full-length structures of Hsp90 reveal dramatic conformational rearrangements, mostly resulting from rigid body rearrangements of the individual domain.*** Shown are the numerous structures of Hsp90 determined in different states with domains colored as in Plate 1. Brackets indicate transient states.

extended state, Hsp90's cleft spans up to 130 Å. Significantly, this large flexibility in cleft size is thought to enable Hsp90 to accommodate the broad range of client sizes from α-synuclein to telomerase (14–290 kDa), and indeed this flexibility is critical for Hsp90 to accommodate binding of a model client protein (Street *et al.*, 2014).

Surprisingly, even without nucleotide, Hsp90 can spontaneously adopt all states, indicating that it exists in a dynamic equilibrium, which is biased but not uniquely determined by nucleotide (Southworth and Agard, 2008). For example, while ATP binding stabilizes the closed state, it does not increase the rate of closure. Furthermore, the conformational equilibrium and closure kinetics are tuned differently for the different homologues (Southworth and Agard, 2008). Some homologues are tuned to respond to client binding and environmental cues such as pH and osmolytes (Krukenberg *et al.*, 2009; Street *et al.*, 2009, 2011a), or temperature (Partridge *et al.*, 2014). Sophisticated single molecule studies suggest even more complex differences in the behavior of the different species (Ratzke *et al.*, 2014, 2012). Despite the relatively complete picture for Hsp90's conformational cycle provided by the extensive structural work, it has remained a major challenge to elucidate how these dramatic conformational rearrangements are utilized for client activation. While for other ATP-driven molecular chaperones nucleotide binding, hydrolysis and release induce conformational changes clearly responsible for controlling the timing of substrate loading and release, this correlation has been far less clear for Hsp90.

Mechanism of ATP Hydrolysis

ATP binding and hydrolysis on Hsp90 is tightly coupled to its conformation changes, with ATP binding stabilizing the closed state and, in turn, the closed state promoting ATP hydrolysis. The rate of ATP hydrolysis by Hsp90 is extremely slow — 1 ATP per minute for yeast and 0.05 for humans (McLaughlin *et al.*, 2002; Panaretou *et al.*, 1998; Richter *et al.*, 2008), the reason being the dependence on very slow conformational changes leading to the closed state (Graf *et al.*, 2009; Hessling *et al.*, 2009). As exemplified by the 20-fold difference in ATP hydrolysis rates between yeast and human Hsp90, hydrolysis rates vary significantly among the different homologues, with the rates reflecting the tuning of the conformational equilibrium for the individual homologues (Southworth and Agard, 2008). This tuning is highly optimized for the specific function of the individual homologue, with both increases and decreases in the ATP hydrolysis rate having detrimental *in vivo* effects (Hawle *et al.*, 2006).

While the rates for converting between the different conformations may differ, the conformational changes required for hydrolysis to occur appear universal. Plate 6.5 illustrates a proposed mechanistic ATP hydrolysis cycle for Hsp90 in which the important structural changes

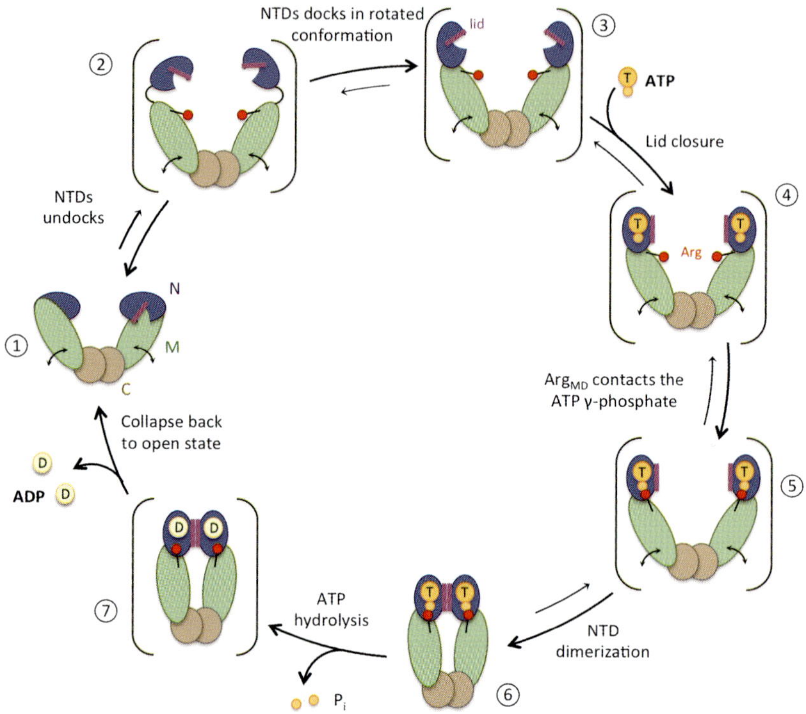

Plate 6.5. ***Mechanistic cycle for ATP hydrolysis on Hsp90***. At step 1, Hsp90's open conformation exhibits flexibility around the MD (green) and CTD (brown) interfaces, and the downward-docked state of the NTD (blue) inhibits the nucleotide lid (purple) from closing. The NTDs transiently undock (step 2), allowing the NTD to dock in an upward-rotated state (step 3, Plate 6), which in turn permits nucleotide lid closure (step 4, Plate 7). In step 3, with the NTD docked in an upward-rotated state, the open state of the lid restricts the movement of the MD such that Arg_{MD} (red) is restricted to a downward conformation (Plate 7). Lid closure allows the MD to rearrange such that Arg_{MD} can flip up to contact the ATP γ-phosphate in the NTD (step 5, Plate 7). Lid closure fully forms the secondary NTD-dimerization interface. Dimerization of the NTDs stabilizes the closed state of Hsp90, in which the MD and CTD orientations become restrained to a compact conformation. ATP hydrolysis breaks the contact between the ATP γ-phosphate and Arg_{MD} (step 7), resulting in Hsp90 relaxing back to the open state. Brackets indicate transient states.

that must occur for Hsp90 to transition from the hydrolysis-inactive open state (step 1) to the closed hydrolysis-competent state (step 6) are depicted. In the open state (step 1), the NTDs are docked in a downward-facing state that prevents the nucleotide lid from closing (Shiau

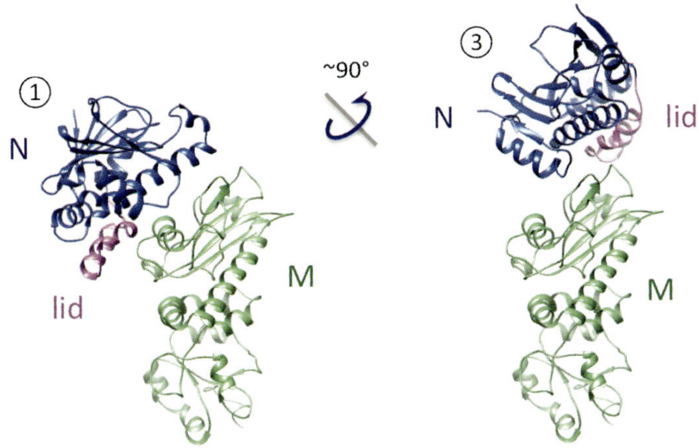

Plate 6.6. ***NTD–MD rotation***. Depicted is a structural model for the rotation of the NTD (blue) about the MD (green) interface based on alignments from the open apo state crystal structure (left) and the closed state crystal structure (right).

et al., 2006). Single molecule pulling experiments indicate that the NTD and MD domains are weakly bound (Jahn *et al.*, 2014), supporting a model where the NTDs transiently undock, allowing the NTD to then dock in an upward-rotated state (step 3, Plate 6.6). Evidence for this transient open state is provided by the cryo-EM structure of Hsp90 stabilized by the cochaperone Hop, in which the NTDs are clearly in an upward-rotated state (Plate 6.17A) (Southworth and Agard, 2011). In the NTD rotated state (step 3), the open state of the lid restricts the movement of the MD such that the Arg_{MD} loop is restrained in a downward conformation (Plate 6.7). Unlike in the unrotated state, the NTD rotated state permits the lid to close over the nucleotide pocket (step 4, Plate 6.3). Closure of the lid in turn allows the MD arginine loop to rearrange such that the Arg_{MD} can flip up and contact the γ-phosphate of the ATP bound in the NTD (step 5, Plate 6.5). The combination of the upward rotation of the NTD and the lid closure aligns the NTD dimerization interfaces such that compaction of the Hsp90 dimer about the MD–CTD interface results in the formation of the secondary dimerization interface (Plate 6.8).

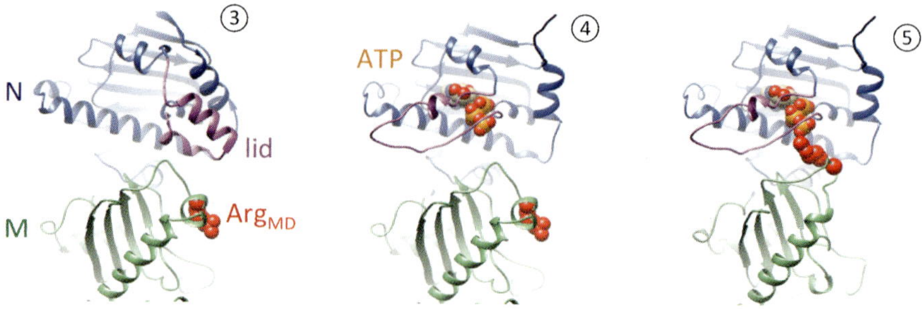

Plate 6.7. ***Nucleotide lid closure and MD rearrangement***. With the NTD docked in an upward-rotated state, the open lid restricts the movement of the MD such that Arg_{MD} is restricted to a downward conformation (left). In the rotated NTD state the nucleotide lid can close. Lid closure (middle) permits restructuring of the Arg_{MD} loop such that Arg_{MD} flips up to make direct contact with the ATP γ-phosphate (right).

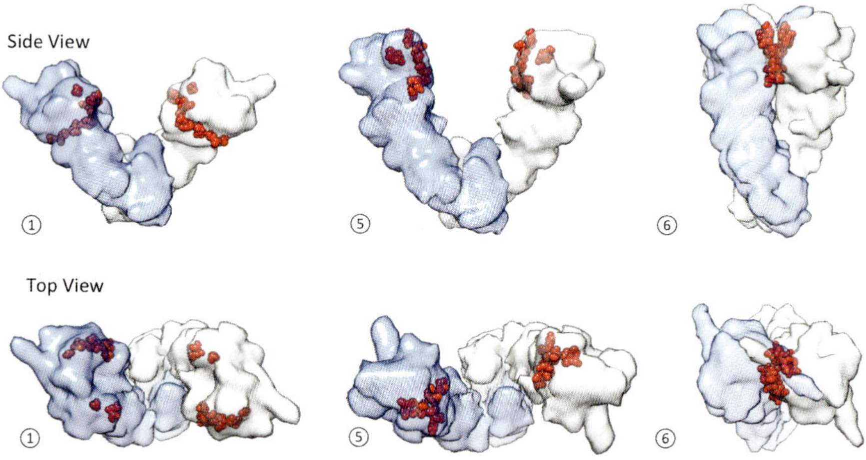

Plate 6.8. ***NTD rotation is required for alignment of the NTD-dimerization interface***. Residues that form contacts in the closed conformation (right) are shown as red spheres, and are significantly out of alignment in the open unrotated state (left) as compared to the rotated NTD state (middle).

NTD dimerization stabilizes both the closed state of the lids and the closed Hsp90 conformation, with the MD and CTD orientations restrained to a compact conformation. Arg_{MD} contributes significantly to the stability of the closed state. Once thought to be a catalytic residue

(Meyer *et al.*, 2003), Arg_{MD} functions more as an ATP sensor that helps stabilize the ATP-hydrolysis-competent conformation, with the extra contact between the MD and the NTD provided by Arg_{MD} and the γ-phosphate stabilizing the intrinsically unstable rotated NTD–MD orientation (Cunningham *et al.*, 2012). In this regard, trans-monomer contacts from the opposing NTD pack against the Arg_{MD} loop in the closed state, stabilizing Arg_{MD} in the upward state that contacts the γ-phosphate (Plate 6.9) (Cunningham *et al.*, 2008). It should be emphasized that the motions converting between the extended open and closed states are highly concerted, resulting in the transient nature of the intermediate states along the pathway between the open (step 1) and closed (step 6) states. Once, in the closed state, biochemical evidence suggests at least two distinct closed states for yeast Hsp90 (Hessling *et al.*, 2009). While one could be explained by the p23-bound closed state (Ali *et al.*, 2006), the exact structure of the other is unclear and is not necessarily the asymmetric closed state described later. Upon ATP hydrolysis, the contacts in both protomers between the ATP γ-phosphate and Arg_{MD} are broken, resulting in relaxation back to the open state. The exact sequence of posthydrolysis events leading to reopening is not well known, but the data are most consistent with reopening primarily occurring from the fully

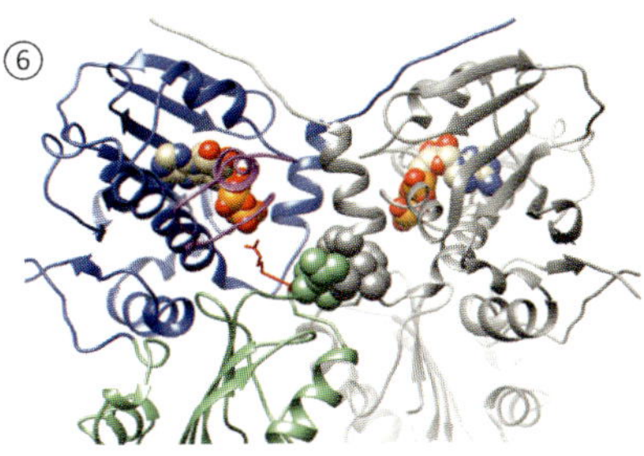

(A)

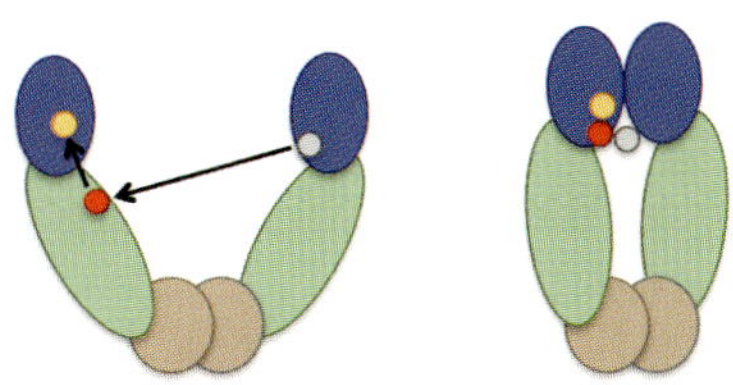

(B)

Plate 6.9. ***Cross-monomer contacts in the closed state stabilize Arg_{MD} and the rotated NTD state*. A**. The crystal structure of Hsp90 in the closed state with the hydrophobic residues involved in the *trans* contact that stabilize the Arg_{MD} loop is depicted as spheres. The Arg_{MD} that makes contact with the ATP in the connected NTD is rendered as sticks. **B**. Schematic depicting the contacts made in **A**.

hydrolyzed state (Partridge *et al.*, 2014). Moreover, the extent to which the transient compact ADP state observed by EM (but not detected by SAXS) is utilized for client release is unclear.

An important feature not depicted in Plate 6.5 is an N-terminal segment of the NTD that forms the first conserved secondary structure across Hsp90s (β-strand 1). This β-strand forms interactions with the NTD of either the *cis* protomer in the apo open state or with the *trans* protomer in the closed state. These interactions contribute strongly to the kinetic barriers to closing and reopening (Plate 6.10). Strand 1 both stabilizes and determines the rotation of the adjacent helix 1,

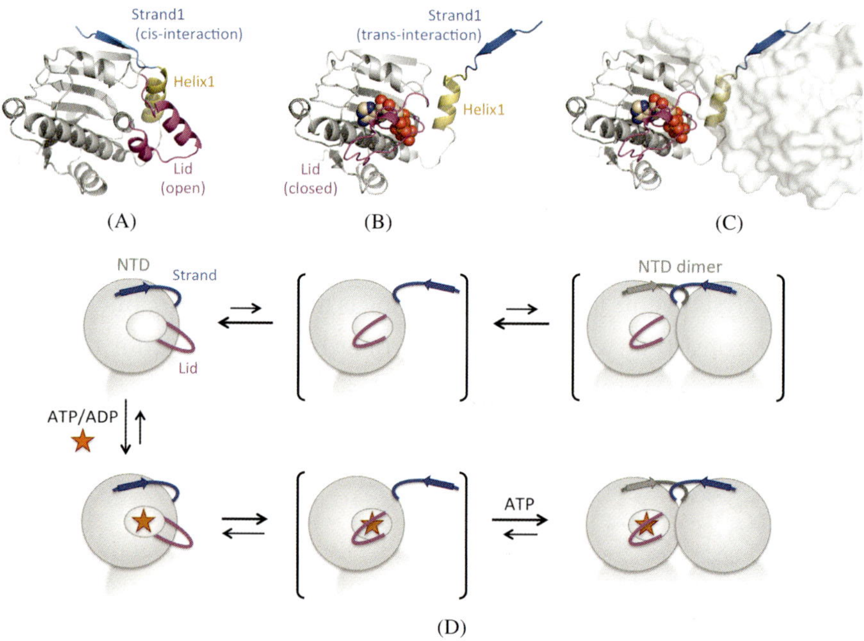

Plate 6.10. ***Anticorrelation between the strand and the lid*.** **A.** In the apo state of the NTD, the lid (purple) is open and β-strand 1 (blue) folds back onto the connected NTD to form the *cis* strap interaction. **B.** In the closed AMPPNP-bound crystal structure, the lid is closed while the β-strand flips away from the connected NTD and instead forms an equivalent interaction with the *trans* NTD monomer **C.**, referred to as the strand swap. **D.** Schematic illustrating the anticorrelation between the lid and the strand swap. In the absence of ATP, the NTD–dimer is unstable due to the inability of Arg_{MD} to stabilize the required N–M rotation required for the closed state, which in turn stabilizes the closed lid.

which interacts with the open lid and occludes the NTD–dimerization interface on helix 1. Consequently, nucleotide binding promotes lid closure, allowing helix 1 to rotate and pack against its *trans* counterpart in the NTD–dimer. Truncation of strand 1 accelerates ATP hydrolysis (Richter *et al.*, 2002), indicating that the release of the strand from the *cis* apo interactions is rate-limiting.

Many homologues contain N-terminal segments that extend beyond strand 1; however, the length and sequence of this region are highly variable (Plate 6.11). This ranges from the lack of N-terminal extensions in bacteria and yeast to 122 residues found in Hsp90α of higher primates (Tripathi and Obermann, 2013). In addition to having a signal sequence for cellular compartment localizations, the features within this N-terminal region are unique to each Hsp90 isoform and may confer homologue specific regulation. The retention of these features throughout evolution hints at the adaptations of Hsp90 in different cellular environments.

In TRAP1, this extra N-terminal domain (termed the "strap") imparts an additional large thermal barrier between the open apo state

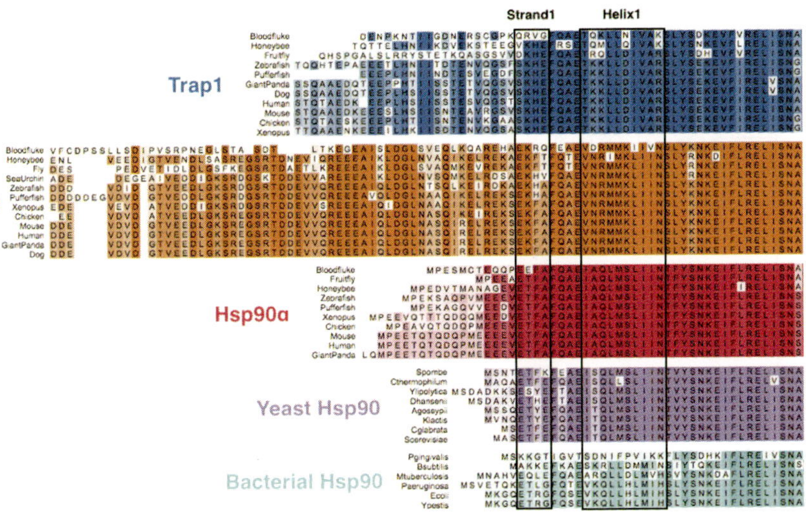

Plate 6.11. ***Evolution of N-terminal extension in Hsp90***. Multiple sequence alignment of the N-terminal of Hsp90 homologues from various eukaryotic and bacterial origins. Sequences of Grp94 and TRAP1 shown are from the mature peptide (excluding the signal peptide).

and the ATP-bound closed state (Partridge *et al.*, 2014). The strap does this by forming extensive interactions along the NTD in both apo and closed states, where it makes *cis* and *trans* NTD interactions, respectively (Plate 6.12). Extra stabilization conferred by the strap makes the transition from *cis* to *trans* much more unfavorable, which explains the extreme temperature dependence of ATPase activity in TRAP1: a nearly 200-fold increase in activity is observed between 25°C and 55°C (Leskovar *et al.*, 2008). A conserved salt bridge in TRAP1 between two residues (a histidine at the strap and a glutamate in the NTD) is required to maintain stabilization via the strap. Although this contact does not appear to be conserved in other Hsp90s, an identical salt bridge appears in Hsp90α at the same position, albeit with swapped identity (histidine in the NTD and glutamate at the strap). Eliminating the putative salt bridge in Hsp90α enhances the ATP hydrolysis rate (Partridge *et al.*, 2014).

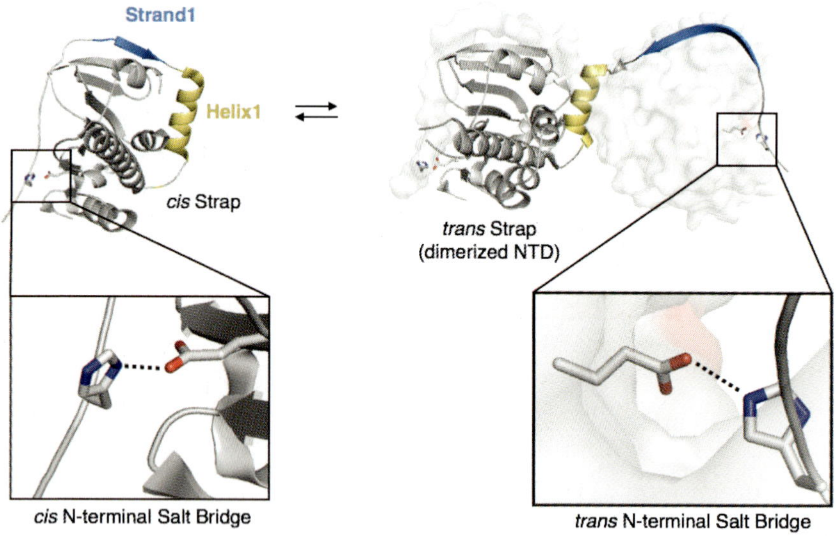

Plate 6.12. ***Two conformations of the N-terminal "strap" observed in mitochondrial Hsp90 (TRAP1).*** In the apo state, the "strap" forms an interaction with its own NTD (*cis* conformation). The NTD–dimer is further stabilized by *trans* strap interaction. The conserved N-terminal salt bridge (boxed) is maintained in both conformations.

Asymmetry

Although Hsp90 homodimers have mostly been observed in symmetric conformations in both open and closed states, ATP-bound mitochondrial Hsp90 (TRAP1) exhibits distinct asymmetry with the largest conformational deviation between monomers localized at the MD:CTD interfaces (Plate 6.4) (Lavery *et al.*, 2014). This asymmetric conformation is thought to arise from strain induced by simultaneously maintaining dimerization at the NTD and the CTD in the closed state. Supporting this, truncating the CTD restores symmetry to the NTD-dimerized TRAP1. In the asymmetric dimer, one protomer maintains a "canonical" MD:CTD interface similar to yeast Hsc82 (Ali *et al.*, 2006), but the opposing "asymmetric" monomer is rearranged such that the MD:CTD interface has 2.5-fold less buried surface area. The ability of TRAP1 to form either one of these interfaces is crucial for its ATPase cycle, given that disruption of either "canonical" or "asymmetric" MD:CTD conformations significantly impairs ATPase activity, despite being 50–60 Å away from the catalytic site. Moreover, orthogonal mutations that target "canonical" and "asymmetric" interfaces do not complement each other to restore wild type activity though they can form the closed state, suggesting that TRAP1 needs to switch between conformations observed in both protomers to progress through its ATPase cycle (Plate 6.13A).

The most exciting hint provided by the TRAP1 structure is the connection between the ATPase cycle and conformational rearrangements that could be utilized for client remodeling. While Hsp90 undergoes large conformational changes between the open and closed states, this transition does not require ATP hydrolysis (Southworth and Agard, 2008), and due to its weak affinity ATP binding is not a strong driver either. This begs the question as to the function of ATP hydrolysis. One possibility is that the strain built up in the closed ATP state is utilized via stepwise hydrolysis to drive client folding. As shown in Plate 6.13, the asymmetry is maximal at the location of key client-binding residues. Thus, in the closed state the client binding

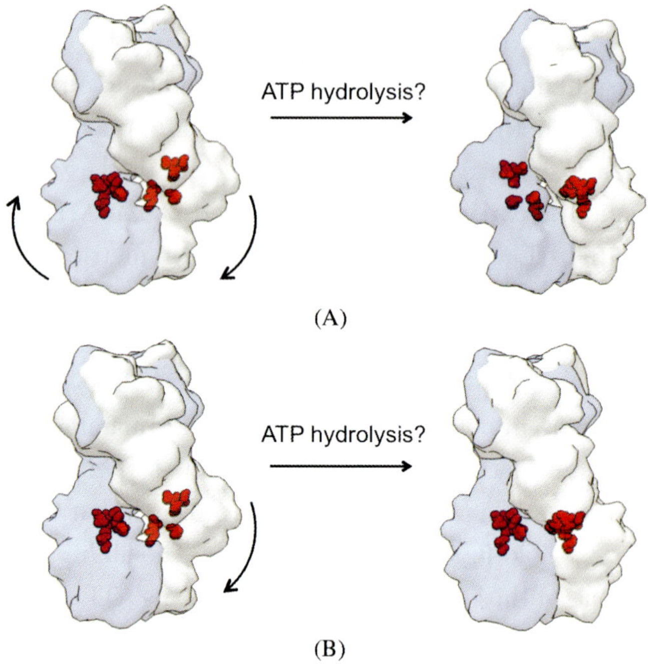

Plate 6.13. ***Proposed mechanism for how asymmetry could be utilized to promote client remodeling***. Differential organization of the client binding residues (red spheres) in the canonical symmetric state versus the noncanonical asymmetric state suggests a mechanism in which ATP hydrolysis either induces switching **A**. or restores symmetry **B**. in the Hsp90 dimer, resulting in reorganization of the client binding regions. Such rearrangement could be coupled to client remodeling.

pocket can have two different arrangements due to the asymmetry. This leads to a sequential ATP hydrolysis model in which the first ATP hydrolysis could be used to alter the asymmetric conformation, directly impacting the structure of a bound client. As suggested in Plate 6.13, hydrolysis could either induce a switch in asymmetry or promote conversion to a symmetric state. While many details are still unknown, the addition of the asymmetric closed state to the ATPase cycle model provides a framework to think about how the energy from ATP hydrolysis can be used to benefit client folding.

Although the asymmetric conformation has been observed only in TRAP1 crystal structure and in solution for the bacterial Hsp90 (Lavery *et al.*, 2014), the presence of strain within the NTD-dimerized closed state exists in other Hsp90 homologues. Yeast Hsp90 undergoes transient CTD opening that coincides with NTD dimerization (Ratzke *et al.*, 2010), suggesting an alternative way to relieve strain in this homologue. While the prevalence of this feature is unclear among other eukaryotic Hsp90s, it hints at the relevance of asymmetry within the Hsp90 family and could be the origin of the asymmetric theme prevalent in cochaperone interactions in the eukaryotic system.

Client Interactions

Generally, a protein is deemed an Hsp90 client if it interacts directly with Hsp90 and displays sensitivity to Hsp90 inhibitors that prevent Hsp90 ATP binding, such as geldanamycin. Treatment with geldanamycin typically results in the immediate loss of a client's activity, and in most cases this is followed by the loss of client protein due to targeted degradation (Schneider *et al.*, 1996; Whitesell and Cook, 1996). It is worth noting that in the case of the steroid hormone receptors, geldanamycin-based inhibitors do not prevent Hsp90 association but instead prevent the essential ATP-hydrolysis-dependent activation step that occurs after client association (Kirschke *et al.*, 2014; Whitesell and Cook, 1996). Moreover, additional investigations with geldanamycin revealed that only the nascent form of many kinases is sensitive to geldanamycin, with a more select class of kinases that continue to depend on Hsp90 in the mature form (Xu *et al.*, 2005). This indicates an important role for Hsp90 in *de novo* folding of its clients and suggests a distinct class within Hsp90's clients whose function is then continuously regulated by Hsp90 throughout the client's cellular lifespan.

While the general consensus is that Hsp90 interacts with proteins in partially folded nonnative states, Hsp90 is also reported to interact with the intrinsically disordered protein tau (Karagöz *et al.*, 2014), and a native state of the glucocorticoid receptor (GR) (Lorenz *et al.*, 2014). In all structural studies so far, large exposed surface areas are utilized for client interactions. In most cases, the client binding region near the MD:CTD interface appears to be a primary interaction site, with the MD:CTD orientation significantly altering the positioning of these residues and thus regulating client recognition (Plate 6.14). Likely for this reason, some clients display a preference for a more compact state of Hsp90. Additionally, secondary low affinity interaction sites span large surfaces of Hsp90, and can form cross-monomer interactions that can regulate Hsp90's conformation. While nucleotide can bias Hsp90 conformation to tune the strength of these interactions, nucleotide binding has a minor-to-modest impact on client interaction affinities, suggesting that client binding and release is not stringently regulated by ATP binding and hydrolysis. This emphasizes very different substrate recognition principles than for other ATP-dependent chaperones that employ some degree of containment that is highly regulated by the nucleotide state.

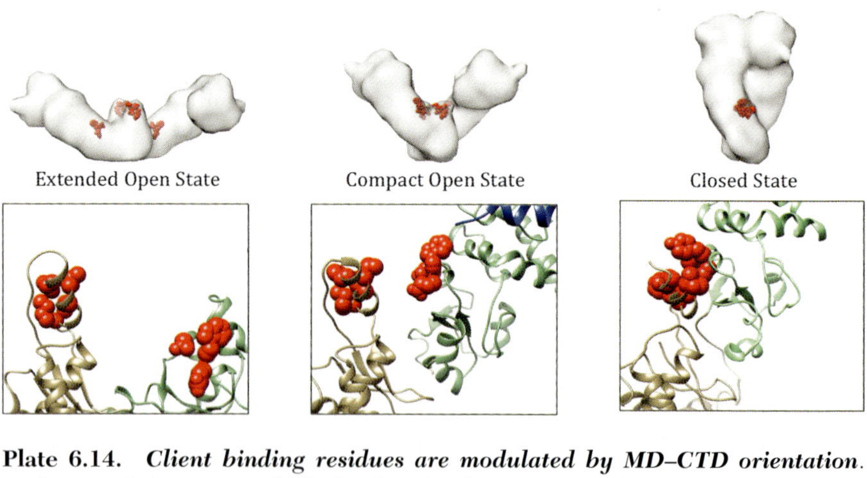

Plate 6.14. ***Client binding residues are modulated by MD–CTD orientation***. In the extended open state (left), the client binding regions (red spheres) in the MD and the CTD are significantly separated. Compaction around the MD–CTD interface brings the client binding regions into closer proximity, as seen in the apo state crystal structure (middle) and the fully closed ATP-bound state (right).

Kinases: A Case Study for Hsp90 Specificity

Despite Hsp90's apparent client specificity, its clients are highly diverse, and lack any unifying sequence or structural determinant that defines a particular Hsp90 recognition site. Kinases are an interesting case, in which 60% of the kinome forms a physical interaction with Hsp90, and yet Hsp90's clients are scattered seemingly randomly across the kinome, showing no correlation with evolutionary relations (Taipale *et al.*, 2012; Wu *et al.*, 2012). A common example of this is the highly homologous v-src and c-src. Despite their shared 98% sequence identity, v-src is significantly more sensitive to Hsp90 perturbations than c-src (Taipale *et al.*, 2012; Xu *et al.*, 1999). Focused attempts at identifying key structural determinants defining the Hsp90 dependence of a particular kinase have suggested that certain surface characteristics at a particular region of the kinase are important (Citri *et al.*, 2006; Xu *et al.*, 2005). However, a more quantitative and systematic approach revealed that the only characteristic correlating with the strength of the Hsp90 interaction is the thermal stability of the kinase, with more unstable kinases likely to have a stronger Hsp90 interaction (Taipale *et al.*, 2012). Furthermore, this study suggested that Hsp90 binds a state distinct from the fully folded active or inactive kinase conformation. This is consistent with a recent study indicating that instead of a distinct structural feature the difference between v-src and c-src in their Hsp90 dependence correlates with compactness, folding cooperativity and lowered energy barriers between states (Boczek *et al.*, 2015). Together, building evidence suggests that Hsp90 preferentially recognizes partially folded nonnative states, with the ability of the client to access these intermediate states manifesting as a major determinant for the strength of the Hsp90 interaction.

Principles Learned from Interactions with a Model Substrate

Structures of Hsp90–substrate complexes have so far eluded crystallization, and biochemical approaches investigating client interactions have also been challenging. The reasons appear rooted in Hsp90's preference for partially folded nonnative states. Stable proteins amenable to *in vitro* investigations rarely sample partially unfolded states required to obtain a strong Hsp90 interaction, while destabilized proteins that do interact are generally prone to aggregation at concentrations required for *in vitro* investigation. That said, a synthetic fragment of staphylococcal nuclease, referred to as Δ131Δ, is globally unfolded yet compact with some local nativelike structure and is fortuitously resistant to aggregation (Alexandrescu *et al.*, 1994; Flanagan *et al.*, 1992). This made Δ131Δ an ideal model client for pioneering rigorous *in vitro* characterization of Hsp90 client interactions.

Conventional binding studies revealed no interactions with full-length native staphylococcal nuclease or with Δ131Δ in the presence of reagents that promote Δ131Δ folding (Street *et al.*, 2011a). However, direct Hsp90 binding is clearly detected for the partially folded Δ131Δ, correlating with hydrogen exchange experiments revealing that Hsp90 interacts with a partially folded state that is transiently sampled in the native state ensemble (Street *et al.*, 2014). As mapped by NMR, within Δ131Δ it is an α-helix and adjacent β-strands that are responsible for Hsp90 binding. Moreover the α-helix remains structured upon Hsp90 binding, explaining the preference for partially folded clients; however, the adjacent strands are distorted, reflecting Hsp90 being able to alter the structure of bound client proteins (Street *et al.*, 2014).

In the absence of nucleotide, a single molecule of Δ131Δ binds per Hsp90 dimer with moderate affinity (K_D of 9 μM for *E. coli*, and 6 μM for yeast) (Street *et al.*, 2011a). The MD–CTD fragment is sufficient for binding a large partially folded region on Δ131Δ, with client interactions spanning both domains (Street *et al.*, 2011b; 2014). Furthermore, using SAXS and NMR, a specific residue on the middle domain (W467 — *E. coli*) was identified as a key determinant for the primary client binding interaction (Street *et al.*, 2011b). This residue was later independently identified in a bacterial genetic screen. This screen together with subsequent *in vitro* characterization, confirmed and expanded this region on the MD and also revealed the importance of the CTD amphipathic helix (Plate 6.2) (Genest *et al.*, 2013). The generality of these regions for client binding *in vivo* was shown by their importance for GR activation in yeast.

Under apo conditions, Hsp90 partially closes around Δ131Δ (Street *et al.*, 2011a). Δ131Δ binding stabilizes changes within the chaperone, leading to a partial closure (extended to V-shaped transition) by rotation about the MD–CTD interface. For Δ131Δ, Hsp90 dimer closure with AMPPNP results in a modest twofold enhancement in binding affinity, whereas an MD–CTD interface mutant severely impairs binding, suggesting that the fully extended state only weakly interacts with this client (Street *et al.*, 2011a, 2014). The ATP binding pocket inhibitor geldanamycin does not inhibit Δ131Δ binding, but prevents the enhancement resulting from AMPPNP. The preference for a specific MD–CTD orientation appears due to the positioning of client binding residues being significantly altered by the MD–CTD orientation (Plate 6.14) (Street *et al.*, 2014). The two client binding regions are significantly separated in the extended apo conformation compared to the more closed Hsp90 states. While these regions form the primary interaction site, Δ131Δ forms additional secondary cross-monomer interactions that promote the NTD rotation, resulting in a fivefold acceleration in dimer closure and a comparable acceleration in ATP hydrolysis for bacterial Hsp90 (but not yeast) (Street *et al.*, 2011b). Validating this principle, a tenfold acceleration in ATP hydrolysis is observed for bacterial Hsp90 by the client protein L2 (Motojima-Miyazaki *et al.*, 2010). Together, this indicates a client recognition mechanism in which the flexibility around the MD–CTD interface is critical for client binding and, in turn, client binding can regulate Hsp90's ATP hydrolysis cycle for some Hsp90 homologues.

Tau: An Unfolded Client Protein That Mimics a Partially Folded Intermediate

While the data discussed so far suggest that interaction with Hsp90 necessitates some degree of secondary and tertiary structure, one of Hsp90's clients, tau, is intrinsically disordered. In the cell, tau binds microtubules and controls microtubule growth (Mandelkow and Mandelkow, 2012). Tau oligomerization and aggregation leads to insoluble tangles present in neurodegenerative diseases referred to as tauopathies, which include Alzheimer's disease. Antagonizing this process, Hsp90, in coordination with Hsp70, plays a decisive role in tau homoeostasis by promoting tau solubility and clearance (Dickey *et al.*, 2007; Dou *et al.*, 2003; Thompson *et al.*, 2012). *In vitro*, tau directly binds Hsp90 independently of nucleotide with moderate affinity (K_D of ~5 μM) (Karagöz *et al.*, 2014). The lack of nucleotide effect is attributed to the fact that these experiments were carried out with human Hsp90, for which the conformational equilibrium is extremely shifted toward the open state, such that even with saturating nucleotide human Hsp90 rarely samples the closed state in the absence of cochaperones. Supporting this, Hsp90 was observed to remain open while bound to tau and ATPγS.

A combination of data from NMR and SAXS was used to structurally characterize this interaction. Splitting of NMR peaks indicates a break in Hsp90's symmetry upon tau binding. Supporting Hsp90's role in aiding tau solubility, Hsp90 binds tau's aggregation-prone repeat regions, which also encompass Hsp70 and microtubule binding sites. In an extended state, tau interacts with a large surface that is 106 Å in length, ranging from the NTD down to the MD, covering 840 Å^2. This extended interaction surface remarkably circumvents known cochaperone binding sites and partially overlaps with GR's client binding site (discussed below) (Karagöz and Rüdiger, 2015). Tau's interaction is driven by numerous low affinity contacts consisting of scattered large hydrophobic and positively charged residues. Intriguingly, the authors argue that the distribution of hydrophobic residues on tau resembles the moderate scattered hydrophobic residues commonly seen on the surface of folding intermediates, suggesting that Hsp90 binds this protein because unfolded tau mimics the surfaces of a partially folded intermediate. Perhaps this mode of interaction, combined with the more localized and structured α-helical binding observed for Δ131Δ, helps explain the remarkable diversity of Hsp90 clients.

Direct Interaction with the Native State of the Glucocorticoid Receptor

Perhaps most surprising, recent work with steroid receptors reveals Hsp90 can interact with a client in its native state (Lorenz *et al.*, 2014). Steroid hormone receptors are a subclass of the nuclear receptor superfamily that, as an evolutionarily related subset, evolved a dependence on chaperones for what appear to be regulatory purposes (Yen, 2015). They are ligand-activated transcription factors that *in vivo* interact with Hsp90 predominately in their apo state and depend on Hsp90 to varying degrees to bind their cognate hormone and become active (Pratt and Toft, 1997). However, *in vivo* observations are convoluted by Hsp70-driven interactions (discussed later). Among the hormone receptors, GR is stringently dependent on Hsp90 *in vivo* (Picard *et al.*, 1990). Hsp90's interaction is mediated through GR's ligand binding domain (GRLBD) (Howard *et al.*, 1990). The apo GRLBD is more unstable and less folded than the hormone bound state (Lorenz *et al.*, 2014), suggesting that Hsp90 helps maintain apo receptor stability *in vivo*.

Surprisingly, but consistent with the observation that hormone does not directly stimulate receptor dissociation from Hsp90 (Smith, 1993), a presumably native ligand-bound state of GRLBD can directly bind Hsp90 with a K_D of 1–5 μM (Lorenz *et al.*, 2014). While GRLBD binds different conformational states of Hsp90 with varying affinity, GRLBD significantly prefers an ATP-bound partially closed state. NMR and SAXS data indicate a large binding surface on Hsp90 (~1500–2000 Å^2) centered on the inner-cleft-facing side of the MD, which spans the MD, parts of the CTD, and a small portion of the NTD. Significantly, this region encompasses the key client binding residues identified in bacteria and with Δ131Δ. The Hsp90 dimer can accommodate two monomers of GRLBD, with GRLBD molecules binding independently on either side of the dimer. Due to overlapping binding sites, direct GRLBD binding is competitive with the cochaperones Hop and p23, which are known to modulate GR's chaperone cycle *in vivo*. Hop's incompatibility is further driven by Hop's preference for the open states of Hsp90 (see below). While p23 and GRLBD are compatible with the same Hsp90 conformation, overlapping binding sites result in simultaneous occupancy of GRLBD and p23, requiring that each occupy opposite sides of Hsp90's dimer, for which biochemical and structural evidence is provided. Unlike Δ131Δ and L2 in the bacterial system, GRLBD decelerates the progression of the ATP hydrolysis cycle by preferentially binding a partially closed state.

While the interaction with the native state of a client seems contradictory to the general evidence put forth so far, GR is a case where Hsp90 has been implicated in having a continued regulatory role for the activated receptor. Hsp90 has been observed to play a role in receptor translocation, nuclear import, and disassembly of GR transcriptional complexes for activated GR (Freeman and Yamamoto, 2002; Galigniana *et al.*, 2001, 2010). Perhaps this reported interaction reflects a distinct mode of interaction specific to ligand-bound GR. Supporting this, Hsp90 was previously proposed to utilize different interaction surfaces to interact with the apo- and ligand-bound states of GR (Fang *et al.*, 2006).

Cochaperones Regulate Hsp90's Chaperone Cycle

In eukaryotes, Hsp90 is found in large complexes with cochaperones, some of which are listed in Table 6.1. Cochaperones provide a means for tightly regulating Hsp90's conformational cycle (Plate 6.15), recruiting specific clients, and additional client specific regulation and functionality. Specific complexes form at different stages of the chaperone cycle, with what appears as a coordinated progression of association and dissociation of specific cochaperones at distinct steps in the cycle (Li *et al.*, 2012). The specific combination of cochaperones utilized is to some extent determined by the client. Many cochaperones compete for overlapping binding sites and/or are only compatible with distinct conformations of Hsp90. However, some cochaperones are more flexible and are compatible with other cochaperones and/or multiple

Table 6.1. Partial List of Select Hsp90-Associated Cochaperones

Cochaperone Vertebrates (Yeast)	Characteristics and Functions	References
Aha-1	Stimulates ATPase activity; competitive with p23, Hop and Cdc37.	*
p23 (Sba1)	Stabilizes closed conformation; inhibits ATPase activity; competitive with Hop and Aha-1.	*
Cdc37 (p50)	Binds kinases and inhibits ATPase activity; competitive with Aha-1 and p23.	*
TPR-Domain-Containing		
Hop (Sti1)	Coordinates Hsp70 interaction and client loading; inhibits ATPase activity; competitive with p23 and Aha-1.	*
CHIP	Also interacts with Hsp70; contains ubiquitin E3 ligase activity and is involved in quality control.	Stankiewicz *et al.*, 2010; Theodoraki and Caplan, 2012
Cyp40 (Cpr6, Cpr7)	Contains peptidyl-prolyl-isomerase.	Prodromou *et al.*, 1999
PP5 (Ppt1)	Contains phosphatase activity; dephosphorylates Cdc37.	*
FKBP52	Potentiates hormone receptor signaling; contains peptidyl-prolyl-isomerase.	Riggs *et al.*, 2003; Storer *et al.*, 2011

(Continued)

Table 6.1. (*Continued*)

Cochaperone Vertebrates (Yeast)	Characteristics and Functions	References
FKBP51	Attenuates steroid hormone receptor signaling; contains peptidyl-prolyl-isomerase.	Reynolds *et al.*, 1999; Storer *et al.*, 2011
GCUNC-45	Positive regulator of PR signaling; competitive with Aha-1.	Chadli *et al.*, 2006
SGT1	Maintains NLR immune sensors; contains domain with homology to p23.	Shirasu, 2009
(Tah1)	Involved in chromatin remodeling and small snoRNP maturation; weakly stimulates ATPase activity; inhibits ATP hydrolysis when in complex with Pih1.	Eckert *et al.*, 2010

*References found in the main text.

For a complete, updated list of Hsp90-related cochaperones, see http://www.picard.ch/downloads/Hsp90interactors.pdf

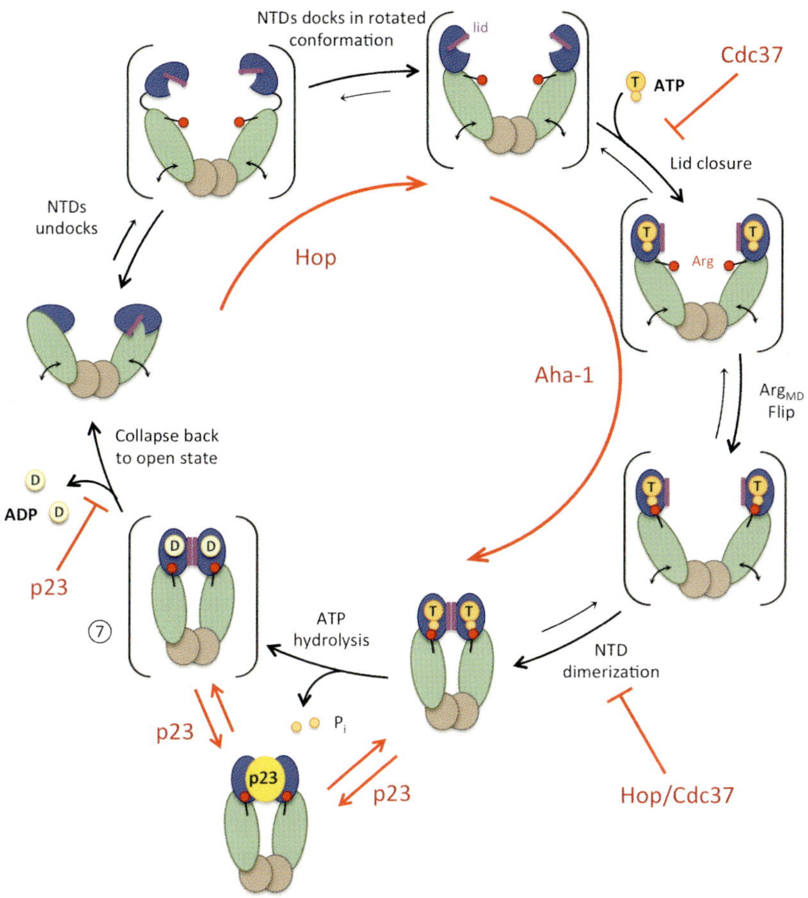

Plate 6.15. ***Cochaperones regulate distinct steps in Hsp90's conformational cycle*.** Mechanistic cycle for Hsp90 ATP hydrolysis as shown in Plate 6.5, but additionally showing the effects of the cochaperones Hop, Aha-1, Cdc-37, and p23.

states of Hsp90, allowing for mixed cochaperone complexes (Li *et al.*, 2010). This complex interplay of cochaperones is thought to both regulate ATP hydrolysis and facilitate a specific order for the assembly and disassembly of different cofactors in the chaperone cycle.

A large number of cochaperones contain TPR domains, which interact with the MEEVD motif on Hsp90's CTD, but some of the most important Hsp90 regulators, including p23, Aha-1 and Cdc37, do not. These three important cochaperones and the TPR containing cochaperone Hop are highly conserved from yeast to humans, and will be discussed in more detail. In addition to the cochaperones, Hsp90 is subject to numerous and varied posttranslational modifications that provide another level of regulation to Hsp90's chaperone cycle (Mollapour and Neckers, 2012); however, unless known to be integral to the specific aspect of the chaperone cycle being discussed, posttranslational modifications are not addressed here.

TPR-Containing Cochaperones Interact with Hsp90's CTD MEEVD Motif

One of the largest definable classes of Hsp90 cochaperones is the TPR (tetratricopeptide repeat) domain–containing proteins. The TPR domain consists of a common structural motif that forms the primary interaction site for recruiting these cochaperones to Hsp90 complexes via binding to Hsp90's C-terminal MEEVD motif. As described for Hop, this primary interaction then promotes additional secondary contacts between the cochaperone and other surfaces on the Hsp90 complex that can have important functional implications in either directly regulating Hsp90's conformation or facilitating client-specific functions.

TPR domains consist of 34-amino-acid repeats folded into antiparallel helical hairpins that bundle together to form a concave cradle-shaped groove. This binding groove accommodates C-terminal EEVD-containing peptides, mostly in an extended conformation (Plate 6.16) (Scheufler *et al.*, 2000; Zeytuni and Zarivach, 2012). While a diversity of sequence and structural specificity occurs for recognition of the N-terminal region of the peptide, the C-terminal interaction is highly conserved, consisting of a two-carboxylate clamp that anchors the peptide through interactions with the main chain carboxylate of peptide's C-terminal Asp (Scheufler *et al.*, 2000; Zhang *et al.*, 2005). Not all TPR domains in Hsp90 cochaperones specifically recognize Hsp90's MEEVD motif. In fact, some also bind to or are specific for the C-terminal end of Hsp70 that contains the closely related IEEVD motif. This allows some cochaperones, such as Hop and CHIP, to coordinate the activity of the two chaperones. The Hsp90 homodimer

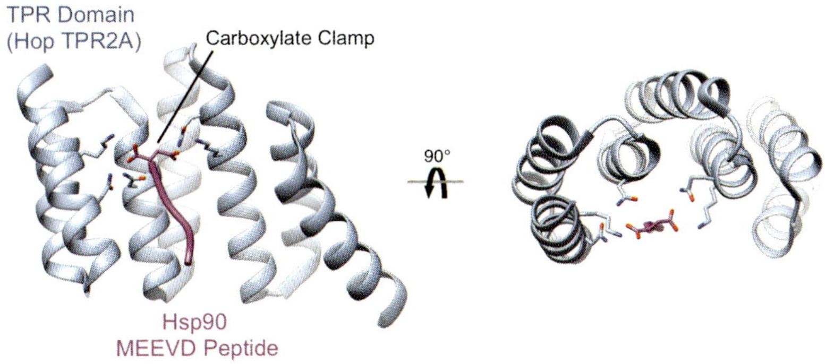

Plate 6.16. ***Hsp90's C-terminal MEEVD motif recruits TPR-domain-containing cochaperones.*** Crystal structure of the Hop TPR2A domain bound to Hsp90's MEEVD-containing peptide (Scheufler *et al.*, 2000).

presents two MEEVD motifs, allowing up to two TPR-containing cochaperones to be bound simultaneously. Some TPR-containing cochaperones are compatible with other TPR-containing cochaperones, allowing mixed cochaperone complexes to form, and in some cases are preferred (Ebong *et al.*, 2011; Li *et al.*, 2010).

Hop Facilitates Hsp90 Client Loading

While nonessential in yeast, the cochaperone Hop (initially named p60) significantly enhances Hsp90-dependent client activation, as seen for the hormone receptor GR and the kinase v-src (Chang *et al.*, 1997; Dittmar and Pratt, 1997; Kirschke *et al.*, 2014). Hop — short for "Hsc70/Hsp90-organizing protein" — facilitates client transfer from Hsp70 to Hsp90. To do so, it provides a physical link between the chaperones by simultaneously binding their C-terminals through separate TPR domains. Beyond interaction with Hsp90's MEEVD motif, Hop makes extensive secondary contacts that regulate Hsp90's conformation in ways that appear to assist client loading and transfer from Hsp70. A cryo-EM reconstruction of an Hsp90–Hop complex reveals that Hsp90 is in an open state with the MD–CTD interface stabilized in a more compact V shape similar to that of the apo state crystal structure (Plate 6.17) (Southworth and Agard, 2011). This MD–CTD conformation brings the

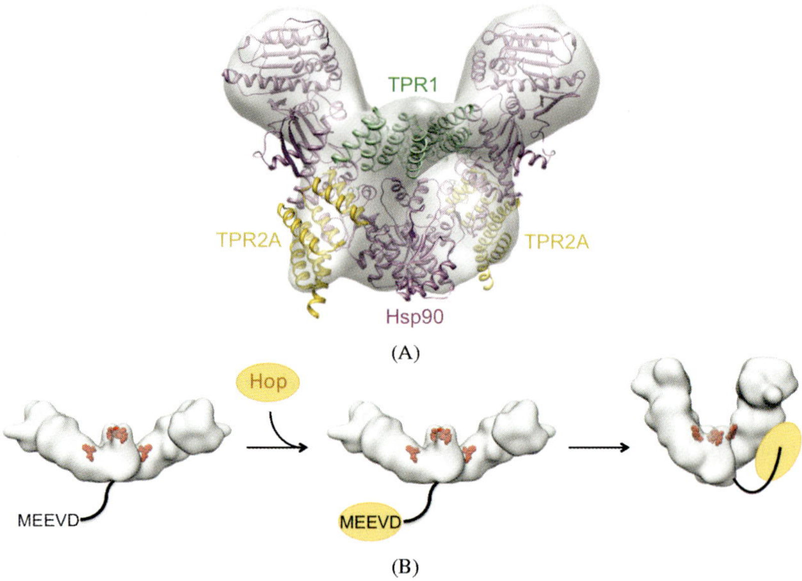

Plate 6.17. ***Hop stabilizes a client loading conformation of Hsp90*.** **A**. Cryo-EM map of Hsp90 bound by two molecules of Hop (Southworth and Agard, 2011). **B**. The interaction with Hsp90 is initiated by the binding of Hop's TPR 2A domain to Hsp90's MEEVD motif. Secondary interactions then stabilize the distinct conformation of Hsp90 shown in **A**., in which the NTDs are rotated up and compaction of the MD–CTD interface, which aligns client binding residues.

two major client binding regions into close proximity, suggesting that Hop organizes Hsp90's client binding region for specific client recognition. Even more striking, Hsp90's NTDs are rotated in the Hop bound structure, such that each arm is primed for NTD dimerization but physically blocked from closure. This supports biochemical data indicating that Hop inhibits ATPase function without effecting nucleotide binding by selectively interacting with an open state and preventing closure (Lee *et al.*, 2012; McLaughlin *et al.*, 2002; Richter *et al.*, 2003). While two molecules of Hop can simultaneously bind Hsp90 (Prodromou *et al.*, 1999; Southworth and Agard, 2011), a single Hop is sufficient to inhibit ATP hydrolysis (Li *et al.*, 2010). Furthermore, the second Hop interaction is noticeably weaker than the first (Ebong *et al.*, 2011; Southworth and Agard, 2011). Combined with the fact that the biological ratios of

Hsp90 to Hop are 10:1, a single Hop-bound complex is likely the most physiologically relevant, leaving the second MEEVD motif accessible for interacting with other TPR-containing cochaperones (Li *et al.*, 2010).

Hop is a modular protein consisting of three TPR domains and two DP domains (rich in aspartate and proline residues) (Plate 6.18). In solution it is dynamic and elongated, with a flexible N-terminal region consisting of the TPR1 and DP1 domains connected by a long flexible linker to the more rigid C-terminal module consisting of the TPR2A,

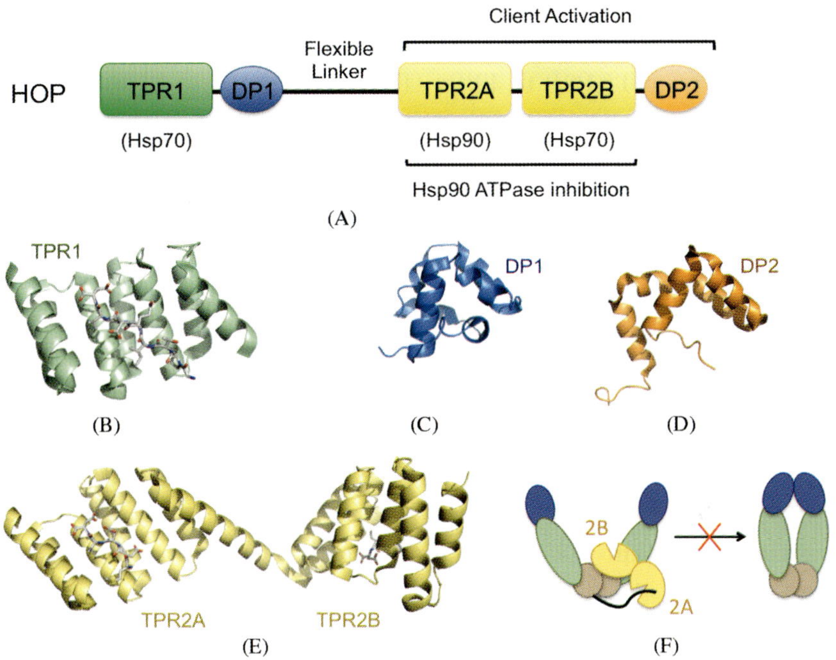

Plate 6.18. ***Hop domain architecture and structure*.** **A**. Domain architecture of Hop. Hsp90's MEEVD motif preferentially binds TPR2A, and Hsp70's EEVD motif binds both TRP1 and TPR2B. The TPR2A–2B fragment is sufficient for Hsp90 ATPase inhibition, but client activation *in vivo* additionally requires the DP2 domain. **B**. Crystal structure of the EEVD peptide-bound TPR1 domain (Scheufler *et al.*, 2000). **C**. NMR structure of the DP1 domain (Schmid *et al.*, 2012; Stankiewicz *et al.*, 2010; Theodoraki and Caplan, 2012). **D**. NMR structure of the DP2 domain (Schmid *et al.*, 2012). **E**. Crystal structure of the peptide-bound TPR2A–2B rigid fragment (Schmid *et al.*, 2012). **F**. Schematic for binding of Hop TPR2A–2B based on NMR data (Schmid *et al.*, 2012). TPR2A initiates the interaction through Hsp90's via Hsp90's MEEVD motif, and TPR2B makes additional interactions that inhibit ATP hydrolysis.

TPR2B and DP2 domains (Onuoha *et al.*, 2008; Röhl *et al.*, 2015). Among the three TPR domains, TPR2A selectively binds to Hsp90's MEEVD-containing C-terminal tail with high affinity, and both TPR1 and TPR2B bind to the EEVD motif at the C-terminal tail of Hsp70 with moderate affinity (Scheufler *et al.*, 2000; Schmid *et al.*, 2012). Genetic studies indicated that TPR1 and TPR2B carry out redundant functions, with mutations in both TPR1 and TPR2B required to disrupt Hsp70 binding *in vivo* (Flom *et al.*, 2007, 2006). Recently, Hsp90 was found to regulate Hop's domain organization, with Hsp90 binding resulting in a less compact Hop state and a shift in Hsp70's binding preference from TRP2B to TPR1 (Röhl *et al.*, 2015).

While TPR2A initiates binding to Hsp90 through the MEEVD motif, TPR2B also interacts directly with Hsp90. In fact, both TPR2A and TPR2B domains are required for Hsp90 interaction *in vivo* (Flom *et al.*, 2007), and together are sufficient to inhibit Hsp90 ATP hydrolysis (Lee *et al.*, 2012; Schmid *et al.*, 2012). A crystal structure of the TPR2A–TPR2B fragment reveals that the two TPR domains are connected by a structured linker (Plate 6.18E), with a defined orientation of the linker with respect to the TPR2B required for Hop function *in vivo* (Schmid *et al.*, 2012). While TPR2A–TPR2B is sufficient for ATPase inhibition and binding to both Hsp90 and Hsp70, the DP2 domain plays an important yet unidentified function *in vivo*, with the DP2 domain required for Hop-enhanced GR and v-src activation *in vivo* (Schmid *et al.*, 2012; Song and Masison, 2005). Based on the NMR structure, the groove formed in DP2, not present in DP1, forms a distinct surface that is suspected to interact directly with clients; however, another explanation for how DP2 might be important in regulating Hsp70 interactions will be discussed later. Consistent with the absence of the TRP1–DP1 region in some homologues (Gaiser *et al.*, 2009; Johnson and Brown, 2008), Hop's second module (TPR2A–2B–DP2) appears to be the more functionally important region, with the TPR1–DP1 module carrying out some accessory function that is not essential for client activation *in vivo*.

Aha-1 Stimulates ATP Hydrolysis

The activator of the *H*sp90 ATPase, Aha1, is unique among Hsp90's cochaperones, in that it potently stimulates Hsp90's ATP hydrolysis activity by directly promoting structural changes required for hydrolysis (Hessling *et al.*, 2009; Panaretou *et al.*, 2002; Retzlaff *et al.*, 2010; Richter *et al.*, 2008). Since Aha-1 increases both thermotolerance, and client (v-src and GR) activity *in vivo*, it is assumed that it generally enhances Hsp90 function through ATPase stimulation (Harst *et al.*, 2005; Lotz *et al.*, 2003; Panaretou *et al.*, 2002). However, it has more recently been shown that Aha-1 may not always be beneficial to Hsp90 activity, with the effects of Aha-1 being context-specific (Sun *et al.*, 2012).

While there is currently no full-length crystal structure of Aha-1, there are structures of its two individual domains, the C-terminal domain (C-domain) and the N-terminal domain (N-domain), which are linked by a coil structure (Plate 6.19). Surprisingly, the two domains appear to have evolved independently as separate functional regulators of Hsp90 and during the course of evolution became a single linked protein with enhanced function (Singh *et al.*, 2014). The N-domain interacts with the middle domain of Hsp90 independently of the nucleotide state, as illustrated in the crystal structure of the Aha-1 N-domain bound to Hsp90's MD (Plate 6.19B) (Meyer *et al.*, 2004). Binding of the Aha N-domain to Hsp90's MD modulates the Arg_{MD} loop, flipping the Arg_{MD} upward closer to the orientation seen in closed state crystal structure. While the Aha-1 N-domain on its own has some stimulation potential, both domains are required and act cooperatively to maximally stimulate ATP hydrolysis (Panaretou *et al.*, 2002).

The binding of the Aha-1 N-domain is the primary binding interaction, with the secondary interaction resulting from the binding of the Aha-1 C-domain at an interface on Hsp90 formed only in the closed state, with the interaction surface spanning the NTD–dimer junction (Plate 6.19C) (Retzlaff *et al.*, 2010). Together, a sequential binding model

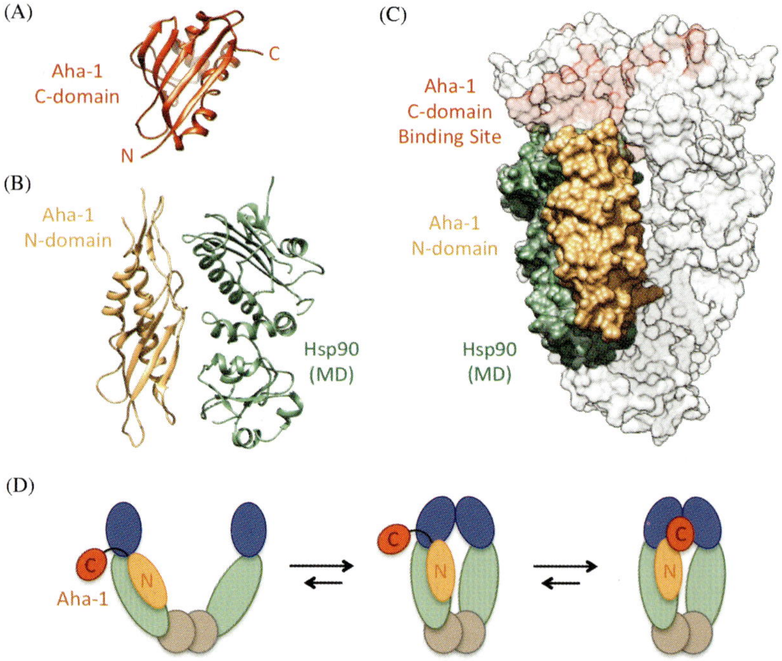

Plate 6.19. ***Aha-1 promotes ATP hydrolysis on Hsp90 through a two-step asymmetric binding mechanism in which cross-monomer contacts stabilize a closed state*.** **A**. NMR structure of the Aha-1 C-domain (pdb 1X53). **B**. Crystal structure of Aha-1's N-domain (yellow) bound to Hsp90's MD (green) (Meyer *et al*., 2004). **C**. Crystal structure from B (green and yellow) aligned to the full-length closed state (transparent gray) with Aha-1's C-domain interaction sites mapped in red (Retzlaff *et al*., 2010). **D**. Schematic of the sequential binding mechanism for Aha-1. The Aha-1 N-domain binds Hsp90's MD independently of Hsp90's conformation. NTD dimerization in the closed state forms the Aha-1 C-domain binding site.

has been proposed in which the N-domain binds the MD in either the open or closed state and, in doing so, localizes the C-domain in proximity to the NTD such that upon transient NTD dimerization the Aha-1 C-domain can rapidly bind and stabilize the catalytically active closed state of Hsp90 (Plate 6.19D). While two molecules of Aha-1 can simultaneously bind one Hsp90 dimer, the activation mechanism is asymmetric, with only one molecule of Aha-1 per Hsp90 dimer required for maximal activation of ATP hydrolysis in both (cis and trans) NTDs (Retzlaff *et al.*, 2010). The ability of a single molecule of Aha-1 to stimulate ATP hydrolysis is likely functionally important, given that the binding of the N-domain to the MD in proximity to the client binding regions is incompatible with the binding of GR and likely many other client proteins (Lorenz *et al.*, 2014). This suggests that Aha-1 stimulates ATP hydrolysis by acting on the dimer interface opposite to that occupied by the client. Furthermore, it should be noted that Aha-1 is not detected in Hsp90 client complexes, suggesting that it interacts only transiently (Sun *et al.*, 2012).

p23 Stabilizes the Closed State of Hsp90

First identified in hormone receptor complexes (Johnson and Toft, 1994), p23 is now well known to play an important role in the later stage of the hormone receptor activation pathway by stabilizing the interaction between Hsp90 and the activated receptor (Dittmar *et al.*, 1997; Kirschke *et al.*, 2014). Beyond steroid receptor activation, p23 is a relatively general cochaperone that regulates Hsp90 by selectively interacting with the ATP stabilized closed state (Grenert *et al.*, 1999; Sullivan and Stensgard, 1997). The reason for this conformational selectivity is clearly illustrated in the closed state crystal structure, in which p23 is bound on both sides of the NTD dimerization interface, with each molecule making interactions that span both NTD monomers (Plate 6.20) (Ali *et al.*, 2006). In this state, each molecule of p23 directly contacts the nucleotide lid, further stabilizing the closed lid state (Plate 6.20B). As a result, p23 inhibits ATP hydrolysis, but only up to 50% for yeast Hsp90 (McLaughlin *et al.*, 2006; Richter *et al.*, 2004). While it is unclear to what extent the exact state seen in the p23-bound closed state crystal structure is catalytically active, inhibition in ATP hydrolysis can be explained by p23 slowing ADP product release through stabilization of both the nucleotide lid and the closed state (Graf *et al.*, 2014). Furthermore, it is notable that while p23 and Aha-1 both preferentially interact with a closed state of Hsp90, their effects on ATPase activity are opposing, and

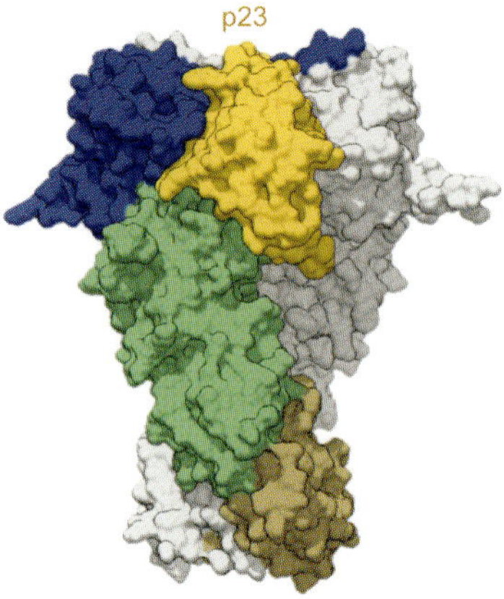

(A)

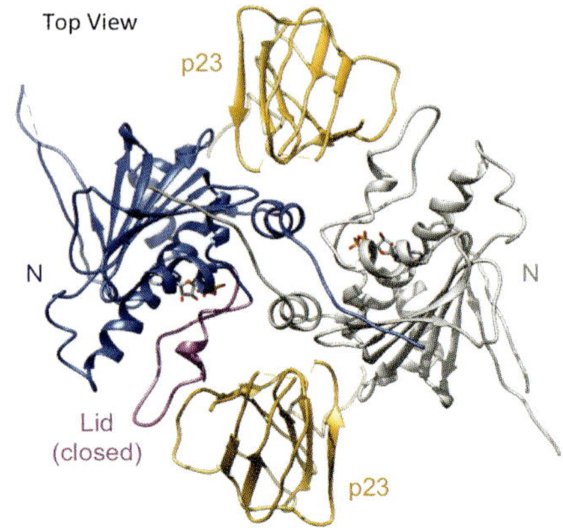

(B)

Plate 6.20. ***p23 stabilizes the closed state of Hsp90*.** **A**. Crystal structure of nucleotide-bound yeast Hsp90 in the closed state bound by the cochaperone p23 (Ali *et al*., 2006). **B**. Top view of the complex from **A**. revealing that p23 stabilizes NTD dimerization by stabilizing the nucleotide lid in the closed conformation and by binding directly to either side of the dimer interface.

have been proposed to interact with distinctly different closed states (Rehn and Buchner, 2015). Supporting this, simultaneous GRLBD binding was compatible with the p23 state but not the Aha-1–bound state (Lorenz *et al*., 2014).

The physiologically relevant binding stoichiometry along with the binding mode of p23 is an unresolved topic. In addition to the 1:1 (Hsp90 monomer: p23) stoichiometry seen in the crystal structure, binding ratios of 2:1 are observed. For the human system, both $Hsp90_2p23_2$ and $Hsp90_2p23_1$ complexes exist under physiological ratios, although the $Hsp90_2p23_2$ state is favored, indicating a cooperative binding mode (Karagöz *et al*., 2011). Conversely, negative allosteric effects have been observed in the yeast system (Richter *et al*., 2004; Siligardi *et al*., 2004). Furthermore, as discussed for GR, client binding to one side of the Hsp90 dimer can occlude p23 binding to the same side of the dimer, while p23 binding to the opposing side of the dimer remains compatible, allowing the formation of an $Hsp90_2p23_1GR_1$ complex (Lorenz *et al.*, 2014).

Perhaps an explanation for these observations lies in the existence of an asymmetric dimeric state as seen in the TRAP1 crystal structure. In the asymmetric state, only one side the Hsp90 dimer would be assessable for p23 binding, with the second site restricted by a steric clash (Lavery *et al.*, 2014). Thus, binding of a second p23 would necessitate conversion to the symmetric closed state. In doing so, p23 would drive remodeling of the client binding sites on Hsp90, suggesting a mechanism for which p23 regulates client interactions (Plate 6.21). Supporting this model, NMR data revealed that binding of two p23s induces

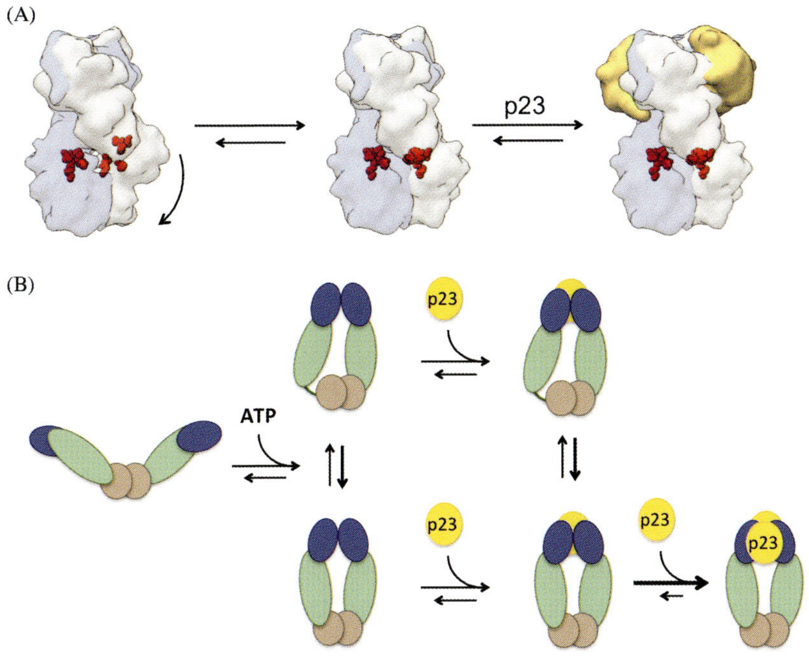

Plate 6.21. ***Stabilization of the symmetric closed state by p23 has implications for regulating client interactions*. A**. Implications for client remodeling and client interactions, client binding residues (red). **B**. Schematic of the proposed p23 binding model incorporating the potential role of asymmetry.

structural changes in Hsp90's MD significantly removed from the p23 binding site (Karagöz *et al.*, 2011); such structural perturbations are consistent with the rotation of the MD from the asymmetric to the symmetric state (Lavery *et al.*, 2014).

Cdc37: A Kinase-Specific Cochaperone

Cdc37 is critical in enabling Hsp90 to maintain a healthy kinase pool. Although it is reported to interact with nonkinase clients (Jinwal *et al.*, 2011; Rao *et al.*, 2001), Cdc37 is mostly kinase-specific, with a near-perfect correlation between the Hsp90–kinase and Cdc37–kinase interaction strengths (Taipale *et al.*, 2012). It is a highly helical, essential protein with three major domains: a flexible N-terminal kinase-interacting domain, a very stable Hsp90-interacting middle domain and a C-terminal domain of unclear function (Plate 6.22A) (Shao *et al.*, 2003). Other than being alpha-helical with a strong prediction for coiled coils (Liu and Landgraf, 2015), the structure of the N-terminal domain remains elusive. A Cdc37 fragment lacking the N-terminal domain has been crystallized in complex with Hsp90's NTD (Plate 6.22). This structure seemed to provide a clear mechanism for how Cdc37 inhibits Hsp90's ATPase activity. This mode of binding would directly block the NTD interface used for dimerization as well as block lid closure, and together these would inhibit ATPase activity (Roe *et al.*, 2004). Importantly, the position of Cdc37 in the crystal structure clashes with essentially all full length Hsp90 structures, implying that Cdc37 binding requires Hsp90 to undergo a partial (but not full) NTD–MD rotation not unlike the rotation seen in the recent Hsp90/Hsp70/GR/Hop EM structure, in which the NTD of the client-bound Hsp90 protomer is in a partially rotated state (Kirschke *et al.*, 2014).

The interaction seen in the crystal structure is supported both by biochemical experiments and by NMR (Sreeramulu *et al.*, 2009), but there is evidence that additional interaction surfaces are involved. Cdc37's interaction with the isolated NTD is weaker than the interaction

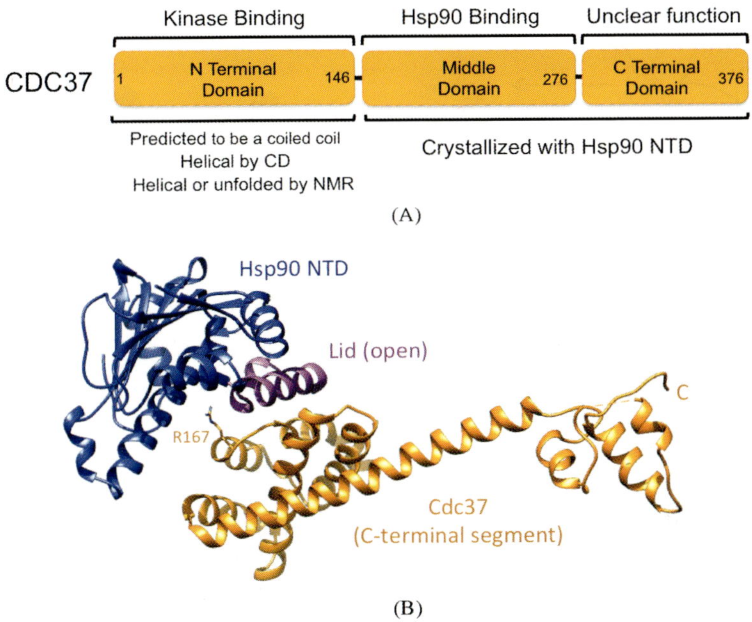

Plate 6.22. ***Cdc37 domain architecture and structure*****.** **A**. Domain architecture of Cdc37. **B**. Crystal structure of the Cdc37 C-terminal fragment bound to Hsp90's NTD (Roe *et al.*, 2004). Cdc37 interacts with the open conformation of Hsp90's nucleotide lid, and Cdc37 R167 protrudes into the nucleotide pocket to prevent proper water coordination required for ATP hydrolysis.

with the full-length chaperone (Eckl *et al.*, 2013; Zhang *et al.*, 2004), and mutations based on the crystal structure give conflicting results *in vivo* (Smith *et al.*, 2015). Finally, in *C. elegans*, Cdc37 seems to mainly interact with the middle domain of Hsp90, and has a higher affinity toward Hsp90's closed state, which would not be accessible based on the crystal structure (Eckl *et al.*, 2013; Gaiser *et al.*, 2010). To reach a consistent interpretation would require that Cdc37 be able to interact with Hsp90 via two distinct interfaces, one on the NTD and the other on the MD, with different species having varying affinities between the two sites. This interpretation also argues for a rearrangement of Cdc37 on Hsp90 during the progression of the cycle, in which Cdc37 dissociates from the Hsp90 NTD and relocates to the MD, allowing Hsp90 closure (Plate 6.23).

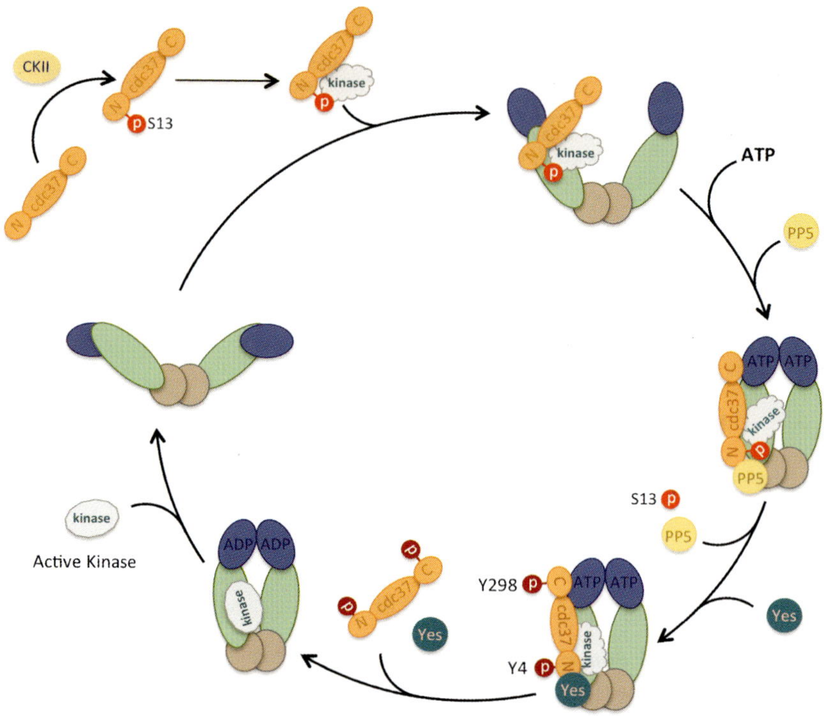

Plate 6.23. ***Proposed mechanism for progression and regulation of Cdc37 through Hsp90's kinase chaperone cycle***. For simplicity, only significant phosphorylations on Cdc37 are shown. Ser13 phosphorylation on Cdc37 by CKII facilitates kinase association with Cdc37. Cdc37 loaded with a pseudofolded kinase binds apo Hsp90 via contacts described in the crystal structure (Plate 22). ATP binding on Hsp90 triggers complex rearrangement, including NTD dimerization and migration of Cdc37 to Hsp90's MD. PP5 binding via Hsp90's MEEVD motif causes Cdc37 Ser13 dephosphorylation. Dissociation of PP5 allows Yes kinase incorporation, which is responsible for phosphorylation of Cdc37 on Tyr 4 and Tyr298, resulting in decreased Cdc37–kinase affinity. PP5 phosphorylation of Hsp90 also promotes Cdc37 dissociation from the Hsp90 (not depicted). ATP hydrolysis by Hsp90 causes rearrangements on the client kinase, followed by release of a folded kinase.

An earlier EM investigation of a kinase-bound Hsp90–Cdc37 complex indicated an asymmetric complex in which a single molecule of Cdc37 and the kinase (Cdk4) bind to a closed-like state of the Hsp90 dimer (Vaughan *et al.*, 2006). Accommodating the interaction seen in the NTD–Cdc37 crystal structure required a significant rotation of one of Hsp90's NTDs. Unfortunately, the low resolution limited by the negative stain made it challenging to unambiguously assign the individual domains. Recent advances in cryo-EM have enabled a high resolution structure of an equivalent kinase bound Hsp90–Cdc37 complex which clearly indicates that Hsp90 assumes the closed state matching that of the ATP-bound yeast crystal structure (Kliment Verba, unpublished results). In this case, the Cdc37 interaction does not match that seen in the NTD–Cdc37 crystal structure, supporting the proposed model for rearrangement of Cdc37–Hsp90 interactions during the cycle.

To help accommodate and coordinate such intricate and precisely timed conformational rearrangements, the system is tightly regulated by phosphorylation. Cdc37 has three known phosphorylation sites — Ser13, Tyr4 and Tyr276, with Ser13 being the best studied. In cells, Cdc37 spends most of its time phosphorylated at Ser13 by the constitutively active CKII (Miyata and Nishida, 2004). This modification strengthens the kinase–Cdc37 interaction by changing the conformation of the very N-terminal of Cdc37 (Liu and Landgraf, 2015; Polier *et al.*, 2013). Dephosphorylation is tightly managed by PP5 phosphatase during the progression through Hsp90's cycle, with perturbations to PP5 having catastrophic effects on kinase stability (Vaughan *et al.*, 2008). Phosphorylation at the other sites is carried out by the Yes kinase, with phosphomimic mutations shown to promote kinase dissociation from Cdc37 (Xu *et al.*, 2012).

Collaboration with the Hsp70 System

For at least some substrate proteins, Hsp90 on its own does not facilitate their refolding, but does so in collaboration with the Hsp70 system through a process dependent on Hsp90 ATP hydrolysis (Genest *et al.*, 2011; Grenert *et al.*, 1999; Wegele *et al.*, 2006). Demonstrated for the refolding of heat-inactivated luciferase, Hsp70 generally acts first as a potent capture device for unfolded proteins in this collaborative refolding process (Wegele *et al.*, 2006). While ATP binding and hydrolysis is not required for a mild Hsp90 effect in the eukaryotic system, ATP hydrolysis is necessary in the bacterial system, and is required for full synergistic stimulation in both systems (Genest *et al.*, 2011; Grenert *et al.*, 1999). Optimal refolding with Hsp70 and Hsp90 in the eukaryotic system necessitates the presence of Hop — consistent with Hop's function in facilitating the interaction between the two chaperones (Johnson *et al.*, 1998). Since Hsp90 and Hsp70 collaboration was similarly observed in the bacterial system that does not possess Hop, it seems that this collaborative interaction between Hsp90 and Hsp70 must be highly conserved and has evolved to be regulated by Hop in the eukaryotic cytosol. Supporting this, a direct interaction between bacterial Hsp70 and Hsp90 can be detected (Genest *et al.*, 2011). Similarly, it seems quite likely that the organellar Hsp90s also interact/collaborate with the Hsp70s within their compartments.

The cooperation between the Hsp70 and Hsp90 systems has long been established for the activation of GR. *In vivo*, Hsp90 is clearly required for GR hormone binding, with lowered Hsp90 levels compromising GR function (Picard *et al.*, 1990). Consistently, purified GR by the Pratt lab was inactive, with Hsp90 association necessary for GR to bind hormone (Bresnick *et al.*, 1989). However, Hsp70 was first required for Hsp90 association. Subsequent progression of the GR-bound Hsp90 to the ATP-bound state correlated with the gain in GR hormone binding, suggesting the correlation between the Hsp90 closed state and client activation (Hutchison *et al.*, 1994; Morishima *et al.*, 2000). In this case, Hop facilitated the progression to the Hsp90 ATP-bound (closed) state and GR remained active while bound to Hsp90 (Dittmar and Pratt, 1997). However, Hsp90's interaction with activated GR was highly transient and required p23 stabilization, which allowed the Hsp90-bound active state of GR to persist (Dittmar *et al.*, 1997). Comparable results were obtained with the heat(~37°C)-inactivated progesterone receptor (Kosano *et al.*, 1998; Smith *et al.*, 1992), in which time course reconstitution experiments suggested that Hsp70 and Hop dissociate from the receptor complex part-way through the activation pathway (Smith, 1993).

In contrast, the purified GR ligand binding domain (GRLBD) can bind ligand without aid or any observable enhancement from Hsp90 (Bledsoe *et al.*, 2002; Kirschke *et al.*, 2014; Lorenz *et al.*, 2014). Nonetheless, when starting with a functional the GRLBD, it was discovered that Hsp70 inhibits GR hormone binding by disrupting regions in the GRLBD important for ligand recognition (Kirschke *et al.*, 2014). Then, in an entirely Hsp90 ATP hydrolysis–dependent process, Hsp90 reverses the Hsp70-induced inhibition, indicating that ATP hydrolysis is utilized to regulate Hsp70 client dissociation, which is a prerequisite for a protein to reach its fully folded active state. In agreement with previous work, this process is highly coordinated by cochaperones, with complete reactivation requiring both Hop and p23. Incorporation of p23 was only detected for hydrolysis-competent Hsp90 client complexes, suggesting that p23's role in the receptor maturation pathway occurs on a posthydrolysis, presumably closed state of Hsp90. Notably, unlike direct interactions between Hsp90 and clients which typically display affinities in the range of 1–10 μM, the avidity provided by the complete system produces an apparent Hsp90 affinity that is in the nM regime.

Together, the emerging model is that ATP hydrolysis on Hsp90 is utilized for collaboration with the Hsp70 system (Plate 6.24), and that Hsp70's propensity to interact with unfolded protein regions appears to have evolved as a regulatory mechanism for inactivation and disassembly of some native proteins, upon which Hsp90 then acts to restore client activity by prompting Hsp70 client dissociation. While Hsp70 nucleotide exchange factors traditionally promote Hsp70 substrate release, comparison between Hsp90 and the human NEF (Bag-1) indicates that Hsp90 provides an important antiaggregation function upon Hsp70 release not provided by the NEF (Kirschke *et al.*, 2014). Moreover, experiments on both the bacterial and human systems suggest a collaboration between Hsp90 and NEFs (Genest *et al.*, 2011; Kirschke *et al.*, 2014), indicating that these are not necessarily different pathways of client release from Hsp70 but that they evolved to work in distinct and collaborative ways within the same pathway.

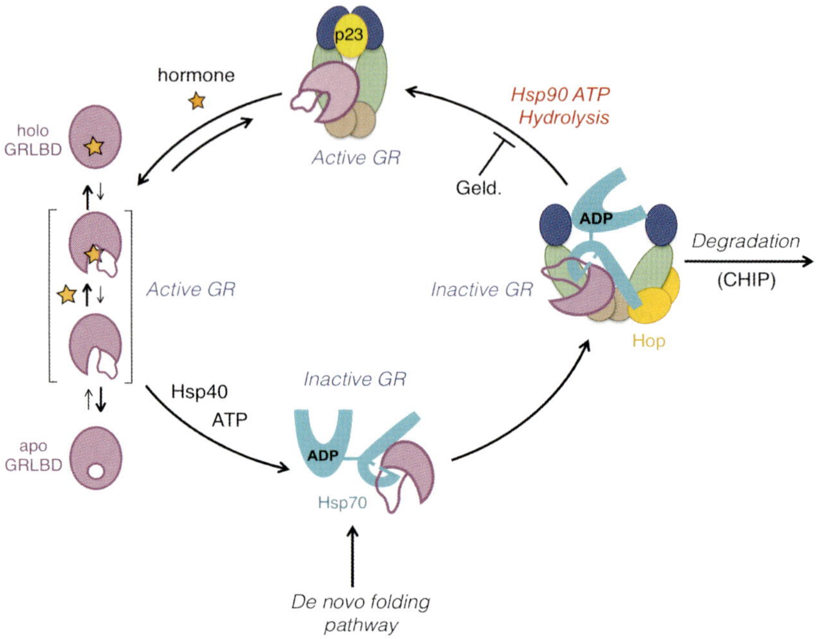

Plate 6.24. ***Schematic of the chaperone regulated GR hormone binding cycle***. The GRLBD binds hormone on its own through a process requiring transient partial unfolding, which allows the hormone binding pocket at the core of the protein to become solvent-accessible for hormone binding and release. In an Hsp40 and ATP hydrolysis–dependent process, Hsp70 binds the GRLBD and promotes hormone release by stabilizing a partially unfolded inactive state. Hop facilitates the loading of the GRLBD-bound Hsp70 complex onto Hsp90, forming an inactive complex that is stalled by geldanamycin (Geld), and promotes targeted degradation of the client. Subsequent ATP hydrolysis on Hsp90 is required to release Hsp70 (and Hop), allowing progression to the closed, p23-stabilized state, in which the GRLBD is held in a state with high affinity for hormone.

Furthermore, constant cycling in the presence of both Hsp70 and Hsp90 results in a higher affinity receptor with more rapid on and off rates. This enhanced GR ligand binding would enable faster physiological response times for hormone signaling and could only be achieved with both chaperones. The mechanism for this enhancement likely lies in the fact that the hormone binding pocket is at the core of the GRLBD, such that the domain has to partially unfold to allow hormone entry and release. Thus, the nuclear hormone receptors likely evolved a regulatory dependence on the Hsp90–Hsp70 chaperones because they are a class of proteins where partial unfolding is potentially beneficial. Together, in addition to aiding in solubility, Hsp70 and Hsp90 provide a mechanism for both precise regulation and enhanced function. Moreover, since CHIP preferentially ubiquitinates clients while bound to Hsp70 (Stankiewicz *et al.*, 2010), the connection between ATP hydrolysis on Hsp90 and client dissociation from Hsp70 provides a mechanistic explanation for how hydrolysis inhibition on Hsp90 leads to client degradation, and additionally implicates client transfer as a critical quality control checkpoint.

Structural insight into client handoff has been provided by a low resolution cryo-EM reconstruction of the Hsp90–Hsp70–GR–Hop complex, which reveals that the biochemical Hsp90–Hsp70 synergy is likely the direct result of a highly coordinated physical interaction between Hsp70's and Hsp90's ATPase domains in the Hop-stabilized GR-bound complex (Plate 6.25) (Kirschke *et al.*, 2014). In this complex, the domains of Hsp90 and Hsp70 are unambiguously assigned, with extra density stretching between the client binding sites of Hsp70 and Hsp90 attributed to the GRLBD, which appears simultaneously bound to the two chaperones. Consistent with the presence of Hop, Hsp90 is in an open V shape similar to the previous Hsp90–Hop EM structure with the defined MD–CTD interface and the NTDs rotated. However, Hsp90 is

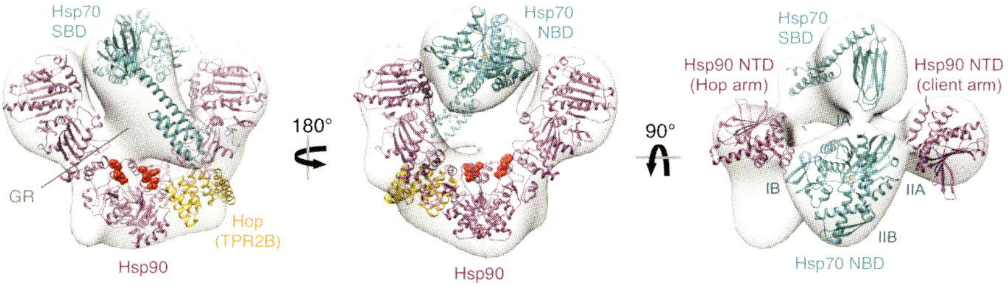

Plate 6.25. ***Cryo-EM reconstruction of the GR-bound Hsp70–Hsp90–Hop complex*****. A.** Front and back views of the EM map fitted with individual domains based on crystal structures of Hsp70 (teal), Hop TPR2B (yellow), Hsp90 (purple) and Hsp90's client binding residues (highlighted red spheres).

slightly asymmetric, with the NTD of the client-bound monomer incompletely rotated, oriented about 23° outward compared to the opposing fully rotated NTD of the Hop-bound monomer. As expected for the client-bound ADP state of Hsp70, the substrate binding domain (SBD) and the nucleotide binding domain (NBD) are dissociated and straddle Hsp90's cleft. On one side, the Hsp70 SBD delivers the GRLBD directly to the client binding region at the MD–CTD interface. On the other side, Hsp70's ATPase domain is nestled between Hsp90 NTDs, providing a structural basis for potentially coupling the two hydrolysis cycles. Importantly, this direct contact requires the rotated NTD state of Hsp90 induced by Hop (Plate 6.26A), indicating that not only does Hop preorganize Hsp90 for receiving clients, but also for interacting with Hsp70.

Here we propose a new model for how Hop's various domains contribute to organizing the Hsp90–Hsp70–client complex which combines previous biochemical and structural evidence with new structural insights from the cryo-EM density map (Elaine Kirschke, unpublished results). In agreement with NMR data (Schmid *et al.*, 2012), the TPR2B fits at Hsp90's MD–CTD interface, positioning TPR2B in proximity to the Hsp70 SBD C-terminal. This suggests that TPR2B is bound to Hsp70's EEVD tail in the complex, enabling TPR2B to recruit and position the Hsp70 SBD for efficient client delivery. If rigidly positioned as observed in the TPR2A–2B crystal fragment, this docking of TRP2B would move TRP2A out of the map density. However, a rotation of TPR2A's six-helix bundle around the linker would allow TPR2A to pack into extra density on Hsp90's MD (Plate 6.26B). This rotation does not perturb the functionally important orientation on the TPR2B side of the linker and supports the observation that the isolated TPR2A domain directly contacts Hsp90's MD at high concentrations (Schmid *et al.*, 2012). The functional implications of this contact are unclear but suggest that Hop could modulate Hsp90's MD conformation. Intriguingly, TPR2B's location positions the Hop DP2 domain in an unoccupied region of density at Hsp90's NTD–MD interface (Plate 6.26B). Fitting

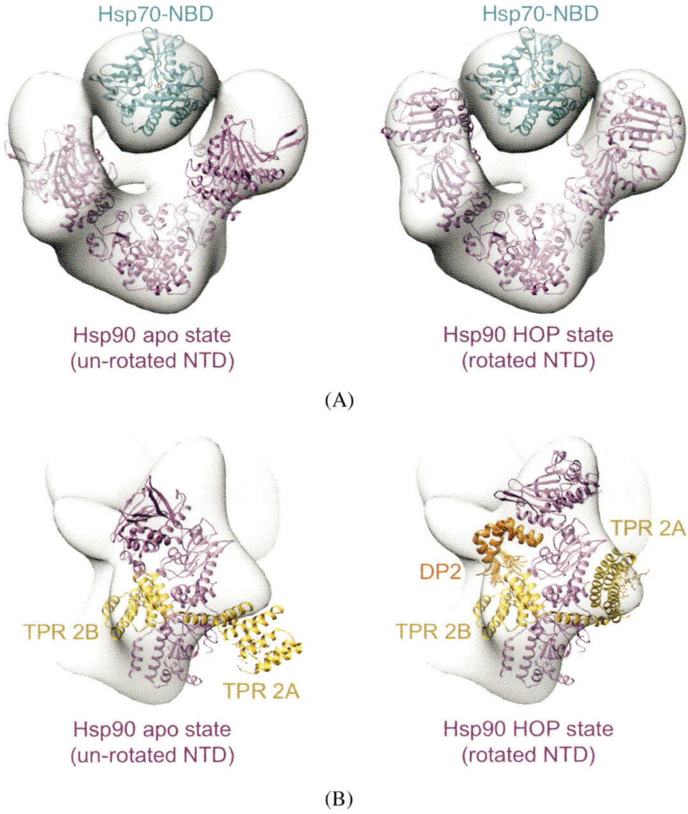

Plate 6.26. ***Hop coordinates a direct interaction between Hsp90 and Hsp70.*** **A**. Contact between the ATPase domains of Hsp90 and Hsp70 requires Hsp90 to be in a Hop-induced rotated NTD conformation (right). **B**. Preliminary fitting comparing the unrotated NTD state (left) versus the rotated NTD state (right) suggests that the DP2 stabilizes the NTD rotation. Additional density centered on Hsp90's MD suggests that Hop TPR2A rotates around the rigid linker to form additional contacts with Hsp90.

the DP2 domain into this density is only compatible with the rotated NTD conformation. This suggests that the DP2 domain could promote the NTD rotation necessary for the direct contact between the ATPase domains of the two chaperones, and thus provides an explanation for the functional importance of the DP2 domain *in vivo*.

It is currently unclear why Hop appears to be arranged differently in the Hsp70–client-bound complex compared to the initial Hsp90–Hop

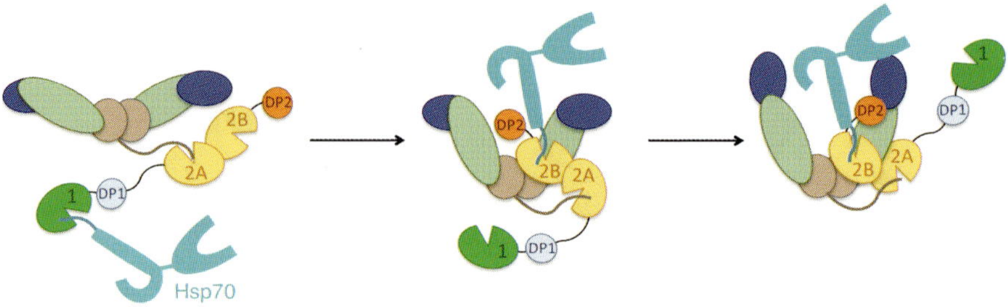

Plate 6.27. ***Proposed model for hop-coordinated client transfer from Hsp70.*** Hop TPR2A initiates the interaction with Hsp90's MEEVD motif, and Hop TPR1 initiates the interaction with Hsp70's EEVD motif. Secondary interactions between Hsp90 and other Hop TPR domains stabilize a client loading conformation of Hsp90. Hsp70 could then be transferred to TPR2B. Hop DP2 docks into Hsp90's MD–NTD interface, stabilizing the rotated NTD Hsp90 conformation required for the direct interaction between Hsp90's and Hsp70's ATPase domains. ATP hydrolysis on Hsp90 then enables faithful client release from Hsp70.

EM complex, for which the interaction appeared to be dominated by TRP2A and TPR1 (Plate 6.17A). Perhaps the explanation lies in the possibility that, while not essential, TPR1–DP1 facilitates the early stage of client transfer, and Hop rearranges during client transfer (Plate 6.27). In this case, Hsp70 could be recruited through TPR1 and then transferred onto TPR2B on progression to the more engaged client bound Hsp90 complex, as recently proposed (Röhl *et al.*, 2015).

Concluding Remarks

Significant structural, biophysical and biochemical contributions from many labs have together provided a near-complete picture for Hsp90's dynamic nature and the important mechanistic details of the conformational changes that occur during Hsp90's ATP hydrolysis cycle. An elaborate system of cochaperones has emerged as a mechanism for precise regulation in the eukaryotic Hsp90 system, in which they stabilize and promote specific conformational states of Hsp90 during the chaperone cycle. The extensive regulation of Hsp90 in eukaryotes is perhaps the most extreme in the human cytoplasm, with human cytosolic Hsp90's high dependence on cochaperones for progression to compact states reflected in its extremely low intrinsic ATP hydrolysis rate. Despite significant progress in our understanding of cochaperone interactions, a complete picture for many of the cochaperones is still lacking, with investigations complicated by the fact that some cochaperones, such as Cdc37 and Hop, could rearrange to form different contacts at distinct stages of the chaperone cycle.

Importantly, while recent investigations have provided insight into the principles and surfaces utilized for recognizing and binding client proteins, it is still not clear how the dramatic conformational changes and ATP hydrolysis are coupled to client protein activation. The differential presentation of the critical client binding residues in the asymmetric closed states observed in TRAP1 provides a compelling mechanism by which ATP-hydrolysis-driven rearrangement of these client binding regions could be directly coupled to client remodeling. However, this still needs to be experimentally demonstrated. Furthermore, if this model is indeed true for TRAP1, it remains to be established how universally conserved the asymmetric closed state is, and the extent to which the asymmetric state is sampled and utilized equivalently in the cytosolic eukaryotic Hsp90 homologues.

Nonetheless, it is now clear that ATP hydrolysis on Hsp90 is critical for collaboration with the Hsp70 system, in which the Hsp70 and Hsp90 chaperone cycles are highly coordinated to cooperatively refold misfolded proteins and to precisely regulate client activity. The coupling of ATP hydrolysis on Hsp90 to client handoff from Hsp70 suggests an important role for cellular Hsp90 in controlling the flux of proteins on and off Hsp70 as well as in critical quality control decisions. Furthermore, constant cycling of mature proteins through the Hsp70–Hsp90 chaperone system, in which transient, partially unfolded intermediates are captured by Hsp70 and transferred to Hsp90, would enable Hsp90 to utilize refoldablility as a metric for protein quality and faithful transfer.

Justifying the energy consumed for continuous cycling with the chaperone systems, this would allow the chaperones to constantly assess the quality of a large portion of the proteome and to expediently identify and target damaged proteins for degradation. Our current understanding of client handoff is very limited. Detailing how this process is coordinated and obtaining insights into how the structure of a client protein changes throughout the chaperone cycle will be key to understanding how chaperone-mediated remodeling and quality control are integrated into determining the client protein fate — either activation or degradation.

Acknowledgments

We thank members of the Agard lab for helpful discussions. Support for this work was provided by the PSI-Biology grant U01 GM098254 (D. A. A.), a Howard Hughes International Predoctoral Fellowship (K. V.), the UCSF Quantitative Biosciences Consortium and the Howard Hughes Medical Institute.

References

Alexandrescu AT, Abeygunawardana C, Shortle D (1994). Structure and dynamics of a denatured 131-residue fragment of staphylococcal nuclease: a heteronuclear NMR study. *Biochemistry* **33:** 1063–1072.

Ali MMU, Roe SM, Vaughan CK, *et al.* (2006). Crystal structure of an Hsp90–nucleotide–p23/Sba1 closed chaperone complex. *Nature* **440:** 1013–1017.

Bledsoe RK, Montana VG, Stanley TB, *et al.* (2002). Crystal structure of the glucocorticoid receptor ligand binding domain reveals a novel mode of receptor dimerization and coactivator recognition. *Cell* **110:** 93–105.

Boczek EE, Reefschläger LG, Dehling M, *et al.* (2015). Conformational processing of oncogenic v-Src kinase by the molecular chaperone Hsp90. *Proc Natl Acad Sci USA* **112:** E3189–E3198.

Borkovich KA, Farrelly FW, Finkelstein DB, *et al.* (1989). hsp82 is an essential protein that is required in higher concentrations for growth of cells at higher temperatures. *Mol Cell Biol* **9:** 3919–3930.

Bresnick EH, Dalman FC, Sanchez ER, Pratt WB (1989). Evidence that the 90-kDa heat shock protein is necessary for the steroid binding conformation of the L cell glucocorticoid receptor. *J Biol Chem* **264:** 4992–4997.

Chadli A, Graham JD, Abel MG, *et al.* (2006). GCUNC-45 is a novel regulator for the progesterone receptor/hsp90 chaperoning pathway. *Mol Cell Biol* **26:** 1722–1730.

Chang HC, Nathan DF, Lindquist S (1997). *In vivo* analysis of the Hsp90 cochaperone Sti1 (p60). *Mol Cell Biol* **17:** 318–325.

Citri A, Harari D, Shohat G, *et al.* (2006). Hsp90 recognizes a common surface on client kinases. *J Biol Chem* **281:** 14361–14369.

Cunningham CN, Krukenberg KA, Agard DA (2008). Intra- and intermonomer interactions are required to synergistically facilitate ATP hydrolysis in Hsp90. *J Biol Chem* **283:** 21170–21178.

Cunningham CN, Southworth DR, Krukenberg KA, Agard DA (2012). The conserved arginine 380 of Hsp90 is not a catalytic residue, but stabilizes the closed conformation required for ATP hydrolysis. *Protein Sci* **21:** 1162–1171.

DeZwaan DC, Freeman BC (2010). HSP90 manages the ends. *Trends Biochem Sci* **35:** 384–391.

Dickey CA, Kamal A, Lundgren K, *et al.* (2007). The high-affinity HSP90–CHIP complex recognizes and selectively degrades phosphorylated tau client proteins. *J Clin Invest* **117:** 648–658.

Dittmar KD, Demady DR, Stancato LF, *et al.* (1997). Folding of the glucocorticoid receptor by the heat shock protein (hsp) 90-based chaperone machinery. The role of p23 is to stabilize receptor.hsp90 heterocomplexes formed by hsp90.p60.hsp70. *J Biol Chem* **272:** 21213–21220.

Dittmar KD, Pratt WB (1997). Folding of the glucocorticoid receptor by the reconstituted hsp90-based chaperone machinery. *J Biol Chem* **272:** 13047–13054.

Dollins DE, Warren JJ, Immormino RM, Gewirth DT (2007). Structures of GRP94–nucleotide complexes reveal mechanistic differences between the hsp90 Chaperones. *Mol Cell* **28:** 41–56.

Dou F, Netzer WJ, Tanemura K, *et al.* (2003). Chaperones increase association of tau protein with microtubules. *Proc Natl Acad Sci USA* **100:** 721–726.

Dutta R, Inouye M (2000). GHKL, an emergent ATPase/kinase superfamily. *Trends Biochem Sci* **25:** 24–28.

Ebong IO, Morgner N, Zhou M, *et al.* (2011). Heterogeneity and dynamics in the assembly of the Heat Shock Protein 90 chaperone complexes. *Pro Nat Acad Sci USA* **108:** 17939–17944.

Eckert K, Saliou JM, Monlezun L, *et al.* (2010). The Pih1–Tah1 cochaperone complex inhibits Hsp90 molecular chaperone ATPase activity. *J Biol Chem* **285:** 31304–31312.

Eckl JM, Rutz DA, Haslbeck V, *et al.* (2013). Cdc37 (cell division cycle 37) restricts Hsp90 (heat shock protein 90) motility by interaction with N-terminal and middle domain binding sites. *J Biol Chem* **288:** 16032–16042.

Edkins AL (2015). CHIP: A co-chaperone for degradation by the proteasome. *Subcell Biochem* **78:** 219–242.

Eustace BK, Sakurai T, Stewart JK, *et al.* (2004). Functional proteomic screens reveal an essential extracellular role for Hsp90 alpha in cancer cell invasiveness. *Nat Cell Biol* **6:** 507–514.

Fang L, Ricketson D, Getubig L, Darimont BD (2006). Unliganded and hormone-bound glucocorticoid receptors interact with distinct hydrophobic sites in the Hsp90 C-terminal domain. *Proc Natl Acad Sci USA* **103:** 1847–18492.

Flanagan JM, Kataoka M, Shortle D, Engelman DM (1992). Truncated staphylococcal nuclease is compact but disordered. *Proc Natl Acad Sci USA* **89:** 748–752.

Flom G, Behal RH, Rosen L, *et al.* (2007). Definition of the minimal fragments of Sti1 required for dimerization, interaction with Hsp70 and Hsp90 and *in vivo* functions. *Biochem J* **404:** 159.

Flom G, Weekes J, Williams JJ, Johnson JL (2006). Effect of mutation of the tetratricopeptide repeat and asparatate-proline 2 domains of Sti1 on Hsp90 signaling and interaction in *Saccharomyces cerevisiae*. *Genetics* **172:** 41–51.

Freeman BC, and Yamamoto KR (2002). Disassembly of transcriptional regulatory complexes by molecular chaperones. *Science* **296:** 2232–2235.

Gaiser AM, Brandt F, Richter K (2009). The non-canonical Hop protein from *Caenorhabditis elegans* exerts essential functions and forms binary complexes with either Hsc70 or Hsp90. *J Mol Biol* **391:** 621–634.

Gaiser AM, Kretzschmar A, Richter K (2010). Cdc37–Hsp90 complexes are responsive to nucleotide-induced conformational changes and binding of further cofactors. *J Biol Chem* **285:** 40921–40932.

Galigniana MD, Radanyi C, Renoir JM, *et al.* (2001). Evidence that the peptidylprolyl isomerase domain of the hsp90-binding immunophilin FKBP52 is involved in both dynein interaction and glucocorticoid receptor movement to the nucleus. *J Biol Chem* **276:** 14884–14889.

Galigniana MD, Echeverría PC, Erlejman AG, Piwien-Pilipuk G (2010). Role of molecular chaperones and TPR-domain proteins in the cytoplasmic transport of steroid receptors and their passage through the nuclear pore. *Nucleus* **1:** 299–308.

Genest O, Hoskins JR, Camberg JL, *et al.* (2011). Heat shock protein 90 from *Escherichia coli* collaborates with the DnaK chaperone system in client protein remodeling. *Proc Natl Acad Sci USA* **108:** 8206–8211.

Genest O, Reidy M, Street TO, *et al.* (2013). Uncovering a region of heat shock protein 90 important for client binding in *E. coli* and chaperone function in yeast. *Mol Cell* **49:** 464–473.

Graf C, Lee C-T., Meier-Andrejszki E, *et al.* (2014). Differences in conformational dynamics within the Hsp90 chaperone family reveal mechanistic insights. *Front Mol Biosci* **1:** 1–15.

Graf C, Stankiewicz M, Kramer G, Mayer MP (2009). Spatially and kinetically resolved changes in the conformational dynamics of the Hsp90 chaperone machine. *EMBO J* **28:** 602–613.

Grenert JP, Johnson BD, Toft DO (1999). The importance of ATP binding and hydrolysis by Hsp90 in formation and function of protein heterocomplexes. *J Biol Chem* **274:** 17525–17533.

Grudniak AM, Markowska K, Wolska KI (2015). Interactions of *Escherichia coli* molecular chaperone HtpG with DnaA replication initiator DNA. *Cell Stress Chaperones* 1–7.

Hainzl O, Lapina MC, Büchner J, Richter K (2009). The charged linker region is an important regulator of Hsp90 function. *J Biol Chem* **284:** 22559–22567.

Harris SF, Shiau AK, Agard DA (2004). The crystal structure of the carboxy-terminal dimerization domain of htpG, the *Escherichia coli* Hsp90, reveals a potential substrate binding site. *Structure* **12:** 1087–1097.

Harst A, Lin H, Obermann WMJ (2005). Aha1 competes with Hop, p50 and p23 for binding to the molecular chaperone Hsp90 and contributes to kinase and hormone receptor activation. *Biochem J* **387:** 789–796.

Hawle P, Siepmann M, Harst A, *et al.* (2006). The middle domain of Hsp90 acts as a discriminator between different types of client proteins. *Mol Cell Biol* **26:** 8385–8395.

Hessling M, Richter K, Buchner J (2009). Dissection of the ATP-induced conformational cycle of the molecular chaperone Hsp90. *Nat Struct Mol Biol* **16:** 287–293.

Howard KJ, Holley SJ, Yamamoto KR, Distelhorst CW (1990). Mapping the HSP90 binding region of the glucocorticoid receptor. *J Biol Chem* **265:** 11928–11935.

Hutchison K, Dittmar KD, Czar MJ, Pratt WB (1994). Proof that hsp70 is required for assembly of the glucocorticoid receptor into a heterocomplex with hsp90. *J Biol Chem* **269:** 5043–5049.

Jahn M, Rehn A, Pelz B, *et al.* (2014). The charged linker of the molecular chaperone Hsp90 modulates domain contacts and biological function. *Proc Natl Acad Sci USA* **111:** 17881–17886.

Jakob U, Lilie H, Meyer I, Buchner J (1995). Transient Interaction of Hsp90 with early unfolding intermediates of citrate synthase. *J Biol Chem* **270:** 7288–7294.

Jinwal UK, Trotter JH, Abisambra JF, *et al.* (2011). The Hsp90 kinase co-chaperone Cdc37 regulates tau stability and phosphorylation dynamics. *J Biol Chem* **286:** 16976–16983.

Johnson BD, Schumacher RJ, Ross ED, Toft DO (1998). Hop modulates Hsp70/Hsp90 interactions in protein folding. *J Biol Chem* **273:** 3679–3686.

Johnson JL, Toft DO (1994). A novel chaperone complex for steroid receptors involving heat shock proteins, immunophilins, and p23. *J Biol Chem* **269:** 24989–24993.

Johnson JL, Brown C (2008). Plasticity of the Hsp90 chaperone machine in divergent eukaryotic organisms. *Cell Stress Chaperones* **14:** 83–94.

Karagöz GE, Rüdiger S (2015). Hsp90 interaction with clients. *Trends Biochem Sci USA*.

Karagöz GE, Duarte AMS, Akoury E, *et al.* (2014). Hsp90–tau complex reveals molecular basis for specificity in chaperone action. *Cell* **156:** 963–974.

Karagöz GE, Duarte AMS, Ippel H, *et al.* (2011). N-terminal domain of human Hsp90 triggers binding to the cochaperone p23. *Proc Natl Acad Sci USA* **108:** 580–585.

Kirschke E, Goswami D, Southworth D, *et al.* (2014). Glucocorticoid receptor function regulated by coordinated action of the Hsp90 and Hsp70 chaperone cycles. *Cell* **157:** 1685–1697.

Kosano H, Stensgard B, Charlesworth MC, *et al.* (1998). The assembly of progesterone receptor–hsp90 complexes using purified proteins. *J Biol Chem* **273:** 32973–32979.

Krukenberg KA, Förster F, Rice LM, *et al.* (2008). Multiple conformations of *E. coli* Hsp90 in solution: insights into the conformational dynamics of Hsp90. *Structure* **16:** 755–765.

Krukenberg KA, Southworth DR, Street TO, Agard DA (2009). pH-dependent conformational changes in bacterial Hsp90 reveal a Grp94-like conformation at pH 6 that is highly active in suppression of citrate synthase aggregation. *J Mol Biol* **390:** 278–291.

Lavery LA, Partridge JR, Ramelot TA, *et al.* (2014). Structural asymmetry in the closed state of mitochondrial Hsp90 (TRAP1) supports a two-step ATP hydrolysis mechanism. *Mol Cell* **53:** 330–343.

Lee C-T., Graf C, Mayer FJ, *et al.* (2012). Dynamics of the regulation of Hsp90 by the co-chaperone Sti1. *EMBO J* **31:** 1518–1528.

Leskovar A, Wegele H, Werbeck ND, *et al.* (2008). The ATPase cycle of the mitochondrial Hsp90 analog Trap1. *J Biol Chem* **283:** 11677–11688.

Li J, Richter K, Buchner J (2010). Mixed Hsp90–cochaperone complexes are important for the progression of the reaction cycle. *Nat Struct Mol Biol* **18:** 61–66.

Li J, Soroka J, Buchner J (2012). The Hsp90 chaperone machinery: Conformational dynamics and regulation by co-chaperones. *Biochimica et Biophysica Acta (BBA) — Molecular Cell Research* **1823:** 624–635.

Liu W, Landgraf R (2015). Phosphorylated and unphosphorylated serine 13 of CDC37 stabilize distinct interactions between its client and HSP90 binding domains. *Biochemistry* **54:** 1493–1504.

Lorenz OR, Freiburger L, Rutz DA, *et al.* (2014). Modulation of the Hsp90 chaperone cycle by a stringent client protein. *Mol Cell* **53:** 941–953.

Lotz GP, Lin H, Harst A, Obermann WMJ (2003). Aha1 binds to the middle domain of Hsp90, contributes to client protein activation, and stimulates the ATPase activity of the molecular chaperone. *J Biol Chem* **278:** 17228–17235.

Mandelkow EM, Mandelkow E (2012). Biochemistry and cell biology of tau protein in neurofibrillary degeneration. *Cold Spring Harb Perspect Med* **2:** a006247–a006247.

McLaughlin SH, Smith HW, Jackson SE (2002). Stimulation of the weak ATPase activity of human Hsp90 by a client protein. *J Mol Biol* **315:** 787–798.

McLaughlin SH, Sobott F, Yao Z-P, *et al.* (2006). The co-chaperone p23 arrests the Hsp90 ATPase cycle to trap client proteins. *J Mol Biol* **356:** 746–758.

Meyer P, Prodromou C, Hu B, *et al.* (2003). Structural and functional analysis of the middle segment of Hsp90: implications for ATP hydrolysis and client protein and cochaperone interactions. *Mol Cell* **11:** 647–658.

Meyer P, Prodromou C, Liao C, *et al.* (2004). Structural basis for recruitment of the ATPase activator Aha1 to the Hsp90 chaperone machinery. *EMBO J* **23:** 1402–1410.

Miyata Y, Nishida E (2004). CK2 controls multiple protein kinases by phosphorylating a kinase-targeting molecular chaperone, Cdc37. *Mol Cell Biol* **24:** 4065–4074.

Mollapour M, Neckers L (2012). Post-translational modifications of Hsp90 and their contributions to chaperone regulation. *Biochimica et Biophysica Acta (BBA) — Molecular Cell Research.*

Morishima Y, Murphy PJ, Li DP, *et al.* (2000). Stepwise assembly of a glucocorticoid receptor.hsp90 heterocomplex resolves two sequential ATP-dependent events involving first hsp70 and then hsp90 in opening of the steroid binding pocket. *J Biol Chem* **275:** 18054–18060.

Motojima-Miyazaki Y, Yoshida M, Motojima F (2010). Ribosomal protein L2 associates with *E. coli* HtpG and activates its ATPase activity. *Biochem Biophys Res Commun* **400:** 241–245.

Nathan DF, Vos MH, Lindquist S (1997). *In vivo* functions of the *Saccharomyces cerevisiae* Hsp90 chaperone. *Proc Natl Acad Sci USA* **94:** 12949–12956.

Obermann WMJ, Sondermann H, Russo A, *et al.* (1998). *In vivo* function of Hsp90 is dependent on ATP binding and ATP hydrolysis. *J Cell Biol* **143:** 901–910.

Onuoha SC, Coulstock ET, Grossmann JG, Jackson SE (2008). Structural studies on the co-chaperone Hop and its complexes with Hsp90. *J Mol Biol* **379:** 732–744.

Panaretou B, Prodromou C, Roe SM, *et al.* (1998). ATP binding and hydrolysis are essential to the function of the Hsp90 molecular chaperone *in vivo*. *EMBO J* **17:** 4829–4836.

Panaretou B, Siligardi G, Meyer P, *et al.* (2002). Activation of the ATPase activity of Hsp90 by the stress-regulated cochaperone aha1. *Mol Cell* **10:** 1307–1318.

Partridge JR, Lavery LA, Elnatan D, *et al.* (2014). A novel N-terminal extension in mitochondrial TRAP1 serves as a thermal regulator of chaperone activity. *Elife* **3:** 1–12.

Picard D, Khursheed B, Garabedian MJ, *et al.* (1990). Reduced levels of hsp90 compromise steroid receptor action *in vivo*. *Nature* **348:** 166–168.

Polier S, Samant RS, Clarke PA, *et al.* (2013). ATP-competitive inhibitors block protein kinase recruitment to the Hsp90–Cdc37 system. *Nat Chem Biol* **9:** 307–312.

Pratt WB, Morishima Y, Osawa Y (2008). The Hsp90 chaperone machinery regulates signaling by modulating ligand binding clefts. *J Biol Chem* **283:** 22885–22889.

Pratt WB, Toft DO (1997). Steroid receptor interactions with heat shock protein and immunophilin chaperones. *Endocr Rev* **18:** 306–360.

Prodromou C, Piper PW, Pearl LH (1999). Regulation of Hsp90 ATPase activity by tetratricopeptide repeat (TPR)–domain co-chaperones. *EMBO J* **18:** 754–762.

Prodromou C, Roe SM, Piper PW, Pearl LH (1997). A molecular clamp in the crystal structure of the N-terminal domain of the yeast Hsp90 chaperone. *Nat Struct Biol* **4:** 477–482.

Pursell NW, Mishra P, Bolon DNA (2012). Solubility-promoting function of Hsp90 contributes to client maturation and robust cell growth. *Eukaryot Cell* **11:** 1033–1041.

Queitsch C, Sangster TA, Lindquist S (2002). Hsp90 as a capacitor of phenotypic variation. *Nature* **417:** 618–624.

Rao J, Lee P, Benzeno S, *et al.* (2001). Functional interaction of human Cdc37 with the androgen receptor but not with the glucocorticoid receptor. *J Biol Chem* **276;** 5814–5820.

Ratzke C, Hellenkamp B, Hugel T (2014). Four-colour FRET reveals directionality in the Hsp90 multicomponent machinery. *Nat Commun* **5:** 4192.

Ratzke C, Mickler M, Hellenkamp B, *et al.* (2010). Dynamics of heat shock protein 90 C-terminal dimerization is an important part of its conformational cycle. *Proc Natl Acad Sci USA* **107:** 16101–16106.

Ratzke C, Minh N, Mayer MP, Hugel T (2012). From a ratchet mechanism to random fluctuations evolution of Hsp90's mechanochemical cycle. *J Mol Biol* **423:** 462–471.

Rehn AB, Buchner J (2015). p23 and Aha1. *Subcell Biochemis* **78:** 113–131.

Retzlaff M, Hagn F, Mitschke L, *et al.* (2010). Asymmetric activation of the Hsp90 dimer by its cochaperone aha1. *Mol Cell* **37:** 344–354.

Reynolds PD, Ruan Y, Smith DF, Scammell JG (1999). Glucocorticoid resistance in the squirrel monkey is associated with overexpression of the immunophilin FKBP51. *J Clin Endocrinol Metab* **84:** 663–669.

Richter K, Reinstein J, Buchner J (2002). N-terminal residues regulate the catalytic efficiency of the Hsp90 ATPase cycle. *J Biol Chem* **277:** 44905–44910.

Richter K, Buchner J (2001). Hsp90: chaperoning signal transduction. *J Cell Physiol* **188:** 281–290.

Richter K, Moser S, Hagn F, *et al.* (2006). Intrinsic inhibition of the Hsp90 ATPase activity. *J Biol Chem* **281:** 11301–11311.

Richter K, Muschler P, Hainzl O, *et al.* (2003). Sti1 is a non-competitive inhibitor of the Hsp90 ATPase. Binding prevents the N-terminal dimerization reaction during the atpase cycle. *J Biol Chem* **278:** 10328–10333.

Richter K, Soroka J, Skalniak L, *et al.* (2008). Conserved conformational changes in the ATPase cycle of human Hsp90. *J Biol Chem* **283:** 17757–17765.

Richter K, Walter S, Buchner J (2004). The co-chaperone sba1 connects the ATPase reaction of Hsp90 to the progression of the chaperone cycle. *J Mol Biol* **342:** 1403–1413.

Riggs DL, Roberts PJ, Chirillo SC, *et al.* (2003). The Hsp90-binding peptidylprolyl isomerase FKBP52 potentiates glucocorticoid signaling *in vivo*. *EMBO J* **22:** 1158–1167.

Roe SM, Ali MMU, Meyer P, *et al.* (2004). The mechanism of Hsp90 regulation by the protein kinase-specific cochaperone p50(cdc37). *Cell* **116:** 87–98.

Röhl A, Wengler D, Madl T, *et al.* (2015). Hsp90 regulates the dynamics of its cochaperone Sti1 the transfer of Hsp70 between modules. *Nat Commun* **6:** 6655.

Rutherford SL, Lindquist S (1998). Hsp90 as a capacitor for morphological evolution. *Nature* **396:** 336–342.

Sato T, Minagawa S, Kojima E, *et al.* (2010). HtpG, the prokaryotic homologue of Hsp90, stabilizes a phycobilisome protein in the cyanobacterium *Synechococcus elongatus* PCC 7942. *Mol Microbiol* **76:** 576–589.

Sawarkar R, Paro R (2013). Hsp90@chromatin.nucleus: an emerging hub of a networker. *Trends Cell Biol.*

Scheufler C, Brinker A, Bourenkov G, Pegoraro S (2000). Structure of TPR domain–peptide complexes: critical elements in the assembly of the Hsp70–Hsp90 multichaperone machine. *Cell.*

Schmid AB, Lagleder S, Gräwert MA, *et al.* (2012). The architecture of functional modules in the Hsp90 co-chaperone Sti1/Hop. *EMBO J* **31:** 1506–1517.

Schneider C, Sepp-Lorenzino L, Nimmesgern E, *et al.* (1996). Pharmacologic shifting of a balance between protein refolding and degradation mediated by Hsp90. *Proc Natl Acad Sci USA* **93:** 14536–14541.

Shao J, Irwin A, Hartson SD, Matts RL (2003). Functional dissection of cdc37: characterization of domain structure and amino acid residues critical for protein kinase binding. *Biochemistry* **42:** 12577–12588.

Shiau AK, Harris SF, Southworth DR, Agard DA (2006). Structural analysis of *E. coli* hsp90 reveals dramatic nucleotide-dependent conformational rearrangements. *Cell* **127:** 329–340.

Shirasu K (2009). The HSP90–SGT1 chaperone complex for NLR immune sensors. *Annu Rev Plant Biol* **60:** 139–164.

Siligardi G, Hu B, Panaretou B, *et al.* (2004). Co-chaperone regulation of conformational switching in the Hsp90 ATPase cycle. *J Biol Chem* **279:** 51989–51998.

Singh M, Shah V, Tatu U (2014). A novel C-terminal homologue of Aha1 co-chaperone binds to heat shock protein 90 and stimulates its ATPase activity in *Entamoeba histolytica*. *J Mol Biol* **426:** 1786–1798.

Smith DF (1993). Dynamics of heat shock protein 90-progesterone receptor binding and the disactivation loop model for steroid receptor complexes. *Mol Endocrinol* **7:** 1418–1429.

Smith DF, Stensgard BA, Welch WJ, Toft DO (1992). Assembly of progesterone receptor with heat shock proteins and receptor activation are ATP mediated events. *J Biol Chem* **267:** 1350–1356.

Smith JR, de Billy E, Hobbs S, *et al.* (2015). Restricting direct interaction of CDC37 with HSP90 does not compromise chaperoning of client proteins. *Oncogene* **34:** 15–26.

Song Y, Masison DC (2005). Independent regulation of Hsp70 and Hsp90 chaperones by Hsp70/Hsp90-organizing protein Sti1 (Hop1). *J Biol Chem* **280:** 34178–34185.

Southworth DR, Agard DA (2008). Species-dependent ensembles of conserved conformational states define the Hsp90 chaperone ATPase cycle. *Mol Cell* **32:** 631–640.

Southworth DR, Agard DA (2011). Client-loading conformation of the Hsp90 molecular chaperone revealed in the cryo-EM structure of the human Hsp90:Hop complex. *Mol Cell* **42:** 771–781.

Sreeramulu S, Jonker HRA, Langer T, *et al.* (2009). The human Cdc37.Hsp90 complex studied by heteronuclear NMR spectroscopy. *J Biol Chem* **284:** 3885–3896.

Stankiewicz M, Nikolay R, Rybin V, Mayer MP (2010). CHIP participates in protein triage decisions by preferentially ubiquitinating Hsp70-bound substrates. *FEBS J* **277:** 3353–3367.

Stebbins CE, Russo AA, Schneider C, *et al.* (1997). Crystal structure of an Hsp90–geldanamycin complex: targeting of a protein chaperone by an anti-tumor agent. *Cell* **89:** 239–250.

Storer CL, Dickey CA, Galigniana MD, *et al.* (2011). FKBP51 and FKBP52 in signaling and disease. *Trends Endocrinol Metab* **22:** 481–490.

Street TO, Krukenberg KA, Rosgen JR, *et al.* (2009). Osmolyte-induced conformational changes in the Hsp90 molecular chaperone. *Protein Sci* **19:** 56–65.

Street TO, Lavery LA, Agard DA (2011a). Substrate binding drives large-scale conformational changes in the Hsp90 molecular chaperone. *Mol Cell* **42:** 96–105.

Street TO, Lavery LA, Verba KA, *et al.* (2011b). Cross-monomer substrate contacts reposition the Hsp90 N-terminal domain prime the chaperone activity. *J Mol Biol* **415:** 3–15.

Street TO, Zeng X, Pellarin R, *et al.* (2014). Elucidating the mechanism of substrate recognition by the bacterial Hsp90 molecular chaperone. *J Mol Biol* **426:** 2393–2404.

Sullivan WP, Stensgard B (1997). Nucleotides two functional states of hsp90. *J Biol Chem* **272:** 8007–8012.

Sun L, Prince T, Manjarrez JR, *et al.* (2012). Characterization of the interaction of Aha1 with components of the Hsp90 chaperone machine and client proteins. *Biochim Biophys Acta* **1823:** 1092–1101.

Taipale M, Jarosz DF, Lindquist S (2010). HSP90 at the hub of protein homeostasis: emerging mechanistic insights. *Nat Rev Mol Cell Biol* **11:** 515–528.

Taipale M, Krykbaeva I, Koeva M, *et al.* (2012). Quantitative analysis of Hsp90–client interactions reveals principles of substrate recognition. *Cell* **150:** 987–1001.

Theodoraki MA, Caplan AJ (2012). Quality control and fate determination of Hsp90 client proteins. *Biochim Biophys Acta* **1823:** 683–688.

Thompson AD, Scaglione KM, Prensner J, *et al.* (2012). Analysis of the tau-associated proteome reveals that exchange of Hsp70 for Hsp90 is involved in tau degradation. *ACS Chem Biol* **7:** 1677–1686.

Trepel J, Mollapour M, Giaccone G, Neckers L (2010). Targeting the dynamic Hsp90 complex in cancer. *Nat Rev Cancer* **10:** 537–549.

Tripathi V, Obermann WMJ (2013). A primate specific extra domain in the molecular chaperone Hsp90. *PLoS ONE* **8:** e71856.

Tsutsumi S, Mollapour M, Prodromou C, *et al.* (2012). Charged linker sequence modulates eukaryotic heat shock protein 90 (Hsp90) chaperone activity. *Proc Natl Acad Sci USA* **109:** 2937–2942.

Vaughan CK, Gohlke U, Sobott F, *et al.* (2006). Structure of an Hsp90–Cdc37–Cdk4 complex. *Mol Cell* **23:** 697–707.

Vaughan CK, Mollapour M, Smith JR, *et al.* (2008). Hsp90-dependent activation of protein kinases is regulated by chaperone-targeted dephosphorylation of Cdc37. *Mol Cell* **31:** 886–895.

Wayne N, Bolon DN (2007). Dimerization of Hsp90 is required for *in vivo* function. *J Biol Chem* **282:** 35386–35395.

Wegele H, Muschler P, Bunck M, *et al.* (2003). Dissection of the contribution of individual domains to the ATPase mechanism of Hsp90. *J Biol Chem* **278:** 39303–39310.

Wegele H, Wandinger SK, Schmid AB, *et al.* (2006). Substrate transfer from the chaperone Hsp70 to Hsp90. *J Mol Biol* **356:** 802–811.

Whitesell L, Cook P (1996). Stable and specific binding of heat shock protein 90 by geldanamycin disrupts glucocorticoid receptor function in intact cells. *Mol Endocrinol* **10:** 705–712.

Wiech H, Büchner J, Zimmermann R, Jakob U (1992). Hsp90 chaperones protein folding *in vitro*. *Nature* **358:** 169–170.

Wu Z, Moghaddas Gholami A, Kuster B (2012). Systematic identification of the HSP90 regulated proteome. *Mol Cell Proteomics* **11:** M111.016675–M111.016675.

Xu W, Mollapour M, Prodromou C, *et al.* (2012). Dynamic tyrosine phosphorylation modulates cycling of the HSP90–P50(CDC37)–AHA1 chaperone machine. *Mol Cell* **47:** 434–443.

Xu W, Yuan X, Xiang Z, *et al.*(2005). Surface charge and hydrophobicity determine ErbB2 binding to the Hsp90 chaperone complex. *Nat Struct Mol Biol* **12:** 120–126.

Xu Y, Singer MA, Lindquist S (1999). Maturation of the tyrosine kinase c-src as a kinase and as a substrate depends on the molecular chaperone Hsp90. *Proc Natl Acad Sci USA* **96:** 109–114.

Yen PM (2015). Classical nuclear hormone receptor activity as a mediator of complex biological responses: a look at health and disease. *Best Pract Res Clin Endocrinol Metab* **29:** 517–528.

Yosef I, Goren MG, Kiro R, *et al.* (2011). High-temperature protein G is essential for activity of the *Escherichia coli* clustered regularly interspaced short palindromic repeats (CRISPR)/Cas system. *Proc Natl Acad Sci USA*.

Young JC (2001). Hsp90: a specialized but essential protein-folding tool. *J Cell Biol* **154:** 267–274.

Zeytuni N, Zarivach R (2012). Structural and functional discussion of the tetra-trico-peptide repeat, a protein interaction module. *Structure* **20:** 397–405.

Zhang M, Windheim M, Roe SM, *et al.* (2005). Chaperoned ubiquitylation — crystal structures of the CHIP U box E3 ubiquitin ligase and a CHIP–Ubc13–Uev1a complex. *Mol Cell* **20:** 525–538.

Zhang W, Hirshberg M, McLaughlin SH (2004). Biochemical and structural studies of the interaction of Cdc37 with Hsp90. *J Mol Biol* **340:** 891–907.

Zhao R, Davey M, Hsu YC, *et al.* (2005). Navigating the chaperone network: an integrative map of physical and genetic interactions mediated by the Hsp90 chaperone. *Cell*.

Index